Composites Science and Technology

Series Editor

Mohammad Jawaid, Lab of Biocomposite Technology, Universiti Putra Malaysia, INTROP, Serdang, Malaysia

Composites Science and Technology (CST) book series publishes the latest developments in the field of composite science and technology. It aims to publish cutting edge research monographs (both edited and authored volumes) comprehensively covering topics shown below:

- Composites from agricultural biomass/natural fibres include conventional composites-Plywood/MDF/Fiberboard
- Fabrication of Composites/conventional composites from biomass and natural fibers
- Utilization of biomass in polymer composites
- Wood, and Wood based materials
- Chemistry and biology of Composites and Biocomposites
- Modelling of damage of Composites and Biocomposites
- Failure Analysis of Composites and Biocomposites
- Structural Health Monitoring of Composites and Biocomposites
- Durability of Composites and Biocomposites
- Biodegradability of Composites and Biocomposites
- Thermal properties of Composites and Biocomposites
- Flammability of Composites and Biocomposites
- Tribology of Composites and Biocomposites
- Bionanocomposites and Nanocomposites
- Applications of Composites, and Biocomposites

To submit a proposal for a research monograph or have further inquries, please contact springer editor, Ramesh Premnath (ramesh.premnath@springer.com).

More information about this series at http://www.springer.com/series/16333

Bhasha Sharma · Purnima Jain
Editors

Graphene Based Biopolymer Nanocomposites

Springer

Editors
Bhasha Sharma
Department of Chemistry
Netaji Subhas University of Technology
New Delhi, Delhi, India

Purnima Jain
Department of Chemistry
Netaji Subhas University of Technology
New Delhi, Delhi, India

Composites Science and Technology
ISBN 978-981-15-9179-2   ISBN 978-981-15-9180-8 (eBook)
https://doi.org/10.1007/978-981-15-9180-8

This Springer imprint is published by the registered company Springer Nature Singapore Pte Ltd.
The registered company address is: 152 Beach Road, #21-01/04 Gateway East, Singapore 189721, Singapore

Preface

On one hand, when research is revolving around the development of biodegradable plastics, new trends have been increasingly shifted towards the development of techniques involving biodegradation of conventional plastics under defined conditions. Biopolymer nanocomposites are the most auspicious aspirant to intercept hazardous situations that can replace conventional plastics. Quest amidst sustainable and green materials have been developed due to their utilization in versatile applications from biomedical to packaging. Our purpose of writing this book is to deliver the idea to work on sustainable development which could be beneficial for researchers, students, and industries. This book will be a virtue for sustainable development to benefit mankind. This book will have a thorough investigation of sustainable biopolymers and their advantages which can reduce conventional plastic wastes and adapt eco-friendly materials.

To proliferate the use of bio-based or renewable resources, readers will get an extensive overview on the ongoing research in the field of biopolymer graphene nanocomposites, as we all know graphene is a wonder material and has outstanding mechanical electrical, thermal properties that can forfeit the existing petroleum-based product. The main emphasis of this book to address and rectify the complications of using plastics which are non-degradable and has an abhorrent impact on the environment. The limitations of the properties of biopolymer can be vanquished by employing graphene as a nanomaterial. Outstanding properties of graphene in accordance with biopolymer can be utilized to develop applications like water treatment, tissue engineering, photo-catalysts, super-absorbents, etc. This book will consist of meticulous chapters contributed by the scientific community who are veterans in field of science and technology in biopolymers and their applications from different universities and countries. This book covers versatile areas such as green biopolymers, bionanoplastics involving graphene and its functionalization, synthesis and properties of graphene-reinforced biopolymers. We hope this book will be useful reference materials for the engineers, chemists, researchers associated with polymers science and technology as we all know the biggest challenge in the current scenerio is the plastic waste predicament. The vital components of this book are environmental-based nanotechnologies which have been studied globally. This book delineates eco-friendly and sustainable products of graphene-reinforced biopolymers in newflanged ways.

With all indebtedness and gratitude, we would like to express our sincere thanks to all the contributors for their dedicated and prompt response. Our wholehearted thank goes to all seniors, juniors, Department of Chemistry and School of Applied Sciences, NSUT, for their continuous support.

In the end, we thank Almighty God for providing us ample potency and the right path to cherish this juncture of my life.

New Delhi, India

Bhasha Sharma
Purnima Jain

Contents

Emerging Trends in Green Polymer Based Composite Materials: Properties, Fabrication and Applications

Partha Pratim Das, Vijay Chaudhary, and Shubhanshu Mishra

Abstract Composite materials showed improved properties compared to metal and polymer materials that made the composites used as parts of the structures. On the other side, fiber-reinforced polymers are primarily manufactured from synthetic fibers, like glass or paper, and petro-chemical thermosetting resin or matrix. An emerging field of high-performance natural fibers, especially Bast fibers (including flax, hemp, and jute), is gaining interest in this context, and is immediately attractive. Therefore, the industry includes many significant innovations at every stage of composite production which extends from the fibers and their precursors or preforms to the manufacturing processes and associated industries. There are plenty of developments on the market across the value chain, with most advancement in the composites industry concentrating on performance enhancement and cost benefits. This chapter includes an overview of emerging trends in fabrication and characterization of green polymer-based composite materials.

Keywords Polymer · Composites · Natural fibers · Graphene oxide

1 Introduction

Most composites used today are at the forefront of materials technology, with performance and costs appropriate for highly demanding applications, and have conquered various sectors such as aerospace industry, aeronautics, automotive industry, manufacturing industries, construction, and the marine industry. Also, more use of composite technologies is emerging. This highly competitive market continues to evolve, with the major emphasis in the past being to produce materials with adequate strength, and high wear resistance. Over a long period, composites enabled us to

P. P. Das · V. Chaudhary (✉) · S. Mishra
Department of Mechanical Engineering, Amity School of Engineering and Technology, Amity University, Uttar Pradesh, Noida 201313, India
e-mail: vijaychaudhary111@gmail.com

P. P. Das
e-mail: das1.parthapratim@gmail.com

B. Sharma and P. Jain (eds.), *Graphene Based Biopolymer Nanocomposites*, Composites Science and Technology,
https://doi.org/10.1007/978-981-15-9180-8_1

make remarkable products with exceptional capacities, but often they are being made in a unique or small series production. Today, large-scale production is needed by end-use sectors, especially the aeronautics and automotive industries. The composites industry must, therefore, follow the demands of a broad series production that the creative industries require [1]. For the composites industry, this is an enormous opportunity that has grown rapidly thanks mainly to innovation that has created new consumer applications and built on existing applications. Growth for various industry would rely on other capabilities in the future. Because of its cost–performance benefits, composites become the material of choice for various industries. Also, major developments are anticipated in the field of composites in the coming years, provided that the composites industry needs to meet the demands of large-scale production to increase broad penetration in several advanced industries. Weight reduction and improved mechanical performance are the key factors in the combination of new materials for automotive and other means of transport. Because of their excellent quality and lightweight, the composite materials have thus become the most attractive candidate. In particular, composite materials with fiber-reinforced polymer (FRP) were used as an alternative to metals and addition of a filler material also influence the mechanical properties. E. T. Thostenson.et.al [2] explained that various filler materials such as calcium carbonate (GCC), precipitated calcium carbonate (PCC), kaolin, talc, carbon black, nano -clay etc. But, one of the most important filler materials is Graphene, which is gaining a lot of importance with polymer composites. As graphene possess low mass density and exceptional mechanical properties, they have been extensively used as engineering materials in various challenging applications.

On the other side, fiber-reinforced polymers are primarily made of synthetic fibers, such as glass or steel, and a thermosetting resin or thermoplastic matrix of petrochemical origin [3]. Graphene, a special hexagon-lattice monolayer, is much more rigid and heavier than the carbon nanotube (CNT). By nanoindentation on a free-standing monolayer graphene, the modulus and intrinsic strength of the Young were estimated at a strain of $\varepsilon_{\text{int}} = 0.25$ to be equal to $\text{E} = 1.0$ TPa and $\sigma_{\text{int}} = 130$ GPa [4, 5].

Amandine Célino et al. [3] explained that composite materials showed improved properties compared to metal and polymeric materials which made the composites used as structural components. In this context, interest is gaining from an emerging field of high-performance natural fibers, especially Bast fibers. Various Bast fibers include flax, hemp and jute and are immediately attractive as they provide comparable or improved tensile strength and stiffness to that of glass fiber, as well as contain favorable vibration dampening and no– abrasive properties [3]. Nonetheless, due to their beneficial properties, low prices, legislative engines, and consumer preferences, natural fibers and resins are on the rise. These forms of fiber composites have some main developments. The industry sectors, users of natural fibers, are now engaged in an eco-design approach. The significance and use of graphene in different host materials was systematically clarified by Kuilla et al. in their detailed review article on graphene-based polymer nanocomposites. They also made a remarkable comparison of different nanofillers, and described in detail their significant applications [6]. Jang.et.al reviewed the processing of graphene nanoplatelets (GNPs) for fabrication of composite materials [7]. Hansma et al. demonstrated the effective manufacture of nanocomposites based on graphene. They optimized the quantity and combination of adhesives and high-strength nanostructures (graphene) required to produce a solid, low-density, lightweight and damage-resistant composite material [8] effectively. Yu et al. [9] observed that few-layer graphene nanocomposites based on epoxy exhibit interesting properties for the electronics industry, appropriate for the development of thermal-interface materials.

The main goal is to upgrade the knowledge of fibers and filler material [10, 11] that must be chosen in order to get the better combination for manufacturing Green Polymers based composite materials and attract the new industries to come up in developing new composites. Various authors used nanofiller to enhance the performance of green composite materials. One of the most important filler is graphene which plays a vital role in enhancement of performance. Various literature related to graphene reinforced composite materials is shown in Table 1.

2 Key Trends in Composite Materials

The industry includes many significant developments at each composite production stage, from the fibers and their precursors or preforms to the manufacturing processes and related industries. Not only do these developments generate opportunities at every stage of product manufacturing for a wide range of companies, but they also produce both improved and new materials that allow this industry to expand and develop applications.

There are plenty of developments on the market across the value chain, with most innovations concentrating on performance enhancement and cost benefits in the composites industry. Innovation generates "business value" in which interest can be extracted from a combination of [27]:

- Fulfilling functional needs in target markets, and/or better or more efficiently satisfying established needs.
- All of these needs highlight the importance of an open-ended approach to innovation in order to meet the medium and long-term needs of the industry.

Main developments in continuous advancement and improvement of composite materials are now as follows:

- increased light weight of automobile, aerospace and industrial components;
- Improved reinforcement (fibre, cloth or particulate matter) and matrix (resin) structures to meet greater mechanical and chemical requirements;
- Reducing the cost of different composite parts;
- Technologies faster and more reliable, applicable to broad series production;
- Reduction in component counts in many applications, developing technology for one-piece or modular products;

Table 1 Shows the literature of graphene reinforced composite materials

Fibres/polymer	Filler material	Fabrication Techniques	Effect on Performance on the fabricate composites	References
Natural fibre/epoxy and carbon fibre reinforced composites	Graphene nanoplatelets (GNPs)	Hot Press molding followed by post curing (pressed at 1500 psi Pressure and at a temperature of 150 °C for around 30 min)	All Mechanical properties like ultimate tensile strength, flexural strength, and flexural modulus were all improved. Furthermore, fatigue life of the developed composites also increased. And increases the overall life span of the composites	Shen et al. [12]
Jute/epoxy	Graphene/graphene oxide	Hand layup process	Enhancement of interfacial strength by facto 236% and tensile strength by 96% more than the untreated fibres i.e. without graphene. Furthermore, it improves the interlocking between fibre and the graphene oxide	Sarker et al. [13]
Jute/epoxy	Nano-engineered graphene	Hand layup process	It enhanced the stress–strain value of jute-epoxy composites by 324% and tensile strength by 110% more than untreated jute fiber composites, by arranging fibers in parallel direction with graphene derivatives by individualization and nano surface engineering	Sarker et al. [14]

(continued)

Table 1 (continued)

Fibres/polymer	Filler material	Fabrication Techniques	Effect on Performance on the fabricate composites	References
Sisal/epoxy	Graphene oxide	Injection molding and extrusion compounding	It shows an enhancement in tensile strength by 36.5%, increase in tensile modulus by 30%, and impact strength also get improved by an increment of 36.27%. Furthermore, it is also mentioned that, there is also increment in thermal stability and water absorption resistance	Chen et al. [15]
Kenaf/polypropylene (PP)	Graphene nanoplatelets (GNPs)	Melt extrusion blending	With increase in concentration of graphene nanoplatelets (GNPs), flexural strength increases, young's modulus also increased from 706 MPa (neat PP) to 1600 MPa with incorporation of GNPs. It is also verified from Thermogravimetric analysis, the thermal stability also improves	Idumah et al. [16]
Flax/epoxy	Graphene	Vacuum infusion	Incorporation of graphene as a filler material, tensile strength and flexural strength increases due to strong interfacial boning. It is also revealed that, incorporation of graphene acts as an effective barrier to flax/epoxy composites water absorption	Kamaraj et al. [17]

(continued)

Table 1 (continued)

Fibres/polymer	Filler material	Fabrication Techniques	Effect on Performance on the fabricate composites	References
Hemp/epoxy	Graphene	Hand layup technique	Enhancement in tensile and other mechanical properties such as young's modulus, flexural strength, etc	Hallad et al. [18]
Kenaf/glass/epoxy	Graphene	Hand layup technique	Incorporation of graphene gives better mechanical properties. With increase in concentration of graphene, both flexural and impact strength increases	Ramesh Kumar et al. [19]
Banana fibre/epoxy	Graphene oxide (GO)	Hand layup technique	With addition of GO, there is and overall increment in the mechanical properties i.e. tensile strength is increased by 275%, value of young's modulus increased by 242%, followed by 239% increment in flexural strength 296% for impact strength in comparison with neat epoxy	Bharadiya et al. [20]
Banana/glass/epoxy	Graphene	Hand layup technique	There is an overall enhancement in the flexural strength of the developed composites	Ramesh Kumar et al. [21]

(continued)

Table 1 (continued)

Fibres/polymer	Filler material	Fabrication Techniques	Effect on Performance on the fabricate composites	References
Cotton/epoxy	Graphene oxide	The cotton/epoxy–GO composite was prepared by dispersing 10.0 g of cotton fiber in 500 mL double distilled water using homogenizer	Improves the mechanical properties and acts as an effective barrier to C/E-GO composites to water absorption	Abd-Elhamid et al. [22]
Coir/fibre/epoxy	Graphene	Hand layup technique	Graphene enriches the overall mechanical properties of the coir-epoxy/graphene composite as compared to near epoxy. The % enrichment in tensile strength is upto 325%, 282% in stress–strain vale, 278% in value of flexural strength and increment in impact strength upto 435% as compared to neat epoxy	Bharadiya et al. [23]
Jute/epoxy	Graphene	Compression molding	With increase in the concentration of graphene content, stress–strain value increases which results in higher strength. It also improves the flexural strength as compared to neat epoxy	Rathinasabapathi et al. [24]
Curaua-plant (ananas arectifolius)/epoxy	Graphene oxide	Hydraulic press	With incorporation of graphene oxide, the thermal degradation was retarded which contributes to a high temperature resistance in comparison with neat epoxy	Costa et al. [25]

(continued)

Table 1 (continued)

Fibres/polymer	Filler material	Fabrication Techniques	Effect on Performance on the fabricate composites	References
Ramie fibre/polypropylene (PP)	Graphene oxide	Film stacking process	It is observed that the interlaminar strength increase by an increment of 40% when graphene oxide is incorporated with ramie fibre/polypropylene composites	Dang et al. [26]

- Environmentally friendly reinforcement (fibre, plastic or particulate matter) and matrix (resin) systems, with a focus on developing renewable materials of high strength.

The use of composite materials varies from simple households to industrial applications that are light to heavy. Consequently, composite materials play an important role in various segments of the industry such as aerospace, automotive, defense and space, marine, consumer goods due to the desirable characteristics of lightweight, corrosion resistance, high strength and flexibility in design among many others [28]. While it continues to replace conventional materials in many industries, composites remain under-represented in several markets and potential applications. The above-mentioned trend in innovations will allow for increased penetration of composite materials across different industries. Aerospace, automotive, transport, wind, and construction segments are expected to grow at a higher rate in the coming years. To meet the higher mechanical and chemical requirements, innovation is expected in the development of higher performance composite materials. Improving stiffness and strength along with developing low-cost carbon fiber or particulate composite parts for various applications in automotive, wind, and industrial applications are also major trends in innovation. Relative to the core materials in layered composites, the latest technologies are being performed to improve strength and rigidity for automotive and other applications, and to obtain the lower density to meet demand from end-use industries [29]. Light-weighting and cost reduction are therefore the main developments in different application types, such as transportation, aerospace, or wind energy.

3 Focus on Graphene as a Filler Green Polymer-Based Composites Material

The development of composite materials has provided an opportunity for the use of new materials, resulting in reduction of cost, increased efficiency and efficient use of available natural resources. In addition, focusing on green materials will give importance to the production of natural fibers of high strength to weight ratio to increase penetration in the automotive, building, and other industries. Even though various filler materials are available, one of the most effective filler used is graphene which plays a potential role in enhancing the properties of green polymer based composite due to its low mass density and exception mechanical properties. The challenges for turning renewable resources into industrial materials are reliability, flexibility, highly efficient and sustainability, biodiversity and environmental impact. In this context, the latest developments regarding the natural fibers in composite materials are [30]:

- Improving strength and rigidity to cope with glass fibre;
- Finding new application areas for widening footprints in various industries;

- Improving efficiency and rigidity for automotive applications;
- Expect the use of modern reinforced composites by advanced and industries to replace parts currently being made with other materials or other composite forms.

A renewed interest in biobased goods has led to concern for the environment. Among them, plant (natural) fibers, particularly in automotive engineering, are considered as an environmentally friendly substitute for composite enhancing glass fibres. Because of their wide availability, low cost, low density, high-specific mechanical properties and eco-friendly picture they are increasingly being used as reinforcements in polymer matrix composites. Increasing environmental issues and depleting oil supplies demand new eco-friendly green materials. Among a variety of natural materials, cellulosic natural fibers or textiles are envisaged as the most suitable means of solving these specific problems. It has been well recognized the ability of cellulosic fibers as reinforcement in composite materials, but the renewed interest in natural fibers has resulted in a large number of modifications to make it equal and even superior to synthetic fibers. We emerged as a replacement for the traditional materials after drastic improvements in the quality of natural fibres, being an environmentally friendly option for the future.

An emerging trend is the implementation of natural fiber reinforced composites with graphene as a filler material as a replacement for synthetic fibre-reinforced composites, taking into consideration the high-performance level of composite materials in terms of durability, maintenance and cost effectiveness [31]. The use of green polymer based composite materials acquired a lot of attention due to the disadvantage of the thermoplastic and fiber glass as reinforcement.

4 Fabrication Techniques for Green Polymers Based Composite Materials

Composite production process selection has emerged as a big challenge in the field composite materials. Fabrication technique for composite materials depends on many factors, such as matrix and reinforcement properties (type of fiber/matrix, fiber material, fiber orientation, fiber length, etc.), product geometry (shape, thickness, etc.), and end-use. A summary of various fabrication techniques used for composite materials based on green polymers is described below.

4.1 Hand Lay-Up Technique

It is the oldest, easiest and most widely used method for producing composite materials. This method uses continuous fibre in unidirectional form, spun, knitted & stitched cloth. The composite materials can be fabricated by mixing layers of different fiber orientations, depending on the type applications. Alternative matrix and reinforcement layers are set above it in this form. In general, the leasing agent is used on the surface of the mold to prevent sticking and to allow easy removal of the finished part [32]. To remove the extra resin from each sheet, a roller moves the matrix onto the reinforcements to ensure a uniform distribution of resin over the surface. The process is repeated for all reinforcing layers until the desired thickness is reached. The entire process is manually hand-made. The whole process is custom-made by custom. The composite material is cured at normal ambient temperatures. After thorough cure, the finished part is removed. Various parameters viz. fiber size, fiber colour, fiber orientation, matrix pressure type and healing time, affect the performance and quality of the composite products produced through the hand lay-up process [33]. The benefits of this approach include less investment in materials, easy-to-change mold/design and virtually no limit on the size of the part to be produced. Nevertheless, some of the drawbacks of this process include: the method is suitable for low volume fraction/reinforcement phase concentration; it is very time consuming; it has high void content and/or porosity; only one finished surface (which is in contact with the mold) is possible; it is labour-intensive; thickness control is not very accurate; it is difficult to achieve uniform fibre. This method is best suited for making blades of wind turbines, tanks, boats, boat hulls etc. (Fig. 1).

Some researchers have carried out a study on composite materials developed using the technique of hand lay-up [34]. Fong et al. [35] performed an experiment on yarn flax fibers for polymer-coated sutures as well as hand lay-up polymer composites. The effect of moisture on the mechanical properties of yarn flax fibers is being studied, as is the possible dependence by Mishra et.al on knot geometry [36]. Garcia et al. [37] performed an investigation into the manufacture and multifunctional properties of in situ grown hybrid materials containing associated carbon nanotubes.

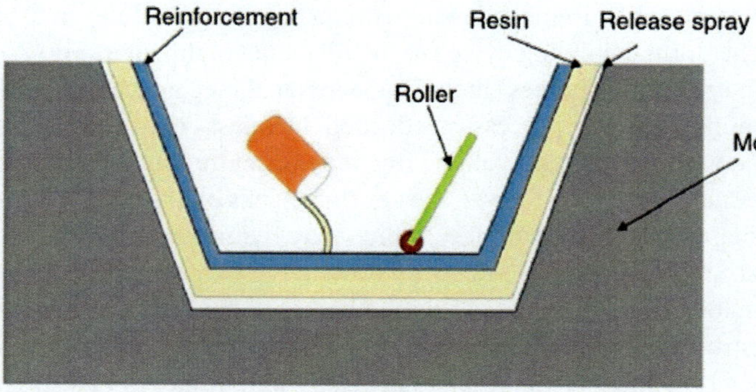

Fig. 1 Schematic diagram of hand lay-up process

4.2 Filament Winding Technique

It is a process in which the composite parts are formed by winding continuous fibers onto a rotating mandrel in different orientations. Filament winding method is the most economical way to manufacture symmetrical composite parts with a high-volume output. This method is primarily used for hollow, usually circular, or oval sectioned sections, such as pipes and tanks. With a tiny gas cylinder to a large cryogenic tank, it has a wide variety of applications. This method is ideal for automation that needs little to no human interaction. Upon winding fibers we are passed into a resin bowl, where the resin wets the fibers. The number of layers and winding thickness depend mainly on the desired properties of the composite parts. To compact the fibers toward the mandrel, the desired tension is given to the fibers. The movement of spinning mandrel and moving carriage will alter the winding pattern. The procedure is usually done at room temperature or at high temperature. The mandrel is extracted from the composite component after curing is finished, and can be reused.

Depending on the desired applications specific winding patterns can be used. Many researchers have been working on the composites made with winding filaments. Lamontia et al. [38] modelled, studied and developed graphite/thermoplastic composite filament wound ring-stiffened pressure hull model as well as hydrostatic testing. Misri et al. [39] performed an investigation into the mechanical actions of kenaf fiber/unsaturated hollow composite polyester shafts created using filament winding technique. Ongoing kenaf fiber rovings were pulled through a drum style resin bath and wounded around an aluminium revolving mandrel. The winding styles most widely used are the hoop winding, polar winding, and helical winding.

Mandrel plays a big part in the filament winding cycle. The mandrel type determines the shape of the generated composite part. The material the mandrel is made with mainly depends on the end-use of the composite parts produced. Mandrels may or may not be removable. Removed mandrels can be marked as fully free, breakable, or soluble. The normal winding cycle for the filaments is shown in Fig. 2.

The advantages of this process include phenomenal mechanical properties due to the use of continuous fibres, process speed, good thickness control, better fiber orientation and material control, high volume fraction/concentration of reinforcing phase, and good internal finishing. However, the key drawbacks of this method are difficulty in winding shapes which are complex, which may require complicated equipment; poor external finish; restriction to convex formed components; high mandrel cost; and need for low viscosity resins. This method also involves the manufacture of open-ended structures such as gas cylinders and pipe systems, and closed end structures such as gas cylinders and piping systems, as well as closed end structure such as pressure vessels and chemical storage tanks.

Mertiny et al. [40] performed an experimental study into the effect of winding multi-angle filaments on the strength properties of tubular composite structures. Cohen et al. [41] examined the effect filament winding parameters have on the strength and efficiency of the composite vessel.

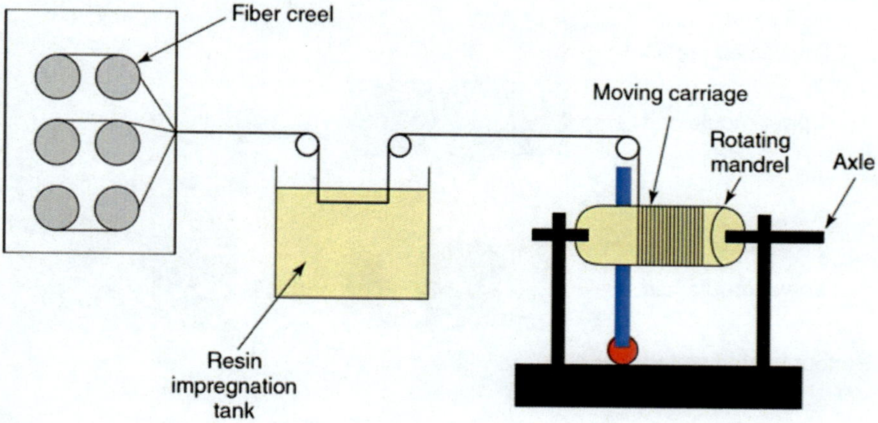

Fig. 2 Schematic diagram filament winding process

4.3 Compression Molding Technique

It is one of the oldest methods of fabricating composite materials; this method gives high mass production capability within a short time as compared to other fabrication methods. In this method, heat and pressure are applied to mold a composite material into the desired shape/form. The final product form depends on the molds used for the matched compression. This method takes place in two stages, namely preheating and pressurizing. The charge or preform is initially positioned in the cavity of matched mold when it is in the open position and the two halves are joined together, thus closing the mold. Then pressure is applied to compress the resin, so it fills the mold's cavity. The material is removed by applying heat when under water. Filho et al. [42] carried out an experimental study on the efficiency of materials with mortier-reinforced compression-molded sisal fiber. Material durability is studied by analysing the effects of accelerated gain on the microstructure and flexural behavior of the composites. After curing, the pressure is released, and the mold is opened to remove the finished part. Pressure, temperature, volume of molding material, type of resin, healing time, etc. are the parameters that affect the performance of the co-produced composites by this method. Figure 3 shows a typical compressive molding process.

The key benefit of this approach is its ability to generate a large variety of parts with minimal size variations. Additionally, short cycle time, finished interior and exterior surfaces, improved surface quality, partial shape uniformity, lower maintenance costs, very little finishing operation needed, and better fiber material control are other advantages. However, the key disadvantages of this method are the high initial capital cost, which is not suitable for very large pieces and economically not viable for limited volume of production, and mold depth limit. Typical items that can

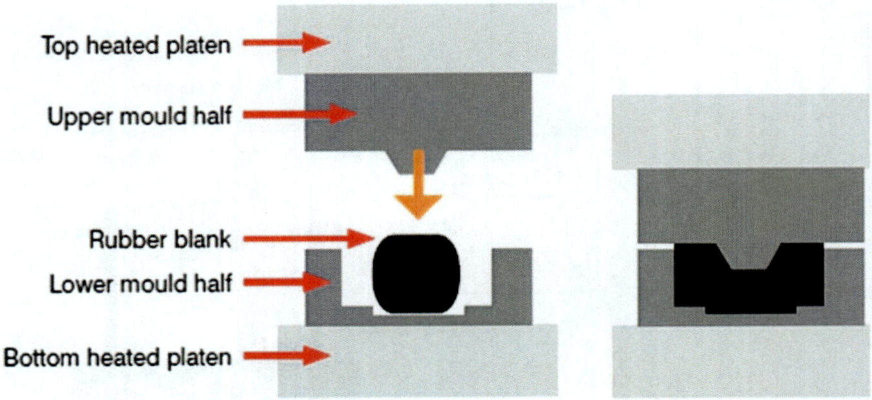

Fig. 3 Diagram showing compression molding process [42]

be manufactured using this process include the bumper boards, road wheels, refrigerator doors, automotive panels, electrical fixtures, machine guards, door panels, hoods, kitchen bowls and trays, control boxes, etc.

Prabu et al. [43] examined the mechanical properties of sisal-filled red mud and polyester-reinforced composites made from banana-fibre compression molding. Chen et al. [44] conducted an experiment on the structure and characteristics of prepolymer polyurethane composites and various soy products made using compression molding. The effect of fiber orientation on mechanical properties such as tensile strength and flexural strength of compression molded sisal fiber reinforced epoxy composites has been investigated [45].

4.4 Vacuum Bagging Technique

It is a method of producing composite materials in which the vacuum pressure is applied during the process of resin cure. To hold the resin and fibers in the desired location atmospheric pressure is used that consolidates the layers within the materials. The components are usually enclosed in an airtight container, and then a vacuum pump evacuates all the air out of the container, resulting in an even ambient pressure over the whole composite material. In the vacuum bagging process, different layers include mold, release agent, composite materials, peel ply, bleeder, release film, respirator and vacuum bag. Peel ply is used for the bonding purposes to create a clean surface. Releasing agent is important to keep the resin from sticking to the mold's surface when laminating a component. Usually, sealant tapes are used on either side of the bag to provide a vacuum-tightened seal between the surface of the mold and the lock. In general, the excess resin from the components is absorbed by a bleeder layer [46]. The release film is a perforated film that lets the trapped air and the volatiles escape. We use the respirator to create uniform pressure around

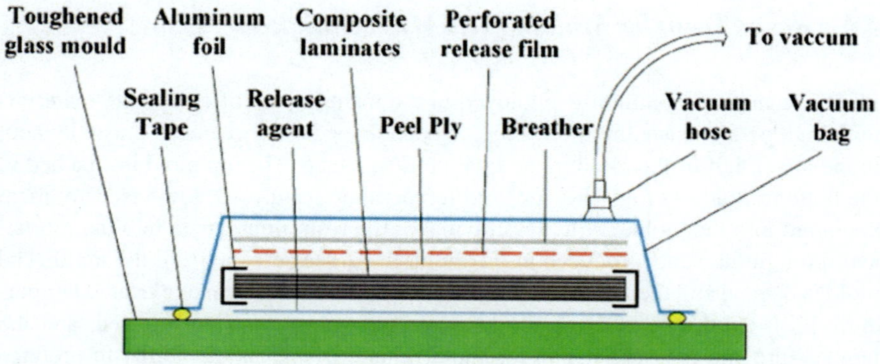

Fig. 4 Schematic diagram of vacuum bagging process

the component. Pressure and heat are applied for a given time. Pressure helps in the even distribution of matrix and enables the relation of matrix-fibers. The resulting composite materials can be cured either at room temperature or at elevated temperature. Many parameters such as type of reinforcement, quality of sheets used, resin viscosity and defined pressure influence the quality of composite materials manufactured by vacuum bagging process (Fig. 4). Advantages of this method such as; strong adhesion between layers leading to higher quality parts; achievable high fiber volume fraction; uniform matrix distribution; very low emissions as all of the materials are packed in a bag; improved component consolidation; and lower molding costs [47]. However, the disadvantages include that the process is not suitable for high production volume that breather cloth needs to be regularly replaced and that costly curing ovens are needed in the process. The vacuum bagging process is primarily used in the manufacture of components including large boat hulls, aircraft frames, race car components, and bathrooms [48].

4.5 Autoclave Molding Technique

With a few modifications, the approach is very similar to the vacuum bagging process. The need for autoclave molding plays a critical role in the manufacture of composite materials due to the high-quality requirements of composite materials for very demanding industries such as aerospace. This advanced process generally produces composite material having lightweight and void-free components. The material most commonly used in autoclave molding is composite prepreg. Autoclave curing offers even consolidation of prepreg laminates with lower voids. In this method the materials used include mold, peel ply, release agent, bleeder, breather and vacuum bag. In this process, prepreg layers are stacked to form the thickness of desired value above the molding plate with different fiber orientation.

4.6 Resin Transfer Molding (RTM) Technique

RTM is a method commonly employed in the manufacture of composite materials with high performance thermosetting., the pre-shaped reinforcement is usually kept in the lower half of the mold in the final product shape. The top mold is attached to the bottom mold. At high pressure and temperature, catalysed, low-viscosity resin is poured into the mold. High pressure makes the resin impregnate into the mortar, and through the vents produced in the mold, the gasses escape from the mold. The mold is opened and the laminate is removed after curing at room or elevated temperature [47]. Curing depends on the laminate size, the form of resin used, and the temperature and pressure within the mold. This process has the ability to produce rapidly large, complex, high-performance composite structures on both sides with good surface finish. Fiber shape, fiber material, viscosity of resin used, applied pressure, temperature of the mold etc., are the parameters affecting the consistency of the final composite component.

This process provides improved laminate density, a high fiber-to-resin ratio and excellent characteristics of strength to weight ratio. Certain advantages include good surface finish on sides, better product thickness control, shorter cycle time, low volatile emissions and better resin and fiber use was explained by Devillard et al. [49]. A few drawbacks of the process, however, are that the tooling cost is high, mold cavity typically limits the part size and the process is limited to low viscosity resins. This method is commonly used in the aerospace, sports, and automotive industries [3].

5 Characterization of Green Polymer Based Composite Materials

Various test methods are needed to check the physical, chemical as well as mechanical properties of the composites that have been manufactured. Standard shapes are necessary to evaluate the properties of reinforcement, matrix material, and laminates. Below are discussed various tests which are adopted for testing the material.

5.1 Reinforcement Testing

Reinforcement plays an important role that greatly affects the overall cumulative behaviour. The reinforcement's physical and chemical properties can be measured with various tests.

5.1.1 Chemical Tests

Various chemical tests that are conducted to determine the chemical properties of the reinforcement. Some of them are discussed below [50]:

(a) **Surface analysis**
For surface analysis, X-ray photoelectron spectroscopy is commonly used. It is also known as ESCA. It gives the total elemental analysis with the exception of hydrogen and helium. It is carried out at debts of (10–200 A°) from the surface and done in vacuum. It is very easy to interpret, highly informative and very sensitive in nature. However, being a sophisticated technique, it is quite expensive also [51].

(b) **Sizing content analysis**

This test is done to determine the weight content of sizing in fiber. Firstly, fibers with sizing are weighed; the sizing is removed and the fibers are weighed again. The weight content of sizing is then calculated. During this test, the sizing is generally removed by dissolving the fibers in a solution, which does not affect the fiber materials. Pyrolysis of sizing can also be done, without affecting the fiber material.

5.1.2 Physical Tests

Physical tests are carried out for determining various physical constants of the reinforced materials used.

(a) **Density**
Archimedes principle is applied to determine the densities of the fibre.

(b) **Weight/unit length**
A known length of fiber is weighed and the weight per unit length is calculated.

(c) **Filament diameter**

This step is relevant only for reinforcement with rounded cross section. A microscope is used to measure the diameter (D) of the fiber directly. It can also be calculated from the density of the fiber material, and the mass M of a known length L of the fiber.

We know that,

$$\text{(density)}, \rho = \frac{Mass}{Volume} = \frac{4M}{\pi.D^2L}$$

or,

$$\mathrm{D} = \sqrt{\frac{4M}{\rho.\pi.L}} \tag{1}$$

5.1.3 Electrical Conductivity

Resistance of each of the known length are measured using a resistance probe. Further, if the graph of resistance with length plotted. The graph obtained will be a straight line, and the slope of the line will give the resistivity. Since resistivity depends on the diameter, its value has to be mentioned.

5.1.4 Surface Roughness

The roughness of the surface is an important parameter for evaluating surface quality and aesthetic interest. The average surface roughness (Ra value) is one of the most commonly used surface roughness parameters, which defines the height of irregularities and provides an approximate measure of surface roughness and depth [52]. This is an absolute average ruggedness over a period of one sample [53]. This Ra value was used by researchers to study the impacts of different process parameters on the surface quality of FRP composites. Davim and Reis studied the impact of CF-reinforced machining parameters on surface quality of plastics [54].

5.1.5 Mechanical Tests

(a) **Tensile strength (MPa)**:

The ability of a material to oppose breaking force under tensile stress. It represents the maximum elongation of material under tension loads and shows a relation between force and elongation. Tensile stress is one of the most utilized and extensively measured properties of materials used in essential applications. Tensile strength avoids unnecessary material costs and achieves specific manufacturing goals [55]. It is measure by a universal testing machine and followed ASTM D3039 standard and the dimension of testing sample is 250 × 25 × 3 mm. tensile strength measures in the form of force per unit area (MPa or psi). Stress is normally measured in N/m^2 or Pa in metric system, such that 1 N/m^2 = 1 Pa. From the experiment the stress value is calculated by dividing the amount of force (F) exerted by the system in the axial direction by its cross-sectional area (A), which is measured before the experiment is performed. It is expressed, mathematically, in Eq. 2. The values of the strain which do not have units can be determined using Equation. In the equation L is the specimen's instantaneous length, and L_0 is the initial length (Fig. 5).

$$\sigma = \frac{F}{A} \tag{2}$$

$$\varepsilon = \frac{L - L_0}{L_0} \tag{3}$$

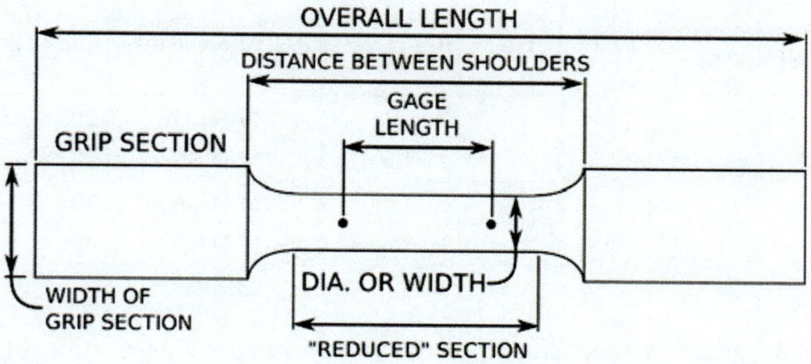

Fig. 5 Standard test specimen nomenclature

(b) **Flexural strength (MPa)**:

The ability of a material to resist deformation under the compression stress. It is the most utilized and widely measured properties of the material used in the essential application. Flexural strength measured by a universal testing machine and follow ASTM D790 standard. The dimension of testing sample 66 × 13 × 3 [56]. The flexural strength and modulus can be calculated using standard relation as follows.

The flexural stress, σ_f is calculated by:

$$\sigma_f = \frac{3FL}{2bd^2} \quad (4)$$

$$\sigma_f = \frac{FL}{\pi R^3} \quad (5)$$

Equations (4) and (5) is used for rectangular cross-section and circular cross section respectively.

The flexural strain, ε_f is calculated by:

$$\varepsilon_f = \frac{6Dd}{L^2} \quad (6)$$

The flexural modulus, E_f is calculated by:

$$E_f = \frac{mL^3}{4bd^3} \quad (7)$$

Here, ε_f.

- F = load at a given point on the load deflection curve, (N)
- L = Support span, (mm)
- b = Width of test beam, (mm)
- d = Depth or thickness of tested beam, (mm)
- D = maximum deflection of the center of the beam, (mm)

Table 2 Dimension of test sample as per ASTM standards

S. No.	Test	ASTM Standard	Dimension of the sample (mm)
1	Tensile	ASTM D3039	250 × 25 × 3
2	Flexural	ASTM D790	127 × 13 × 3
3	Impact	ASTM D256	80 × 10 × 4
4	Hardness	ASTM D785	20

- m = The gradient (i.e., slope) of the initial straight-line portion of the load deflection curve, (N/mm)
- R = The radius of the beam, (mm).

(c) **Impact strength**:

The ability of a material to resist a sudden applied force or resist the fracture under shock loading. It represents the material behaviour during impact load at high speed. Impact strength is tested through ASTM D256 dimension of testing sample 127 × 13 × 3 [57]. The unit of impact strength is J.cm. The impact test performs into two types Izod and Charpy methods. Impact behaviour is the most important factor in mechanical properties.

(d) **Hardness properties**:

The ability of a material to resist indentation load or scratching is evaluated as hardness. Hardness is the resistance of a material to scratch or localized plastic deformation. It is widely used in mechanical engineering and reduces the chance of failures. The hardness test is performing three types Brinell hardness test, Rockwell hardness test and Vicker hardness test.

The ASTM D785 specifications are generally followed during harness testing of composite materials based on polymers. Hardness test are conducted more frequently as compared to other mechanical testing for various reason as follows: Table 2 displays the dimension of various mechanical tests specimen.

a. They are simple and cheaper to perform- usually no special specimen needs to be prepared.
b. It is a non-destructive- the specimen is neither fractured nor excessively deformed.

The various hardness tests may be divided into three categories:

a. Elastic hardness.
b. Resistance to cutting or abrasion.
c. Resistance to indentation.

6 Application of Green Polymers Based Composite Materials

The polymer matrix nanocomposites with graphene and its derivatives as fillers have demonstrated tremendous potential for various important applications, such as electronics, renewable energy, aerospace, and the automotive. As mentioned above, 2-D graphene exhibits better electrical, mechanical and thermal properties and other unique features including higher aspect ratio and greater specific surface area compared to other reinforcements such as CNTs and carbon as well as Kevlar fibres. Below are some of the main applications discussed [58, 59]:

- *Aerospace industry*—About 50% of the portion of airspace is composite. The key advantages of weight reduction and simplification of assembly by composite parts. In the new helicopter production programme, the large-scale use of composites, military fighter jets, small and large civil transport aircraft, rockets, launch vehicles, and missiles. Composites such as rudder, spoilers, airbrakes, elevators, LG doors, engine cowlings, keel beams, rear bulkheads, wing ribs, main wings, turbine engine fan blades, propellers, internal parts, etc. are used to make various aircraft parts.
- *Automotive industry*—Composites are considered lighter in weight, heavier and more fuel-efficient automobiles. A composite consists of a high-resistance fiber (carbon or glass) in a matrix material (epoxy polymer) which, when combined, magnifies properties in comparison to the individual materials by itself. Many components are made from composite materials such as steering wheel, windshield, bench, roof, doors, walls, power absorber, instrument cluster, interior and exterior frame, leaf spring, wheels, engine cover, etc.
- *Medical applications*—A composite is a non-viable material that is used in medical devices and intended for biological system communication. The advances in synthetic materials, surgical practice, and methods of sterilization have, over the decades, allowed the use of composite material in many ways. Today a large variety of technologies and instruments are used in medical practice. Composites in the form of sutures, bone and joint replacements, vascular grafts, cardiac valves, intraocular lenses, dental implants, pacemakers, biosensors, artificial hearts etc. are commonly used for repairing and/or restoring the function of damaged or degenerated tissues or organs, improving function, helping to heal, rebuild or restore.
- *In sports equipment*—lightweight materials are used because they have ease of movement, resistance, low weight, low maintenance and longevity. Owing to its high shock resistance, natural materials, such as wood, were initially used but these materials had some disadvantages. The anisotropic architecture resulted in low resistance, and varying properties and high absorption of moisture attribute various deformations. The composite material has properties of resistance to fatigue breakage, superior thermal stability, friction resistance, abrasion resistance and vibration attenuation, light weight, high strength and high shape flexibility, and can be easily manufactured and shaped.

Fig. 6 Applications of green polymer based composite materials

- *Others application*—Composites have been used for industrial supports, homes, long span roof frameworks, tanks, refrigerator parts, etc. With composites which exhibit excellent marine resistance. With the aid of composite, for domestic and construction purposes, we render lightweight doors, windows, furniture, house, bridge etc. Figure 6 shows the application of green polymer based composite materials in different sectors.

7 Conclusions

To date, several advances have been made on substitute materials and natural resources such as plant fiber have been promoted as matrix material reinforcement. Most fibers have excellent basic strength and low density, such as sisal, jute, banana fibres etc. Because we know that natural fibers are biodegradable, the safe disposal of these fibers and their adverse impact on the environment would also occur. Natural materials reinforced by fibre have some excellent mechanical properties such as low density, stronger thermal insulation etc. In recent years, graphene comes as a potential filler material in comparison with other filler material that is available. Various researches has been going on using graphene as a filler material incorporate with polymer-based composites material in order to obtain better properties. These do have certain intrinsic properties, such as stronger damping and acoustic properties, due to their porous nature, which allows them to have a major role to industries which are dealing with musical instruments.

References

1. Summerscales J, Dissanayake NPJ, Virk AS, Hall W (2010) Fibers as reinforcements composites Part A 41:1329–1335
2. Thostenson ET, Ren Z, Chou T-W (2001) Compos Sci Technol 61(13):1899–1912
3. Célino A, Fréour S, Jacquemin F, Cesari P (2014) Frontiers in chemistry, vol 1, pp 1–12
4. Lee C, Wei X, Kysar JW, Hone J (2008) Science 321. https://www.sciencemag.org/content/321/5887/385
5. Scharfenberg S, Rocklin DZ, Chialvo C, Weaver RL, Goldbart PM, Mason N (2011) Appl Phys Lett 98
6. Kuilla T, Bhadra S, Yao D, Kim NH, Bose S, Lee JH (2010) Progr Polym Sci 35(11):1350–1375
7. Jang BZ, Zhamu A (2008) J Mater Sci 43(15):5092–5101
8. Hansma PK, Turner PJ, Ruoff RS. Nanotechnology 18(4)
9. Yu A, Ramesh P, Itkis ME, Bekyarova E, Haddon RC (2007) J Phys Chem C 111(21):7565–7569
10. Todor M-P, Bulei C, Heput T, Kiss I (2018) IOP Conf Ser Mater Sci Eng 294:1–9
11. Todor M-P, Bulei C, Kiss I (2018) IOP Conf Ser Mater Sci Eng 393:1–6
12. Shen M-Y, Chang T-Y, Hsieh T-H, Li Y-L, Chiang C-L, Yang H, Yip M-C (2013). J Nanomat. https://doi.org/10.1155/2013/565401
13. Sarker F, Karim N, Afroj S, Koncherry V, Novoselov KS, Potluri P (2018) ACS Appl Mater Interf 10:34502–34512
14. Sarker F, Potluri P, Afroj S, Koncherry V, Novoselov KS, Karim N (2019) ACS Appl Mater Interf 11(23):21166–21176. doi: https://doi.org/10.1021/acsami.9b04696
15. Chen J, Huang Z, Lv W, Wang CC (2017) Polym Compos. doi: https://doi.org/10.1002/pc.24433
16. Idumah CI, Hassan A (2017) J Nat Fib doi: https://doi.org/10.1080/15440478.2016.1277817
17. Kamaraj M, Anush Dodson E, Datta S (2019) Adv Compos Mater. doi: https://doi.org/10.1080/09243046.2019.1709679
18. Hallad SA, Banapurmath NR, Patil V, Ajarekar VS, Patil A, Godi MT, Shettar AS (2018) IOP Conf Ser Mater Sci Eng 376:012072. doi: https://doi.org/10.1088/1757-899X/376/1/012072
19. Ramesh Kumar SC, Shivanand HK, Vidayasagar HN, Nagabhushan V (2018) AIP conference proceedings 1943, 020115. doi: https://doi.org/10.1063/1.5029691
20. Bharadiya PS, Singh MK, Mishra S (2018) JOM. Miner Metal Mater Soc. https://doi.org/10.1007/s11837-018-3239-8
21. Ramesh Kumar S, Shivananad HK, Madegowda B (2018) Nanjundaswamy: (IJERMCE) 3(5). ISSN (Online) 2456-1290
22. Abd-Elhamid AI, Nayl AA, El Shanshory AA, Solimana HMA, Alyc HF (2019) RSC Adv 9:5770

23. Preetam SB, Satyendra M (2019) Res Dev Mater Sci 10(5). RDMS.000749.2019. doi: https://doi.org/10.31031/RDMS.2019.10.000749
24. Rathinasabapathi G, Jason Ronaldo M, Manoj SP, Mohamed Arif H, Mohamed Asif MI (2019) IRJET 6(4)
25. Costa UO, Nascimento LFC, Garcia JM, Monteiro SN, da Luz FS, Pinheiro WA, da Costa Garcia Filho F (2019) Polymers 11:1356: doi: https://doi.org/10.3390/polym11081356
26. Danga C-Y, Shena X-J, Niea H-J, Yanga S, Shena J-X, Yanga X-H, Shao-Yun Fu (2019) Compos B 168:488–495. doi: https://doi.org/10.1016/j.compositesb.2019.03.080
27. Bharath KN, Basavaraj S (2016) Sci Eng Compos Mater 23:2
28. Cherif C (2016) Textile materials for lightweight constructions. Technol Methods Mater Proper
29. Hausding J, Diestel O (2016) Chapter: Pre-impregnated textile semi-finished Products (Prepregs), pp 361–379
30. Zimniewska M, Myalski J, Koziol M, Mankowski J, Bogacz E (2012) J Nat Fib 9(4)
31. Hinrichsen J (2000) CAES. 19th European conference on materials for aerospace applications, Munich, pp 6–8
32. Avila AF, Morais DTS (2005) Compos Sci Technol 65(6):827–838
33. Otaka M (1997) Hand lay-up molding process. Japanese Patent JP 09,314,686
34. Patnaik A, Biswas S, Deo B, Satapathy A (2012) Composite polymer 32(4):665–674
35. Saba N, Fong TC, Liew CKE et al (2015) Development of reinforced polymer composites for natural fibre. Springer, Cham, pp 155–175
36. Mishra BV (2016) Composite polymer 37(1):270–278
37. Garcia EJ, Wardle BL, John Hart A, Yamamoto N (2008) Compos Sci Technol 68(9):2034–2041
38. Lamontia MA, Gruber MB, Smoot MA et al (1995) J Thermoplast Compos Mater 8(1):15–36
39. Misri S, Sapuan SM, Leman Z, Ishak MR (2015) Mater Des 65:953–960
40. Mertiny P, Ellyin F, Hothan A (2004) Compos Sci Technol 64(1):1–9
41. Cohen D (1997) Compos a Appl Sci Manuf 28(12):1035–1047
42. Toledo Filho RD, Silva FdA, Fairbairn EMR, de Almeida Melo Filho J (2009) Constr Build Mater 23(6):2409–2420
43. Prabu VA, Manikandan V, Uthayakumar M, Kalirasu S (2012) Mater Phys Mechan 15(2):173–179
44. Chen Y, Zhang L, Du L (2003) Ind Eng Chem Res 42(26):6786–6794
45. Chapman MR, Watson RM, Anderson DA (2006) US Patent 2,006,108,058
46. Liu L, Wang W, Liu Z (2006) Chinese Patent 1(775):498
47. Otten V, Siebrecht D (2006) US Patent 2,006,186,580
48. Sharma S, Kumar P, Chandra R (2016) J Compos Mater. 51:3299–3313
49. Devillard M, Hsiao KT, Gokce A et al (2003) J Compos Mater 37(17):1525–1541
50. Kokta BV, Maldas D, Daneault C, Wand P (1990) J Vinyl Technol 12(3):146–153
51. Lv C, Xue Q, Xia D (2012) Appl Surf Sci 258:2077–2082
52. Zahavi E, Torbilo V (1996) CRC Press, Boca Raton
53. Graham TS (2002) Industrial metrology: surfaces and roughness. Springer, Berlin
54. Davim JP, Reis P (2005) J Mater Process Technol 160:160
55. Termonia Y (1990) Tensile strength of discontinuous fibre-reinforced composites. J Mater Sci 259(11):4644–4653
56. Hill RG (1968) Exp Mech 8:75
57. Lantz RB (1969) Boron epoxy laminate test method. J Compos Mater 3:642
58. Stankovich S, Dikin DA, Dommett GHB, Kohlhaas KM, Zimney EJ, Stach EA, Piner RD, Nguyen ST, Ruoff RS (2006) Nature 442. doi: https://doi.org/10.1038/nature04969
59. Xu Y, Hong W, Bai H, Li C, Shi G (2009) Carbon 47. https://www.sciencedirect.com/science/article/pii/S0008622309005296.

Graphene and Its Derivatives: Fundamental Properties

Rukmani Sharma, Shreya Sharma, and Anjana Sarkar

Abstract The emerging science of graphene based engineered materials as functional surfaces for mechanics, electronics, sensing and other myriad applications is growing. This chapter brings state of the art graphene based materials to study the structure, preparation, properties, and implications. Derivatives studied in this chapter include graphene, graphene oxide, and reduced graphene oxide. The potential industrial implementation of the graphene synthesis methods is reviewed using the key criteria of cost, process condition, yield, scalability, product quality and environmental impact. In recent years nano graphene and its derivatives are finding their vast applications. Although, the properties are tremendous still there are some limitations as their toxicity is a major issue. Regardless of specific irregularities in various detailed test results and theories of toxicity mechanisms, results infer that the physicochemical properties, for instance, surface functionalities, coatings, charges, and structural imperfections of graphene may influence its in vitro/in vivo conduct just as its toxicity in biological frameworks. Henceforth, toxicity of graphene and its functionalized derivatives as a major issue is also dealt in later part of the chapter.

Keywords Graphene · Super capacitors · Toxicity · dentistry · Derivatives of graphene

1 Introduction

Graphene has found its unique position in numerous fields due to its astonishing properties. Graphene is a monolayer structure of graphite held by Vander Waal forces. Graphite is an abundant mineral, allotrope of carbon which is made of carbon atoms arranged in hexagonal shape. Carbon atoms in graphite are arranged in horizontal sheets. Graphite is dark grey to black in color, opaque and very soft due to the arrangement of carbon atoms. Graphite has high melting point therefore finds it application in making substance which requires high heat resistance. It is used in

R. Sharma (✉) · S. Sharma · A. Sarkar
Department of Chemistry, Netaji Subhas University of Technology, New Delhi, India
e-mail: rukmanis13@gmail.com

B. Sharma and P. Jain (eds.), *Graphene Based Biopolymer Nanocomposites*,
Composites Science and Technology,
https://doi.org/10.1007/978-981-15-9180-8_2

making electric motor brushes, batteries and nuclear reactor cores. Graphene is a sp^2 hybridized monolayer material that finds tremendous application in the development of nanocomposites, supercapacitors, optoelectronic devices, etc. [1–5]. Tahriri et al. in 2019 have reported that graphene and its derivatives can play an important role in the field of dentistry i.e. like using graphene many applications like tooth whitening, membrane for bone regeneration, dental implants can be done [6]. Le et al. propose the sandwich-like structure of Ni, Co incorporated with rGO. All the result recommends that sandwich like LDH/RGO composites auspicious electrode for practical operation in supercapacitors [7]. A high-performance red QLEDs have been synthesized by Lei et al. UV ozone treated GO (RGO) and a PEDOT: PSS using stepwise bilayer hole injection layer [8].

1.1 Graphene

The basic elemental atom present in graphene is 15th most abundant metal in earth crust 'carbon'. Carbon has four electrons in the outermost shell so it can share its outermost electron through covalent bond with different elements. Carbon can be hybridized in three different states which are *sp*, sp^2 and sp^3. Graphene is a two dimensional structure having sp^2 hybridization. In graphene, orbitals involved in sigma bonding are p_x, p_y and in plane while which is responsible for the (π) bonding is p_z and is perpendicular to the plane. The orbitals which in plane form sigma bond having interatomic length ~1.42 Å which make the C–C bond stronger and results in the strong mechanical resistance to graphene as shown in Table 2 [9]. Additionally, the π band is half-filled causing zero band gaps between the valence and conductive band which results in the free movement of electron making it highly conductive. p_z orbital which is perpendicular to the plane makes weak π bond resulting in weak Vander Waals interaction between the monolayers of graphene which allows them to move under very weak share [10, 11]. In past it was believed that the two dimensional structure cannot exist as it is thermally unstable but Geim et al. invented a single layer of graphene in their lab in 2004 who proposed that it is a carbon monolayer that is arranged in the shape of honeycomb as shown in Fig. 1 [10]. The invention of graphene has unfastened the new dimensions to the field of material science. An ideal structure of graphene is highly conductive with zero band gap [12], high thermal conductivity [13] and good tensile strength [14] at room temperature. In spite of all these tremendous properties graphene cannot be utilized alone which limits its potential application. So, there is need for incorporating graphene with different materials like functionalized thin films [15], fibres [16], and coatings [17] for the better use of graphene at industry level.

Graphene can also be used as the nanofiller based carbon which enhances the mechanical and thermal properties for the development of high-performance polymer matrix based nanocomposites [20]. Due to the high electric conductivity and mechanical flexibility, it can find the substitution of metal conductors. Graphene has these outstanding properties due to its high aspect ratio and single thin layer. It is known

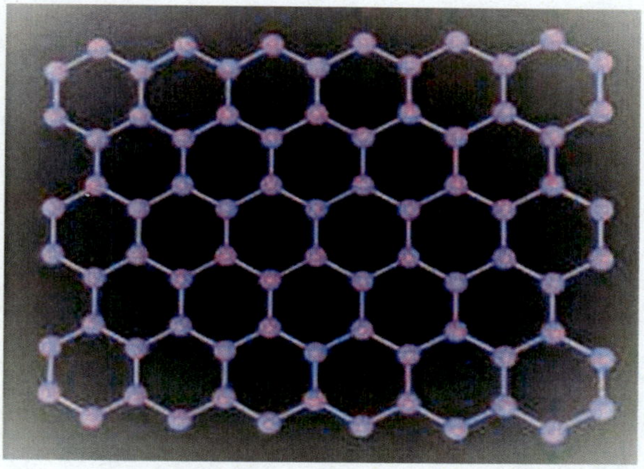

Fig. 1 Depicting the 3-D structure of graphene (reprinted with the permission of Bhasha et al. [18, 19]

that graphene agglomerates due to its high surface area when coming in contact with the polymer. Therefore, for the reduction of this agglomeration functionalization of graphene is required which can be performed by oxidizing or by reducing agents.

1.1.1 Fabrication Methods of Graphene

Different fabrication techniques are utilized for graphene including the mechanical exfoliation of graphite and chemical vapor deposition (CVD), Chemical reduction, etc. discussed in subsequent sections.

a. **Chemical Vapor Deposition (CVD) Method**

   One of the most appropriate methods for the preparation of graphene is CVD method (Fig. 2). In this technique graphene is fabricated by graphite target or catalytic decomposition of hydrocarbon on the surface of metal catalyst. In this, the metal debris is very low that is the major advantage of this technique [21]. This fabrication is best for the preparation of single- layer graphene and can also be used for heteroatoms doped graphene nanostructure in which graphene can be doped with sulphur, nitrogen, phosphorus, fluorine, or bromine, etc. which will improve the catalytic activity enzymatic application, energy conversion [22].

b. **Mechanical Exfoliation**

   It is one of the most common method in which graphene is mechanically exfoliated from graphite. In this method, the graphene is to be peeled from the bulk of graphite layer by layer. To overcome the resistance offered by the Van der Waal attraction in between the adjacent flakes of graphene. There are two paths for mechanical exfoliation one is normal and the other one is lateral force. In normal force the Van der Waal force can be overcome by Scotch tape [23, 24]. Graphite has ability of self-lubricating in lateral direction so lateral force can also use to for peeling the two layer of graphene.

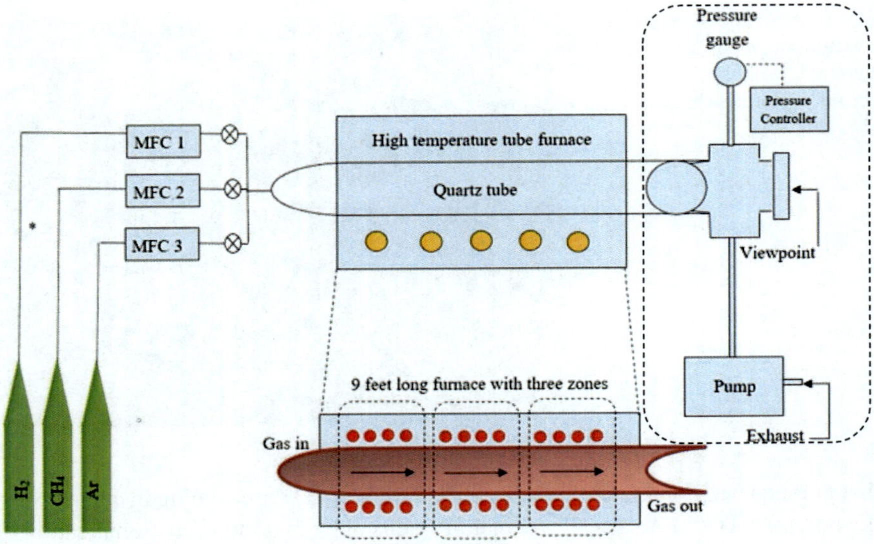

Fig. 2 Apparatus set up for chemical vapor deposition of graphene

c. **Chemical Reduction**
100 mg GO was taken in 250 ml of round bottom flask and 100 ml water is added into it which yields into an inhomogeneous dispersion of yellow brown color. Then the solution was dispersed in a sonicated until the solution becomes clear. Then a particular amount of reducing agent is added and then the solution is heated in the oil bath under cool bath condenser for 24 h. GO will start reducing and precipitates out as black solid. The solution was filtered and washes thoroughly with water and methanol. The product is dried in the air [19].

d. **Unzipping Carbon Nanotubes**
Graphene nano ribbons which are thin and elongated strips of graphene can be fabricated by using this method. In this approach initial material is multilayer or single-layer nanotubes. This technique is known as CNTs unzipping. An electric field is applied to the nanotubes using tungsten electrode. From non-contact end nanotubes start unzipping and graphene nanoribbons are formed. From this method highly pure, defect free graphene is produced [25].

e. **Liquid Phase Exfoliation**
This technique is introduced in 2008 and is one of the widely used fabrication technique [26]. In LPE technique three main steps are there for fabrication: (i) dispersion of graphite in a suitable solvent, (ii) exfoliate and last (iii) purification of filter product [27]. Figure 3 depicts the liquid phase exfoliation of graphene. In some of the discussed methods of fabrication mechanical exfoliation of graphite gives very low yield while the CVD technique is used for the production of graphene at large scale industries. Figure 4 gives a summary of the fabrication techniques discussed in the chapter (Table 1).

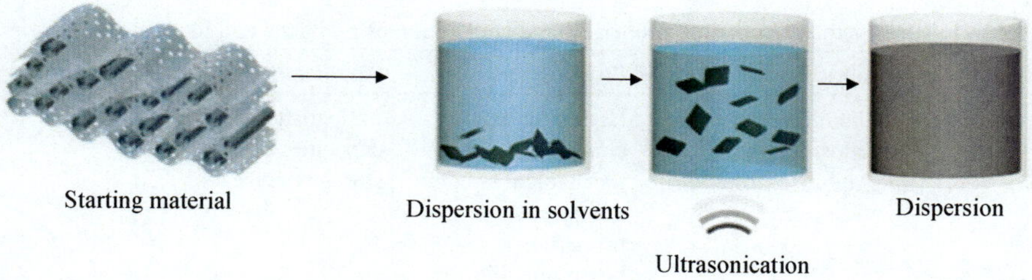

Fig. 3 Liquid phase exfoliation of graphene

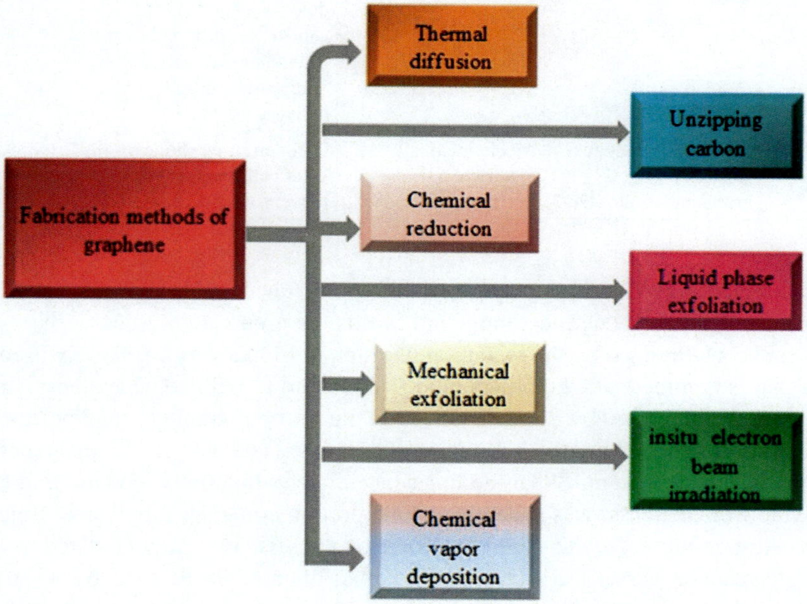

Fig. 4 Flowchart depicting fabrication methods of graphene

1.2 Graphene Oxide (GO)

GO is functionalized graphene containing chemical groups of oxygen, has recently intrigued researchers due to its extraordinary properties like mechanical stability, ambulatory properties in electrical and optical, huge surface area. The functional group is attached to the surface like epoxy, carboxyl and hydroxyl group which makes it easily available for reacting with the different molecules and gives extraordinary properties to the material [28]. Graphite exfoliation is the most trivial approach followed by the addition of strong oxidizing agents to yield GO which is a non-conductive hydrophilic carbon [23, 29–31]. Graphene's previously contiguous aromatic lattice is distorted for GO by alcohols, epoxide, ketones, carboxylic group

Table 1 Reports the preparation, properties and application of graphene and its derivatives

Compound	Fabrication methods	Properties	Applications	References
Graphite	Chemical vapour deposition and graphite exfoliation	High electrical conductivity, high mechanical strength, good optical properties, biocompatible	In dentistry, super capacitor, energy storage devices	[6, 12]
GO	Oxidation of graphite followed by exfoliation	Antimicrobial, magnetic properties, highly hydrophilic, insulator	Anti-static coatings, automotive industry, corrosion resistant devices, super capacitors, solar cells, fillers in polymer composites, drug delivery	[7, 13, 14]
RGO	Thermal or chemical reduction of grapheme oxide (GO)	Mechanical and electrical performance	Light emitting diodes, solar cell devices, biosensors	[8, 15]

and carbonyl [32–34]. GO contain hydroxyl and epoxy functional group on the basal plane of carbon while at edges carbonyl and carboxylic group are present.

GO can be synthesized by the oxidation of graphite from a very commonly used method that is hummers and hummers modified method [23, 33]. The synthesis of GO was first demonstrated by Brodie in 1859. To the slurry of graphite small portion of $KMNO_4$ (potassium chlorate) in fuming nitric acid was added [34]. Staudenmaier has improved this method in 1898 by adding concentrated sulphuric acid with fuming nitric oxide while chlorate was added in a small portion at regular intervals of time in the reaction mixture. By this slight modification the yield of highly oxidized GO in a single reaction vessel [35]. Hummer reported the method in 1958 by which graphite can be oxidized to GO which is commonly used today (Fig. 5). Graphite is oxidized in a strongly acidic medium for this Potassium permanganate ($KMnO_4$) and sodium nitrate ($NaNO_3$) in concentrated sulphuric acid (H_2SO_4) [23]. All three

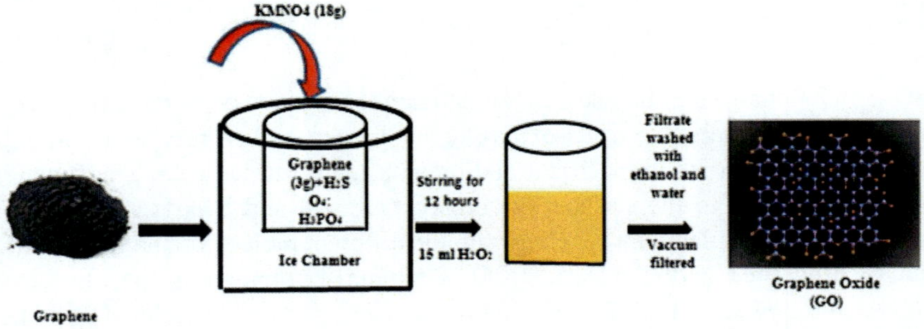

Fig. 5 Synthesis of GO from graphene using Hummer's methods

processes indicated above generated toxic gases like ClO_2, N_2O_4, NO_2, and the latter is also explosive. Oxidation of graphite drastically distorts the graphite's network and interspacing is increased. For graphite, it is 0.335 nm while for GO is more than 0.625 nm [36]. Dispersion of GO in an organic solvent and water aids due to the presence of hydroxylic and carboxylic groups [37]. In GO the main reactive present site is epoxy and the functionalization takes place from the most reactive site.

1.3 Reduced Graphene Oxide (RGO)

RGO is the reduced form of GO in which different oxygen groups are reduced. Different approaches have been used for the reduction process like chemical reduction, thermal reduction, microwave-assisted reduction, solvothermal reduction, and photo-reduction. GO is reduced to enhance the honeycomb hexagonal lattice which has been distorted due to the oxidation of graphene and the electric properties enhance [38, 39]. GO is an insulator but after reduction, the conductivity can be increased. Several agents have been reported which can reduce GO chemically which are hydrazine, sodium borohydride, alcohol, hydroxides, metals, redox active sulfur species, reductive acids, or even enzymatic reduction.

1.3.1 Fabrication Method of RGO

RGO can be fabricated by different methods but mostly used are chemical functionalization of GO and thermal reduction of GO.

a. **Chemical Functionalization of Graphene Oxide**
   Chemical functionalization many reducing agents like hydrazine [18, 40], alcohol [41], sodium borohydride [42, 43], sodium hydroxide [44] etc.

By Using Hydrazine
By using hydrazine hydrate GO can be reduced chemically. Figure 6 depicting the formation of RGO from GO. 1 g GO was taken with 100 ml deionized water in a beaker. For uniform dispersion of the solvent, the mixture was sonicated for 2 h. After ultra-sonication, 0.4 ml of hydrazine hydrate is added and kept for magnetic stirring at 60 °C for 12 h. The mixture was filtered using Vaccum suction and washed with ethanol and water several times. The filtrate was kept in a vaccum oven for 24 h at 40 °C to obtain dry products [45].

By Using Alcohol
Su et al. described the process of fabricating RGO alcohol as reducing agent. In this, GO is heated with 20% H_2/Ar under high temperature with alcohol. In this method the RGO is prepared is highly conducting in nature [46].

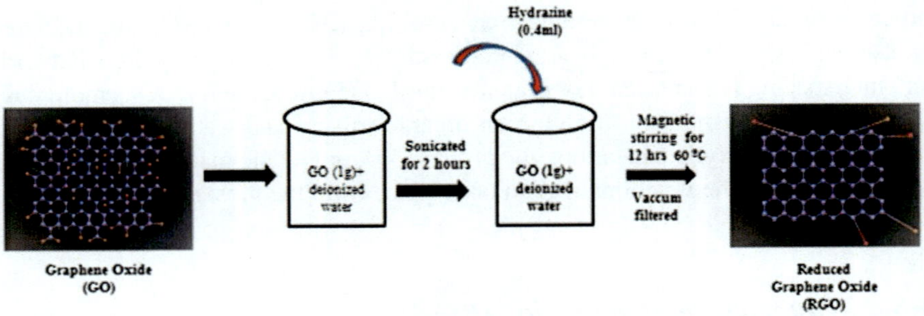

Fig. 6 Synthesis of RGO from GO using Hummer's method

By Using Sodium Borohydride

A two-step reduction process is involved in this reduction as described by Geo et al. in first step deoxygenation in RGO structure using sodium borohydride while in other step dehydration using conc. Sulphuric acid [47].

1.3.2 Thermal Reduction of GO

Another fabricating method of RGO is thermal reduction of GO. In this GO can be heated in different atmosphere like Argon (Ar), hydrogen (H_2), Ammonia (NH_3) and high vaccum or different heating methods like electric heating, heated AFM tip, laser heating, plasma heating, etc. [38]. RGO obtained by heating has high electrical conductivity. Although the reduction level of RGO can be controlled by duration, atmosphere of gas and heating temperature. Hydrazine gives good yield after reduction but is very toxic and its limits its use in mass production and the application in field of biomedical. GO can also be reduced thermally at the temperature of 1000 °C, while researchers have also opted for the less temperature process. For the eradication and removing hazardous chemical researchers has also shifted their focus on natural reducing like same green agents are being explored for the functionalization of graphene like green tea polyphenol [48], bacteria [41], amino acid [49], alcohol [50], serum albumin [51], vitamin C [51], etc. Zhu et al. have used glucose, fructose for the reduction of graphene oxide in the aqueous solution of ammonia [52].

There is no specific chemical reduction method which can precise for the production at large scale as they are toxic in nature while in thermal reduction level can be controlled so by combing chemical and thermal method high reduction level can be achieved (Table 2).

Table 2 Depicting the properties of graphene, GO, RGO

Characteristics	Graphene	Graphene oxide	Reduced grapheme oxide
Young's modulus (TPa)	1–2	2–4	2–4
Thermal conductivity (W/mK)	4.84–5.30 × 10^{-3}	1500–5800	0.1–2
Electrical conductivity(S/m)	7200	1–5 × 10–5	2–7 × 10–3
Electron mobility (cm^2 v^{-1} s^{-1})	10,000–50,000	Insulator	0.05–200
C: O ratio	No oxygen	2–3	8–246
Production cost	High	Low	Low

2 Applications

Graphene and its derivatives find its application in many fields due to its outstanding properties like huge surface area, magnificent optical and mechanical properties, etc. They are used in sensors, supercapacitors, energy storage devices. Functionalized graphene may find its application in nerve conduit, biosensors and biomedical

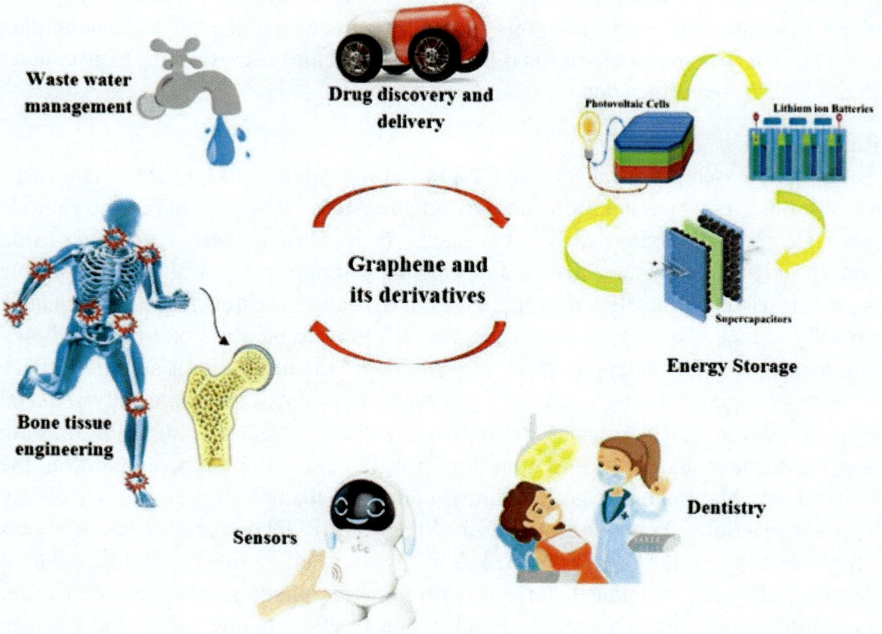

Fig. 7 Depicting applications of graphene and its derivatives

devices [53]. Figure 7 illustrates different implications of graphene and its derivatives. Detailed properties are described as:

Solar Cells

Nowadays, reduced graphene oxide films are being applied for many technological purposes, even in renewable energy sources. This material can be considered less dangerous to health then others commonly used in photovoltaic modules, like indium (In), cadmium (Cd) and others. Graphene and its derivatives are being considered good candidates to replace traditional materials in thin films solar cells especially because of their adequate electrical and optical properties, which enable the application of graphene films as transparent and conductive layers and also in back contact in these devices. In this work, two different approaches were used. The first one was to produce graphene oxide films by a spray-coating automatized technique and then thermally reducing, resulting in thermally reduced graphene oxide (trGO). The second was to produce coatings from a dispersion of chemically reduced graphene oxide (rGO), using the same spray method. The films were characterized with respect to thickness, morphology, transmittance and resistivity, and the best candidates to work as front contact and back contact in CdS/CdTe solar cells were considered candidates for solar cell application. To improve the properties of front contact samples [3, 4], the best approach was the reduction in H_2 atmosphere, while for the back contact it was to cover rGO films by spray with graphene flakes produced by liquid-phase exfoliation of graphite. In this way, rGO films were adequate to back contact (65.6 Ω/□ of sheet resistance), and rGO films showed the best results for front contact (5323 Ω/□ of resistance and 61.9% of transmittance) [54].

Batteries

A graphene-based porous electrode for a lithium–oxygen (Li–O_2) battery is investigated for use in next generation energy storage systems. The porosity of the cathode electrode in Li–O_2 batteries is a key factor in increasing their oxygen diffusion rate and electrochemical activity, and enables a longer cycle life [55]. A simple approach has been developed to increase the SIB anode performance by preparing partially reduced holey graphene oxide through low temperature reduction of holey graphene oxide. The enlarged interlayer spacing (0.434 nm) ensures sufficient space for energy storage through sodium ions, while the pores on the partially reduced graphene oxide sheets enhance the insertion/extraction rate by shortening the ion transfer distance and providing more ion insertion/extraction sites. In addition, the residual oxygen-containing groups further contribute to the energy storage capability by redox reactions. As anode materials for SIBs, the HRGO prepared at 300 °C shows a high capacity of 365 mAh g^{-1} at 0.1 A g^{-1} with excellent rate capability and long cycle stability, much better than the control intact graphene sample prepared under the same temperature. This work provides a facile and scalable strategy to produce high-performance anode materials for sodium ion-based energy storage devices.

Super Capacitors

Super capacitors find their unique application in industries, transport and communication. They are formed by ions accumulating on the surface of solid, liquid and ionic liquid. So for this reason electrode should be inert chemically, open porous structure and must possess great surface area. Minimum resistance should be offered and mechanical durability should be providing the maximum number of charge and discharge cycle at highest value current. Yun et al. performed experiments to enhance the performance of super capacitor which can be fabricated with modification of graphene and its derivatives which can be further used energy storage devices [56].

Dentistry

Tehiri et al. described the use of graphene and its derivatives in dentistry [6]. Implantation of teeth is done by using titanium but a coating of GO on titanium can improved the whole property with biocompatibility. Graphene material is being used as the material for whitening of teeth. Graphene improves the capability of the membrane for preventing soft tissue cells from infiltrating into the new bones [57].

Biosensors

Graphene and its derivatives are use fluorescence resonance energy transfer (FRET) characterization which works precisely as biosensors [58].

Water Waste Treatment

Graphene and its derivatives have a different type of functional group which offers the application water waste treatment as it can absorb various organic and inorganics pollutants and dyes. GO membranes are a potential candidate for the desalination and sewage purification as it acquires great mechanical strength, hydrophilic property and exceptional flexibility [59].

Heavy Metal Removal

As the industrialization has grown recently, has generated many toxic heavy metals into the environment which has created the problem to the environment. So, the removal and separation of heavy metals are required for the safety of human and environment safety. Graphene and its derivatives have shown a tremendous physio-chemical property having high surface area, good absorbent property due to various oxygen groups present which helps in the elimination of heavy metals ions [60] such as lead (Pb(II)) [61], copper (Cu(II)) [62] etc.

Bone Tissue Engineering

The essential need for bone tissue engineering is to refine and repair defects in the bone based on bioactive biomaterials having different properties that can activate and bolster the formation of bone tissues. Graphene and its derivatives find their places in this very application. They support the cell growth and are cell growth and procreation, attachment of cell, bone development pathway [63]. Heshmatpour et al. have modified GO composites with polyvinylpyrrolidone, chitosan and polyethylene glycol. The studies show that cell growth, proliferation, and viability have enhanced after 14 days [64].

Energy Storage Devices
Electronic gas sensors and supercapacitors have been fabricated with the CA-rGO and show good performance, which demonstrates the potential of CA-rGO for sensing and energy storage applications.

Drug Delivery
Graphene and its derivatives are being used in the delivery of drugs. They are being used as the drug carriers. Wei et al. has prepared a drug carrier RGPD which is non-cytotoxic, has a good stability, possesses appropriate photothermal performance [65]. The functionalized graphene mainly acts as nanofillers in polymer composites for improving the properties and even a small amount of functionalized graphene can improve the optical, electrical, thermal, mechanical and magnetic properties assuredly.

3 Toxicity of Graphene and Its Derivatives

As it has discussed the material used for the fabrication is toxic. Many researchers are being conducted to diminishing the biological toxicity. Although the studies have not provided an authenticated result of material toxicity as different techniques are involved in the fabrication that has led to momentous difference in size, density, functional groups of the graphene and its derivatives [66]. In GO surface reactivity give rise to toxicity [67]. GO is toxic to bacteria, fauna and cells in human [68–70]. They also show a high potential of bioaccumulation in the organisms and also in the food chain in aquatic ecosystem.

4 Effect of Nanosize on Graphene and Its Derivatives

Nano graphene and its derivative are advanced class in which the properties of the material take a marvelous change in its properties. They are mainly used for the biomedical application as they possess intrinsic optical properties, larger surface area, functionalization is easy and most important its small size.[71]. These are being used as drug carriers. Nano sheets of GO and RGO are being used for the surface coating the antimicrobial activity increases and also attributes to enhancement of oxidative mechanism which is linked to higher deficiency of smaller nanosheets [72, 73].

5 Limitations of Graphene and Its Derivatives

Researcher are trying to explore a feasible route for the fabrication of graphene [68–70] and its derivatives mainly RGO at mass production and at effective course so, at industrial level it can be easily available. As it is discussed in every synthesis that the product obtained is acidic in nature so it is one of the limitation that the products obtained is washed every time with water and methanol or ethanol for achieving neutral product. Products which are used in the fabrication are toxic for environment and also for the application like biomedical as it is directly related to the drug delivery, bone tissue engineering so that is one of the major limitation. Researchers are trying to replace the toxic reagents with the eco-friendly reagents.

6 Conclusion

Graphene and its derivatives are a boon for the material science. The application of materials incorporated with graphene and its derivative is really vast. From super capacitors to dentistry in every field it has been used. The nano material of graphene and its derivatives find their applications in biomedical field so, toxicity of the material matters. Although, for the reduction of toxicity greener approach are being practiced in lab. Scientists are also trying to approach fabrication method of RGO for the production at large scale without creating harm to the environment and the application regarding biomedical field. It is very important to maintain the acidity of graphene and its derivatives which limits its use.

References

1. Ramanathan T, Abdala AA, Stankovich S, Dikin DA, Herrera-Alonso M, Piner RD, Adamson DH, Schinepp HC, Ruoff RS, Nguyen ST, Aksay IA, Prud-Homme RK, Brinson LC (2008) Functionalized graphene sheets for polymer nanocomposites. Nat Nanotechnol 3(6):327–331
2. Sharma B, Gautam S, Shekhar S, Sharma R, Rajawat DS, Jain P (2019) Facile synthesis of poly (vinyl alcohol) nanocomposite & its potential application to enhance electrochemical performance. Polym Test 74:119–126
3. Wu S, He Q, Tan C, Wang Y, Zhang H (2013) Graphene-based electrochemical sensors. Small 9(8):1160–1172
4. Sharma B, Shekhar S, Gautam S, Jain P (2018) Dynamic shear rheology behavior and long term stability kinetics of reduced graphene oxide filled poly (vinyl alcohol) biofilm. Polym Test 69:583–592
5. Sharma B, Shekhar S, Gautam S, Sarkar A, Jain P (2018) Nanomechanical analysis of chemically reduced graphene oxide reinforced poly (vinyl alcohol) nanocomposite thin films. Polym Test 70:458–466
6. Tahriri M, Del Monico M, Moghanian A, Yaraki MT, Torres R, Yadegari A, Tayebi L (2019) Graphene and its derivatives: opportunities and challenges in dentistry. Mater Sci Eng C 102:171–185
7. Le K, Wang Z, Wang F, Wang Q, Shao Q, Murugadoss V, Wu S, Liu W, Liu J, Gao Q, Guo Z (2019) Sandwich-like NiCo layered double hydroxide/reduced graphene oxide nanocomposite cathodes for high energy density asymmetric supercapacitors. Dalton Trans 48(16):5193–5202
8. Lei Y, Zhao Y, Zhang Q, Xiong Z, Chen L (2020) Highly efficient and bright red quantum dot light-emitting diodes with balanced charge injection. Organ Electron 81:105683

9. Sharma B, Malik P, Jain P (2018) Biopolymer reinforced nanocomposites: a comprehensive review. Mater Today Commun 16:353–363
10. Geim AK, Geim AK, Novoselov KS (2010) The rise of graphene. Nanosci Technol Collect Rev Nat J, 11–19
11. Yang G, Li L, Lee WB, Ng MC (2018) Structure of graphene and its disorders: a review. Sci Technol Adv Mater 19(1):613–648
12. Balandin AA, Ghosh S, Bao W, Calizo I, Teweldebrhan D, Miao F, Lau CN (2008) Superior thermal conductivity of single-layer graphene. Nano Lett 8(3):902–907
13. Thompson TE, Falardeau ER, Hanlon LR (1977) The electrical conductivity and optical reflectance of graphite-SbF5 compounds. Carbon 15(1):39–43
14. Novoselov KS, Geim AK, Morozov S, Jiang D, Katsnelson MI, Grigorieva I, Dubonos SV, Nguyen ST, Firsov AA (2005) Two-dimensional gas of massless Dirac fermions in graphene. Nature 438(7065):197–200
15. Stankovich S, Dikin DA, Piner RD, Kohlhaas KA, Kleinhammes A, Jia Y, Wu Y, Nguyen ST, Ruoff RS (2007) Synthesis of graphene-based nanosheets via chemical reduction of exfoliated graphite oxide. Carbon 45(7):1558–1565
16. Novoselov KS, Jiang D, Schedin F, Booth TJ, Khotkevich VV, Morozov SV, Geim AK (2005) Two-dimensional atomic crystals. Proc Natl Acad Sci 102(30):10451–10453
17. Novoselov KS, Geim AK, Morozov SV, Jiang D, Zhang Y, Dubonos SV, Grigorieva IV, Firsov AA (2004) Electric field effect in atomically thin carbon films. Science 306(5696):666–669
18. Bhasha S, Singh S, Jain P, Malik P (2015) Synthesis and characterization of nanocrystalline zinc oxide thin films via green chemistry. J Nanoanal 2(1):10–16
19. Sharma B, Shekhar S, Malik P, Jain P (2018) Study of mechanism involved in synthesis of graphene oxide and reduced graphene oxide from graphene nanoplatelets. Mater Res Exp 5(6):065012
20. Sato S, Kakushima K, Ahmet P, Ohmori K, Natori K, Yamada K, Iwai H (2011) Structural advantages of rectangular-like channel cross-section on electrical characteristics of silicon nanowire field-effect transistors. Microelectron Reliab 51(5):879–884
21. Dong Z, Jiang C, Cheng H, Zhao Y, Shi G, Jiang L, Qu L (2012) Facile fabrication of light, flexible and multifunctional graphene fibers. Adv Mater 24(14):1856–1861
22. Higginbotham AL, Lomeda JR, Morgan AB, Tour JM (2009) Graphite oxide flame-retardant polymer nanocomposites. ACS Appl Mater Interfaces 1(10):2256–2261
23. Hummers WS Jr, Offeman RE (1958) Preparation of graphitic oxide. J Am Chem Soc 80(6):1339–1339
24. Lerf A, He H, Forster M, Klinowski J (1998) Structure of graphite oxide revisited. J Phys Chem B 102(23):4477–4482
25. Bhuyan MSA, Uddin MN, Islam MM, Bipasha FA, Hossain SS (2016) Synthesis of graphene. Int Nano Lett 6(2):65–83
26. Hernandez Y, Nicolosi V, Lotya M, Blighe FM, Sun Z, De S, McGovern IT, Holand B, Byrne M, Gun'Ko YK, Boland JJ, Niraj P, Duesberg G, Krishnamurthy S, Goodhue R, Hutchison J, Scardaci V, Ferrari AC, Coleman JN (2008) High-yield production of graphene by liquid-phase exfoliation of graphite. Nat Nanotechnol 3(9):563
27. Ciesielski A, Samorì P (2014) Graphene via sonication assisted liquid-phase exfoliation. Chem Soc Rev 43(1):381–398
28. Li F, Jiang X, Zhao J, Zhang S (2015) Graphene oxide: a promising nanomaterial for energy and environmental applications. Nano Energy 16:488–515
29. Dreyer DR, Park S, Bielawski CW, Ruoff RS (2010) The chemistry of graphene oxide. Chem Soc Rev 39(1):228–240
30. He H, Klinowski J, Forster M, Lerf A (1998) A new structural model for graphite oxide. Chem Phys Lett 287(1–2):53–56
31. Uhl FM, Wilkie CA (2004) Preparation of nanocomposites from styrene and modified graphite oxides. Polym Degrad Stab 84(2):215–226
32. Stankovich S, Piner RD, Chen X, Wu N, Nguyen ST, Ruoff RS (2006) Stable aqueous dispersions of graphitic nanoplatelets via the reduction of exfoliated graphite oxide in the presence of poly (sodium 4-styrenesulfonate). J Mater Chem 16(2):155–158

33. Marcano DC, Kosynkin DV, Berlin JM, Sinitskii A, Sun Z, Slesarev A, Lawrence B, Lu W, Tour JM (2010) Improved synthesis of graphene oxide. ACS Nano 4(8):4806–4814
34. Brodie BC (1859) XIII. On the atomic weight of graphite. Philos Trans R Soc Lond 149:249–259
35. Staudenmaier L (1899) Verfahren zur darstellung der graphitsäure. Ber Dtsch Chem Ges 32(2):1394–1399
36. Hontoria-Lucas C, López-Peinado AJ, López-González JDD, Rojas-Cervantes ML, Martin-Aranda RM (1995) Study of oxygen-containing groups in a series of graphite oxides: physical and chemical characterization. Carbon 33(11):1585–1592
37. Bourlinos AB, Gournis D, Petridis D, Szabó T, Szeri A, Dékány I (2003) Graphite oxide: chemical reduction to graphite and surface modification with primary aliphatic amines and amino acids. Langmuir 19(15):6050–6055
38. Mao S, Pu H, Chen J (2012) Graphene oxide and its reduction: modeling and experimental progress. RSC Adv 2(7):2643–2662
39. Priyadarsini S, Mohanty S, Mukherjee S, Basu S, Mishra M (2018) Graphene and graphene oxide as nanomaterials for medicine and biology application. J Nanostruct Chem 8(2):123–137
40. Saravanan N, Rajasekar R, Mahalakshmi S, Sathishkumar TP, Sasikumar KSK, Sahoo S (2014) Graphene and modified graphene-based polymer nanocomposites—a review. J Reinf Plast Compos 33(12):1158–1170
41. Dreyer DR, Murali S, Zhu Y, Ruoff RS, Bielawski CW (2011) Reduction of graphite oxide using alcohols. J Mater Chem 21(10):3443–3447
42. Si Y, Samulski ET (2008) Synthesis of water soluble graphene. Nano Lett 8(6):1679–1682
43. Wang R, Sun J, Gao L, Xu C, Zhang J, Liu Y (2011) Effective post treatment for preparing highly conductive carbon nanotube/reduced graphite oxide hybrid films. Nanoscale 3(3):904–906
44. Fan X, Peng W, Li Y, Li X, Wang S, Zhang G, Zhang F (2008) Deoxygenation of exfoliated graphite oxide under alkaline conditions: a green route to graphene preparation. Adv Mater 20(23):4490–4493
45. Sharma B, Malik P, Jain P (2019) To study the effect of processing conditions on structural and mechanical characterization of graphite and graphene oxide-reinforced PVA nanocomposite. Polym Bull 76(8):3841–3855
46. Jiang L, Shen XP, Wu JL, Shen KC (2010) Preparation and characterization of graphene/poly (vinyl alcohol) nanocomposites. J Appl Polym Sci 118(1):275–279
47. Gao W, Alemany LB, Ci L, Ajayan PM (2009) New insights into the structure and reduction of graphite oxide. Nat Chem 1(5):403
48. Akhavan O, Kalaee M, Alavi ZS, Ghiasi SMA, Esfandiar A (2012) Increasing the antioxidant activity of green tea polyphenols in the presence of iron for the reduction of graphene oxide. Carbon 50(8):3015–3025
49. Gao J, Liu F, Liu Y, Ma N, Wang Z, Zhang X (2010) Environment-friendly method to produce graphene that employs vitamin C and amino acid. Chem Mater 22(7):2213–2218
50. Salas EC, Sun Z, Luttge A, Tour JM (2010) Reduction of graphene oxide via bacterial respiration. ACS Nano 4(8):4852–4856
51. Liu J, Fu S, Yuan B, Li Y, Deng Z (2010) Toward a universal "adhesive nanosheet" for the assembly of multiple nanoparticles based on a protein-induced reduction/decoration of graphene oxide. J Am Chem Soc 132(21):7279–7281
52. Lehtola J, Hakala M, Hamalainen K (2010) Structure of liquid linear alcohols. J Phys Chem B 114(19):6426–6436
53. Layek RK, Nandi AK (2013) A review on synthesis and properties of polymer functionalized graphene. Polymer 54(19):5087–5103
54. Stauber T (ed) (2019) Handbook of graphene, volume 2: physics, chemistry, and biology. Wiley
55. Jung IS, Kwon HJ, Kim M, Kim D, Kim JH, Lee H, Yun D, Byun S, Yu D, An H, Im D, Lee H, Jang J (2019) Rapid oxygen diffusive lithium–oxygen batteries using a restacking-inhibited, free-standing graphene cathode film. J Mater Chem A 7(17):10397–10404
56. Yun XW, Tang B, Xiong ZY, Wang XG (2020) Understanding self-assembly, colloidal behavior and rheological properties of graphene derivatives for high-performance supercapacitor fabrication. Chin J Polym Sci, 1–12

57. Cucchi A, Ghensi P (2014) Vertical guided bone regeneration using titanium-reinforced d-PTFE membrane and prehydrated corticocancellous bone graft. Open Dent J 8:194
58. Obodo RM, Ahmad I, Ezema FI (2019) Introductory chapter: graphene and its applications. In: Graphene and its derivatives-synthesis and applications. IntechOpen
59. Kumar R, Singh RK, Kumar V, Moshkalev SA (2019) Functionalized nanosize graphene and its derivatives for removal of contaminations and water treatment. In: A new generation material graphene: applications in water technology. Springer, Cham, pp 133–185
60. Mao L, Liu C, Lu K, Su Y, Gu C, Huang Q, Petersen EJ (2016) Exposure of few layer graphene to Limnodrilus hoffmeisteri modifies the graphene and changes its bioaccumulation by other organisms. Carbon 109:566–574
61. Zhang T, Jiang Y, Song Z, Huang H, He Y, Fang Z, Weng H, Fang C (2019) Catalogue of topological electronic materials. Nature 566(7745):475–479
62. Fu M, Huang Z, Zang X, Pan L, Liang W, Chen J, Qian H, Xu W, Jiang P, Zhang X (2018) Long noncoding RNA LINC 00978 promotes cancer growth and acts as a diagnostic biomarker in gastric cancer. Cell Prolif 51(1):e12425
63. Dinescu S, Ionita M, Ignat SR, Costache M, Hermenean A (2019) Graphene oxide enhances chitosan-based 3D Scaffold properties for bone tissue engineering. Int J Mol Sci 20(20):5077
64. Heshmatpour F, Haghbin S (2020) Nanohydroxyapatite/graphene oxide nanocomposites modified with synthetic polymers: promising materials for bone tissue engineering applications. Int J Polym Mater Polym Biomater, 1–7
65. Wei M, Lu T, Nong Z, Li G, Pan X, Wei Y, Yang Y, Wu N, Huang J, Pan M, Meng F, Li X (2019) Reductive response and RGD targeting nano-graphene oxide drug delivery system. J Drug Deliv Sci Technol 53:101202
66. Han S, Sun J, He S, Tang M, Chai R (2019) The application of graphene-based biomaterials in biomedicine. Am J Transl Res 11(6):3246
67. Hondroulis E, Zhang Z, Chen C, Li CZ (2012) Impedance based nanotoxicity assessment of graphene nanomaterials at the cellular and tissue level. Anal Lett 45(2–3):272–282
68. Liao KH, Lin YS, Macosko CW, Haynes CL (2011) Cytotoxicity of graphene oxide and graphene in human erythrocytes and skin fibroblasts. ACS Appl Mater Interfaces 3(7):2607–2615
69. Liu S, Zeng TH, Hofmann M, Burcombe E, Wei J, Jiang R, Kong J, Chen Y (2011) Antibacterial activity of graphite, graphite oxide, graphene oxide, and reduced graphene oxide: membrane and oxidative stress. ACS Nano 5(9):6971–6980
70. Fu C, Liu T, Li L, Liu H, Liang Q, Meng X (2015) Effects of graphene oxide on the development of offspring mice in lactation period. Biomaterials 40:23–31
71. Gonçalves G, Vila M, Bdikin I, De Andrés A, Emami N, Ferreira RA, Carlos LD, Gracio J, Marques PA (2014) Breakdown into nanoscale of graphene oxide: confined hot spot atomic reduction and fragmentation. Sci Rep 4(1):1–8
72. Shekhar S, Sarkar A, Sharma B, Jain P (2020) Electrochemical evaluation of functionalized graphene oxide filled PVA-chitosan biohybrid for supercapacitor applications. J Appl Polym Sci 137(17):48610
73. Shekhar S, Sharma R, Sharma S, Sharma B, Sarkar A, Jain P (2020) An exploration of electrocatalytic analysis and antibacterial efficacy of electrically conductive poly (D-glucosamine)/graphene oxide bionanohybrid. Carbohydr Polym 240:116242

Graphene-Based Biopolymer and Nano Composites: Fabrication, Characterization, and Applications

Ankit Manral, Rahul Joshi, and Pramendra Kumar Bajpai

Abstract Composites materials are most favourable used over conventional materials. Incorporation of varying fillers in bio polymers improves the properties of its associated composites. Now days different types of graphene phases are used as fillers in varying bio-polymers. In this book chapter different fabrication, characterization and application of graphene-based biopolymer nano composites are discussed. Apart from these the chapter also discussed about the type of graphene used in biopolymers. This chapter also discussed about the thermal and electric properties of graphene-based nanocomposites. Besides it how fabrication techniques and concentration of graphene fillers affect the properties of graphene based nano composites. The morphological study of graphene based nano composites clearly indicates that the dispersion of graphene nano fillers depends upon the fabrication technique. As graphene is used in a nano fillers form so it can be dispersed and mold easily with biopolymer in any complicated shape very easily.

Keywords Graphene · Biopolymer · Nano composites · Applications · Characterization

1 Introduction

Graphene nano particles are used as reinforcement material in biopolymer instead of natural and synthetic fiber reinforcement in composites. Graphene derived from graphite, consists of hexagonal crystalline structure of 2D carbon atoms with bonded with sp^2 bond [1]. Graphene have high thermal conductivity, electron mobility and mechanical strength. Low density of graphene fillers marks its application in various fields, biomedical and electronics are few of them [1–3]. Graphene are used as fillers

A. Manral (✉)
MPAE Division, Netaji Subhas Institute of Technology, Delhi, India
e-mail: ankitmanralnsit@gmail.com

R. Joshi · P. K. Bajpai
Mechanical Engineering Department, Netaji Subhas University of Technology, New Delhi 110078, India

B. Sharma and P. Jain (eds.), *Graphene Based Biopolymer Nanocomposites*, Composites Science and Technology,
https://doi.org/10.1007/978-981-15-9180-8_3

in neat polymer and many times graphene nano fillers coated on carbon by chemical process and used this coated carbon fiber as reinforcement material in composites [1]. These coated graphene particulates help to enhance the properties of developed composites by increasing interfacial adhesion between reinforcement and matrix material. But graphene-based coating process is very complicated and not cost effective, on contrary various author reported a cost effective and simple processes for coating of graphene fiber over fiber. Hung et al. [1, 4] used electrophoretic deposition method for coating of graphene oxide on fiber to enhanced the properties of composites. Graphene oxide (GO) is very popular as a coating material in field of fiber reinforced composites. It is highly oxygenated and appropriate for bio polymer [5–7]. Therefore, GO fillers are widely used as a reinforced material for biopolymer nanocomposites. GO has nonconductive properties and work as an electric insulating material in biopolymer nanocomposites. Incorporation of graphene in biopolymers increased its application spectrum. Apart from basic applications, graphene in nano form is used in various technology field applications such as in supercapacitor, single molecule sensor, high frequency analog electronics and batteries [8, 9]. Further, on increasing the application of graphene-based nanocomposites, graphene sheets are incorporated in polymers to develop nanocomposite. Incorporation of graphene nanoparticle as a reinforced material in resin biopolymer is done easily by normal mixing or by ultrasonic dispersion, for proper dispersion of graphene particles in bio polymer resin. Proper dispersion of graphene in biopolymer makes composite a near isotropic material that exhibits better properties in comparison to neat biopolymer.

Graphene is used as a reinforcement and additives material with both thermoplastic and thermoset polymers. There is a wide availability of thermoplastic and thermoset polymer in both synthetic and natural form. But due to the adverse effect of synthetic polymers on environment and depletion of petroleum products, material engineers have laid emphasis on developing some new type of polymers that are ecofriendly called as biopolymer. These biopolymers are fully biodegradable in nature and can easily be extracted from plants. Although biopolymers have lower strength and lower thermal stability as compared to synthetic polymer, but its recyclability and easily degradability marks it as a better polymer. Further on the incorporation of reinforcement material (mats, particulates, fillers etc. form of fiber and additives materials) in biopolymer enhances its properties compared to its neat form. The various biopolymers that are used as matrix material in graphene-based nanocomposites are shown in Fig. 1.

In neat biopolymer only 1% of polymer volume of graphene is required or sufficient for development of composites for various electrical applications. Apart from electrical conductivity of material, graphene also improves the thermal stability of neat polymer due to the higher thermal conductivity of graphene that is around 5000 $Wm^{-1}\,k^{-1}$ [10]. Due to high thermal conductivity of graphene-based polymer composites, it can be used in various thermal applications. Lesser thermal conductivity biopolymer has lower thermal stability due to marginal transfer of heat, but incorporation of graphene increases the heat transfer and improves the thermal

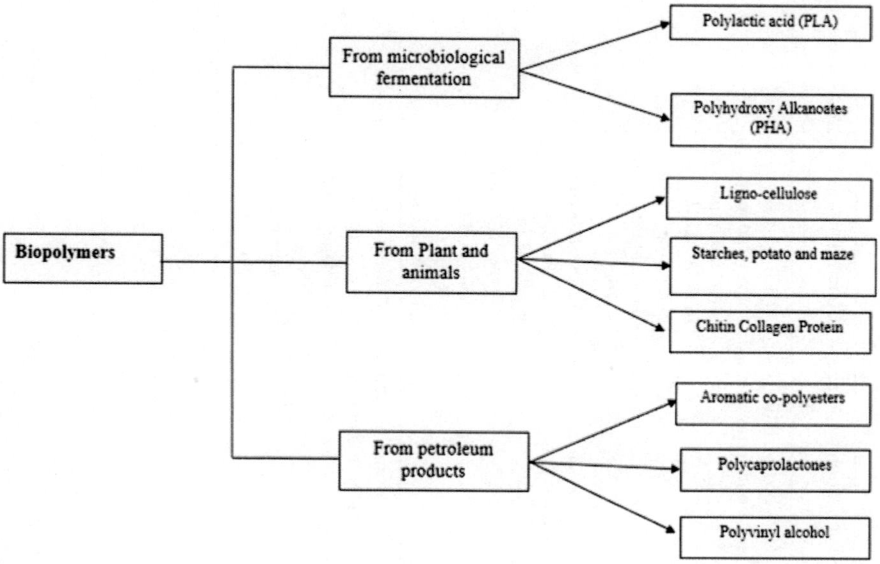

Fig. 1 Classification of biopolymers

stability of biopolymers. For small products like sensor, high frequency analog electronic etc., graphene-based biopolymers composites are introduced. In the applications where high strength and stiffness are required, graphene is used as filler or additive in composites laminates. Various factor influences the properties of graphene-based nanocomposites like particulate size of graphene, fabrication technique, processing parameters. Various studies have been performed in which graphene was reinforced with different biopolymers for the development of nanocomposites. Table 1 shows the fabrication and dispersion technique of graphene into biopolymer. In most of the studies, graphene was incorporated with polylactic acid and least with bio-epoxy and Poly (hydroxyal-kanoate). Dispersion of graphene in liquid resin is done by normal mixing or by sonication techniques. Though for granules form of polymer, first polymer is converted into resin form with the help of chloroform or by other chemicals dissolver. After diluting the biopolymers, graphene is added by normal mixing for development of composites. During fabrication, concentration of graphene in bio polymer and processing parameter influence the properties of nanocomposites. In some studies where high strength and low electric conductivity is required, glucose was incorporated with graphene to reduce the electric conductivity of its associated composites [11]. This chapter highlights the various fabrication technique used for the fabrication of graphene-based biopolymer nanocomposites. Apart from fabrication techniques, effect of processing parameters on developed composites are also discussed. As graphene has high electric and thermal conductivity, so various applications related to these properties are discussed in this chapter.

Table 1 Graphene based biopolymer nanocomposites

S. No	Graphene type	Graphene concentration	Biopolymers	Graphene dispersion	Fabrication technique	Conclusion	References
1	Graphene powder (GP) (1–44 μm)	0.01, 0.1, 0.2, 0.5 1.0, and 2.0 wt%	Poly lactic acid (PLA)	Sonication	Chloroform dilution + solution mixing method	At 1.0 wt % concentration of graphene reinforced in PLA based nanocomposites shows higher tensile strength	[11]
2	Graphene oxide (GO)	0.1, 0.2, 0.3 wt %	Bio-epoxy	Sonication	Solution mixing method	SEM results shows that, dispersion of graphene in polymer altered the tensile properties of developed nanocomposites	[12]
3	Graphene oxide (GO)	2.5 Vol. %	Poly (hydroxyal-kanoate) -PHA	Sonication	Chloroform dilution + solution mixing method	In developed composites incorporation of graphene improved the electrical conductivity	[13]

(continued)

Table 1 (continued)

S. No	Graphene type	Graphene concentration	Biopolymers	Graphene dispersion	Fabrication technique	Conclusion	References
4	Graphene platelets	5 wt %	Poly lactic acid (PLA)	–	Co-rotating twin-screw extruder + compression molding	Multiple reprocessing of graphene-PLA improves the dispersion and improved the crystallinity phase of the developed composites	[14]
5	Graphene nanoplatelets (1 μm)	5, 7, 10, 13 and 15 wt%	Poly lactic acid (PLA)	Normal mixing	Mini-extruder + compression molding	5 wt% is the optimized limit for improving the properties of developed nano composites. On further increasing the wt% of graphene makes composites brittle	[15]
6	Graphene oxide (GO)	0.5, 1 and 2 wt %	Poly L-Lactic acid (PLLA)	Sonication and stirring	N, N-dimethylformamide dilution + dried at 80 °C in vacuum	Incorporation of GO 1 wt% improved the overall isothermal melt crystallization rates as compared to neat PLLA	[16]

(continued)

Table 1 (continued)

S. No	Graphene type	Graphene concentration	Biopolymers	Graphene dispersion	Fabrication technique	Conclusion	References
7	Graphene oxide (GO) + Glucose	–	Poly lactic acid (PLA)	Normal mixing	N, N-dimethylformamide dilution + compression molding	Addition of glucose in Go reduces the electric conductivity of its associated composites with PLA	[17]
8	Expanded graphene (EG)	3 and 6.75 wt %	Poly lactic acid (PLA)	twin-screw extruder	Compression molding	Incorporation of EG in PLA improved the thermal conductivity and elastic modulus of developed composites	[18]
9	Exfoliated graphite nanoplatelets	1, 3 and 5 wt %	Polylactic acid (PLA) + Kenaf fiber reinforcement	Twin screw extruder	Injection molding	Coating of graphite nanoplatelets improved the interfacial adhesion between kenaf and PLA	[19]

2 Graphene

Graphene is the simple structural element of carbon allotropes, fullerenes and graphite. Graphite is available in nano and sheet (*sp^2* bonded carbon atom) form [20]. It has excellent thermal, electrical properties and is excessively used as filler in polymer-based composites. Its high electric conductivity and outstanding properties makes it suitable for many electrical based small applications such as nano sensors and batteries. Graphene based composites is also used in manufacturing of semiconductor. Apart from these properties, there are some potential properties of graphene such as effective moisture barrier, lower density than steel and high Young's modulus [20].

2.1 Synthesis of Graphene

Graphene is a non-metallic material but it has high thermal and electric conductivity. At room temperature graphene has charge carrying mobility of 1×10^4 $cm^{2\,V-1}$ s^{-1}. Although the mobility increases with increase in temperature. The synthesis of Graphene, classification based on two basic techniques shows in Fig. 2.

Graphene is also classified on the basis of its type like Graphene films, Graphene oxide (GO), Reduced Graphene oxide (rGO) and Graphene nanoplatelets (GNPs). Every graphene has its distinct synthesis technique. Chemical vapor deposition (CVD) method is used for producing of graphene films. In this technique, volatile gas

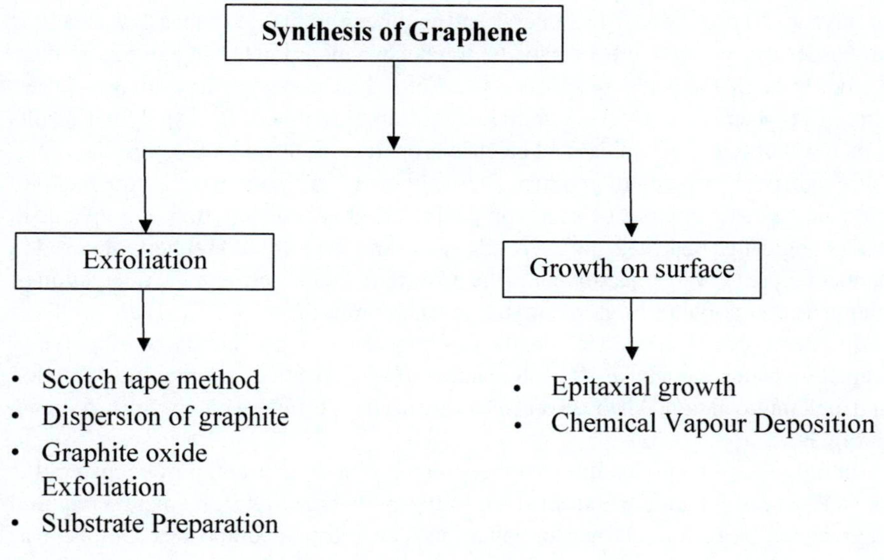

Fig. 2 Synthesis of graphene

on the surface of substrate in a chamber is deposited and waste gases are pumped out. CVD develops high quality of graphene sheets but the by-products extracted during processing are very harmful and toxic. There are two steps in CVD for the development of graphene sheet, in the first step high heat required to debone the carbon bond from raw form of graphene in metal catalyst. Further, in the second step, intensive heat is required for assembling the carbon atoms and with the help of roll- to roll process graphene film developed on copper foils [21]. Apart from graphene sheet, GO is produced by oxidation of graphene. Oxidation functionalizes the surface of graphene and forms multiple type of functional group on the surface and improves hydrophilicity. Improved hydrophilicity helps GO to properly disperse in different solvents, GO has lower electric conductivity or act as an electric insulator. Whereas, GNPs were obtained by mechanical cleavage or plasma exfoliation of bulk graphite.

3 Graphene Based Biopolymer Nano Composites

A variety of biopolymers are available for the development of composites. But every individual biopolymer has distinct method of fabrication for incorporation with reinforced material. Nowadays variety of fillers is incorporated with biopolymer to enhance the properties of its associated composites. Natural fiber fillers are incorporated with neat polymers and with reinforcement and matrix material to improve the properties of final developed composites. Though this improvement in properties is not up to the mark or which can challenge the requirement of current applications spectrum. Therefore, advanced fillers such as minerals and carbon-based allotrope such as graphene are being used in composites to meet the challenging requirement of advanced applications. In mineral and graphene fillers, graphene has excellent properties compared to others. Mostly graphene is incorporated in polymer in nano form, although for better properties of developed nanocomposites, proper dispersion of graphene is required. A good distribution of graphene in biopolymer results in better thermal, mechanical and electric properties of developed composites [22]. Although, concentration of graphene in biopolymer can also be a governing factor in influencing the properties of nano composites. Higher concentration of graphene in nano composites makes its highly brittle, so optimized level of graphene are used in nanocomposites. Whereas solvent is the governing factor between the interaction of graphene and polymer of graphene-based nanocomposites [22, 23]. There are basically three types of techniques for the development of graphene-based biopolymer nanocomposites namely in-situ intercalative polymerization, solution intercalation and melt intercalation. All these processes are used for different biopolymer as shown in Fig. 3.

In only specific applications graphene-based biopolymer composites material is used. Whereas, for wide spectrum of applications fiber mats reinforcement is required in graphene-based biopolymer for better load capability of composites. Graphene as fillers improves the thermal and electric properties of developed composites. Apart from improving the interaction between matrix and reinforcement material, graphene

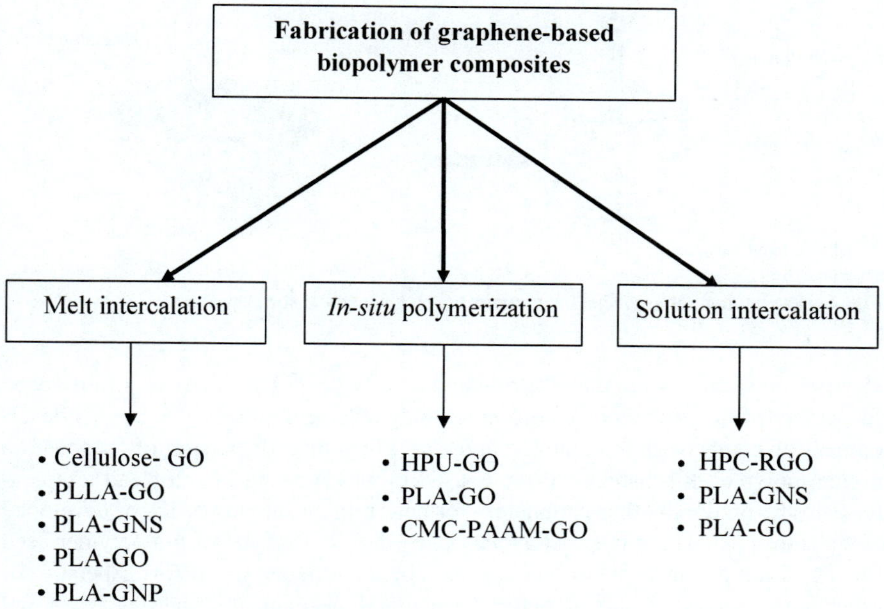

Fig. 3 Fabrication techniques for graphene-based biopolymer nano composites. PLLA- Poly-L-lactic acid, CMC- Carboxy methyl cellulose, PAAM- poly (methyl methacrylate), HPC-hydroxypropyl cellulose, RGO- reduced graphene oxide, GNS-graphene nano sheets

improved the thermal stability and viscoelastic properties of developed composites. The most common method for the development of graphene-based biopolymer composites is solution intercalation method in which thermoplastic biopolymer granules were first dissolves in solvents. After dilution of biopolymer in solvent, graphene is added in an adequate amount. Further for proper dispersion, ultrasonic or magnetic stirrers are used and after proper mixing solution pour in petri dishes and leave it for 24 h in a room temperature. After 24 h the solvent gets evaporated and desired graphene-based biopolymer sheet are obtained. Sridhar et al. [24] developed graphene-based poly (3-hydroxybutyrate-co-4-hydroxybutyrate) nanocomposites. Author concluded that incorporation of graphene in polymer improved the crystallinity phase. Apart from crystallinity phase substantial improved the modulus and thermal degradation behavior of the developed nanocomposites.

3.1 Melt Intercalation

This process is same as casting method used in conventional material. In this method graphene in nano form is added to molten form of bio polymer to get desired nano composite. But addition of graphene in molten polymer in presence of oxygen reduces

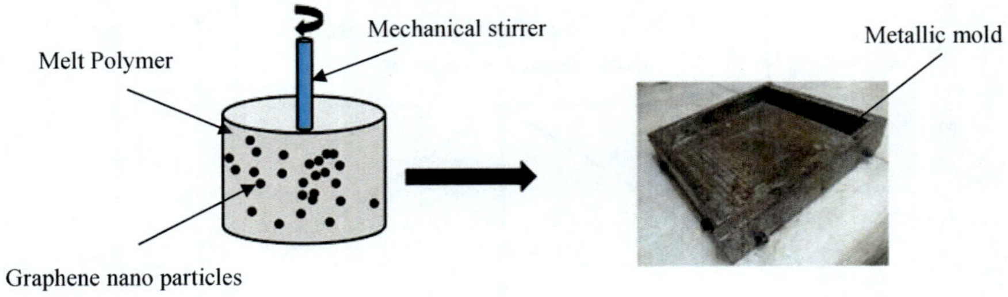

Fig. 4 Development of graphene bio composites by mechanical stirrer process

electron flow and lowers the electric conductivity [22]. This process is introduced for variety of biopolymers like Bio-polyamide (PA), cellulose, PLA etc. for development of graphene-based nano composites. For proper dispersion of graphene in molten polymer, mechanical stirrer is used in which speed of rotor and temperature are the predominating parameters for determining the properties of developed composites [22]. These parameters vary according to the type of bio polymer used for the development of nano composites. Higher rotating speed for dispersion of graphene in polymer results in better thermal and mechanical properties. In normal method, molten mixture of graphene and biopolymer are poured in mold where it is cured in a room temperature, the schematic diagram of this process is shown in Fig. 4.

This is a very simple method for development of nanocomposites, but there are some defects associated with it such as formation of voids etc. These associated defects with this process make it unsuitable for the development of nano composites. Therefore, closed mold process such are injection and extrusion molding are mostly preferred to overcome these defects [25–28]. In injection molding, proper dispersion of biopolymer and graphene are achieved, as it is a closed mold process resulting in less formation of voids in developed products. In this process, neat granules of polymers or hybrid granules of polymer and graphene (By extrusion) are fed into injection molding barrel through hopper. Screw rotation in barrel intermix the constituents of composites; circular heaters are attached at outer periphery of barrel that helps to convert granules of polymer into molten form. This intermix form of constituents is fed into mold cavity to get the desired shaped of the products. The schematic diagram of injection molding machine shown in Fig. 5.

Similar to any process, injection molding process has various processing parameters on which the properties of developed products depend. Screw speed, feed rate, heat temperature are the parameters which are need to optimized to get better properties of fabricated products. Low barrel temperature reduces the viscosity of polymer that restricts the flow of molten polymer and it cannot properly fill the mold cavity. Unfilled polymer in cavity increases the formation of crack in products. So, every processing parameters need to be optimized to enhance the properties of fabricated products. Increase in graphene percentage in nanocomposites up to certain

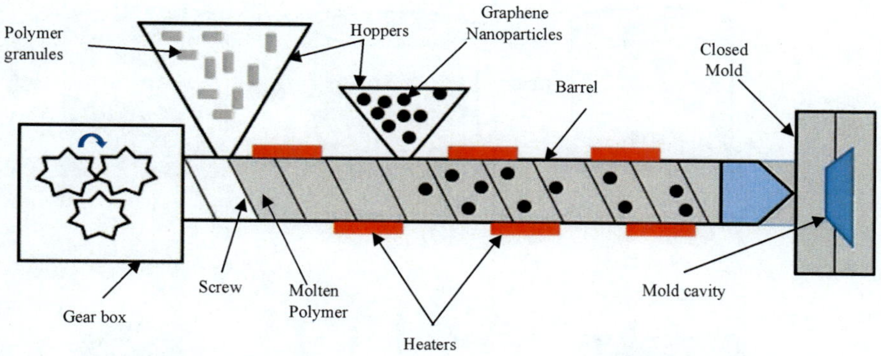

Fig. 5 Fabrication of graphene-based nano composites by injection molding

limit increases the properties of developed composites. Asmatulu et al. [29] developed a recycled high-density polyethylene graphene-based nanocomposite. Author concluded that at 8 wt% of graphene concentration in composite has high thermal stability. Whereas as for higher elastic modulus 4 wt % is the optimized concentration of graphene in composites.

3.2 *Solution Intercalation (SI)*

It is quite easy method than melt intercalation and others used for the fabrication of graphene based biocomposites. This method is only applicable for thermoplastic biopolymer such are PLA, PHB etc. In this method, a simple and shearing stirring of biopolymer in a solvent (chloroform) takes place. During stirring, sometimes graphene nano flasks are added and dispersed with biopolymer with normal stirring. Whereas, sometime ultrasonic stirrer is used for better dispersion of graphene in diluted biopolymer. After dispersion of graphene in biopolymer the solution is kept in the petri dish where solvent evaporates and form a sandwich layer of graphene and biopolymer, resulting in the development of nanocomposites [30]. The schematic diagram of SI techniques shown in Fig. 6.

For most of the biopolymer, there were little changes in chemical structure with increasing concentration of graphene in it. But there was an incredible change in hydrogen bonding or interaction between the graphene and biopolymers. Graphene based nanocomposites such as polyvinyl alcohol (PVA) = graphene [31], polypropylene (PP) = graphene [27] polystyrene (PS) = graphene [26] were prepared by this method. Various researches were reported related to development of graphene-based nanocomposites fabricated with SI technique. Mu et al. [32] performed the comparative studies on effect of various fabrication process of graphene-based composite on its thermal properties. Author concluded that the sample prepared by SI technique have shown better thermal conductivity compared to the sample

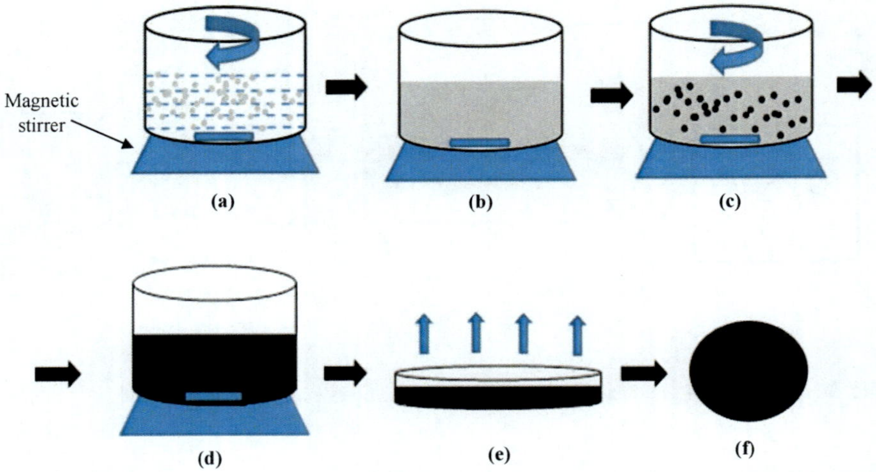

Fig. 6 **a** Stirring of polymer granules in solvent, **b** dilution of polymer in solvent, **c** stirring of graphene nano flasks in dilute polymer, **d** dispersion of graphene in dilute polymer, **e** evaporation of solvent, **f** developed graphene biopolymer sheet

prepared by melt intercalation technique. This statement supports that the dispersion of graphene nanoparticles in polymer is better in solution intercalation techniques than other fabrication techniques. Similarly, Shen et al. [33] fabricated expanded graphite/polypropylene nano composites by solution intercalation technique. Author concluded that at 3.90 volume % of expanded graphene, the developed composite achieved six order magnitude of higher conductivity than that of latter. Author also concluded using SEM, TEM results, that multiple network structures were formed in nanocomposites using SI technique. The mechanical properties influenced by varying concentration of graphene on nano composites shown in Table 2.

3.3 *In-Situ Polymerization*

As earlier discussed, there is another method for fabrication of graphene based nano composites called In-situ polymerization shown in Fig. 7. In this method graphene nanofillers dispersed in monomers solution, then followed polymerization process as followed for polymer to get nano composites. When monomers were bind up through polymerization process, nanofillers were also bind up with monomers unit and formed a composition of polymer and nano fillers. In this process, no exfoliation step is required for proper dispersion of fillers [35]. The formation of nanocomposites is the result of formation of covalent bond between monomers and graphene nanofillers with varying condensation reaction. Although, it is suitable for non-covalent bonded composites such as poly-methyl-methacrylate-GO nanocomposites [36]. Various researchers used situ-polymerization method for incorporation of graphene with

Table 2 Properties of nano composites developed form solution intercalation

S. No.	Graphene type (wt %)	Bio-polymer	Fabrication technique	Tensile strength (MPa)	Tensile modulus (GPa)	Other properties	References
1	GP (1)	PLA	SI	51.14	–	At 0.01 wt % developed nano composites show higher toughness	[11]
2	GO (0.3)	Bio-epoxy	SI	73.55	3.5	–	[12]
3	GO (2.5 vol%)	PHA	SI	–	–	At given concentration of graphene, the associated composites have increase young's modulus by 180–590%, whereas elongation at break decrease by 41–89%	[13]
4	Thermally reduced graphene (TRD)	PHA	SI	–	–	TRG improved the crystallinity phase of the developed composites	[34]

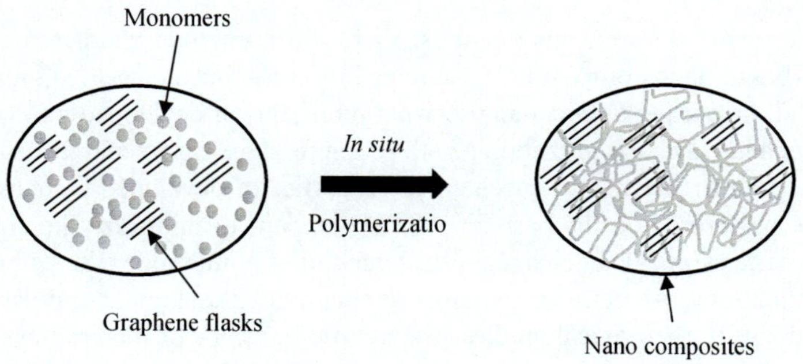

Fig. 7 In situ-polymerization

varying biopolymers like alginate, cellulose, hyperbranched polyurethane (HPU) for development of nanocomposites. In addition, in some cases the formation of hybrid composition of polymer and graphene has further uses in melt intercalation process to get desired shape of nanocomposites. But as compared to other fabrication technique this technique changes intensive properties of developed composites for some extent.

Fabrication of lactide based nanocomposites the dispersion of graphene oxide nano fillers at varying concentration to form thermally reduced graphene (TRG) based nanocomposites. For confirmation of conversion of GO to TRG various technique are used XRD, FTIR and Raman. These techniques confirm the transformation of lactide into PLLA and reduction of GO to TRG [22]. In some research there was an improvement in electric conductivity of nanocomposites after fabrication using situ-polymerization. Similar to electric and thermal conductive properties, this process also highly influenced the mechanical properties of developed nano composites. Dispersion of graphene fillers in monomers units occurs at almost microscopic levels which results in formation of strong covalent bonding between the constituents. As a result of strong covalent bonding, the mechanical properties of developed nanocomposites get enhanced. The compressive properties of 1.6 wt% of CMC-PAAM-GO improved to 260% with respect to the neat polymer compressive strength and this highlights the improvement in mechanical properties with little addition of graphene oxide [22]. Although in other studies, the fabrication of biobased thermosetting graphene oxide based nano composites showed minimal increase in tensile strength and modulus value [37]. Although, apart from tensile strength and modulus, the elongation at break, increased 1.3 times.

4 Characterization Technique for Graphene Based Nano Composites

According to the different applications, there are numerous characterizations of graphene-based nano composites. Characterization has been classified on the basis mechanical, physical, chemical and thermal properties of developed nano composites. In mechanical characterization tensile, compressive, impact and flexural testing were done to characterize the mechanical properties of developed composites. All tests were performed on the basis of ASTM standard to minimize the chances of error. These tests under mechanical characterization come under the categories of static applications, whereas for dynamic applications, rheological and tribological tests are done. In rheological studies viscoelastic behavior of material are studied. In this test, material is subjected to cyclic load with varying temperature conditions, to evaluate the visco-elastic behavior of materials. In tribological studies, materials wear and tear rate are analyzed at varying load and speed rates. Figure 8 shows the classification of characterization on the basis of mechanical, physical and chemical testing. In mechanical characterization, tensile test tells about the tensile load ability

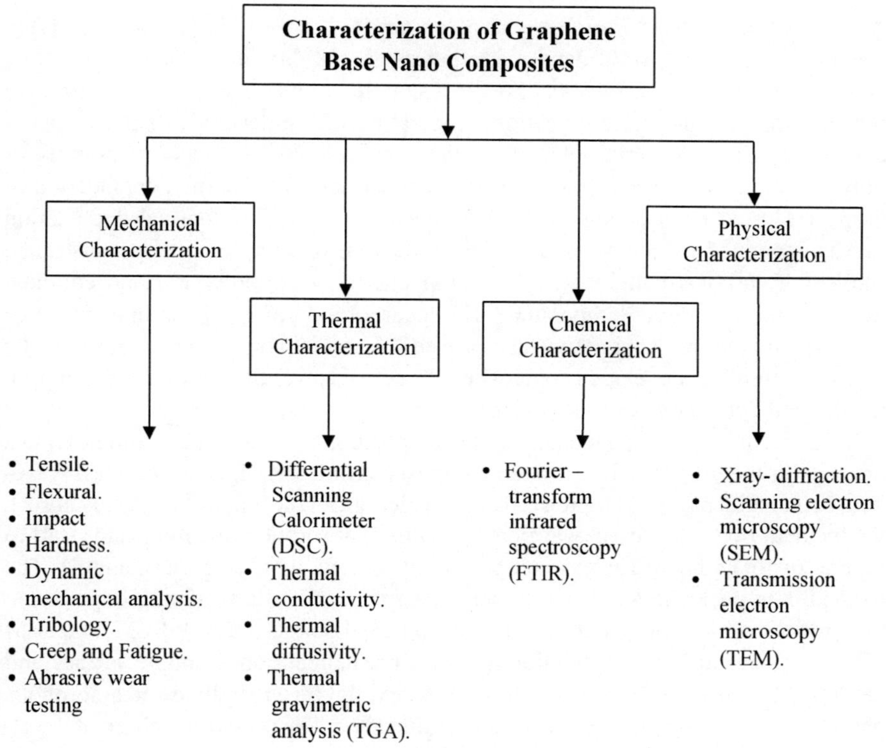

Fig. 8 Characterization of graphene based nano composites

of material and about the stiffness of material under a given tensile load. Whereas, in flexural test, bending strength and flexural modulus of material is checked. It is generally 3-point bending test. In impact test, energy absorption capability of material due to sudden impact of load is studied. Due to presence of nano form of graphene in biopolymers, chances of crack propagation due to sudden impact of load is reduced thereby enhancing the impact properties of nano composite materials.

The tribological study of material is done to predict the wear and tear rates with different abrasives particles. In tribological test, a simple pin and disc rotatory arrangement, in which nano composites samples are mounted on a pin that exerts pressure over a disc. Different abrasive materials are used on the disk to check the wear and tear behavior of different composition of graphene based nano composites. There is one more category of characterization that comes under mechanical testing namely, rheological study. In this study, the composite material is tested under cyclic loads with varying temperature ranges to check its visco-elastic behavior. Storage modulus, loss modulus and damping factor are measured for examining the visco-elastic behavior of material. The nature of cyclic load may be applied torque or by 3-point bending arrangements.

In thermal characterization, thermal properties such as thermal conductivity, specific heat, thermal diffusivity and thermal stability are studied. In DSC, endothermic and exothermic heat flow in a sample is studied. DSC results also give the specific heat and glass transition temperature of the developed nano composites. Incorporated nano particles of graphene nano fillers, restricts the mobility of polymer that enhances the glass transition temperature. The thermal degradation of graphene based nano composites is examined by TGA. The thermal degradation rate is examined on the basis of weight reduction of samples. Every constituent (moisture, carbons, cellulose etc.) of graphene-based biopolymer nano composites degrades at varying temperature. The incorporation of graphene in composites improves its thermal conductivity. Improved in thermal conductivity improved the thermal stability of developed composites. A special apparatus is required for finding the thermal conductivity of nano composites.

FTIR technique is used to cheque the composition of different functional groups presents in the graphene based nano composites. In FTIR, light of different wavelengths is incident on a sample surface in which different functional groups absorb the incident rays at different wavelengths. During absorption, the molecules start to vibrate or rotate about its axis and peaks are formed at varying wavelengths. The formation of peaks height helps in identifying the type of functional group present in a given sample. For physical and morphological studies, X-RD, SEM and TEM are performed. In X-ray diffraction, the rays are incident on a sample surface and varying 2 θ peaks are developed. These peaks exhibit the crystalline and amorphous phase of developed graphene based nano composites. The crystalline phase of developed composites depends upon the dispersion of graphene nano fillers in developed composites. Improper dispersion reduces the crystalline phase of composites. The proper dispersion of graphene makes a material a near isotropic material or one can simply say that the material has a crystalline phase. SEM called as scanning electron microscopy, is used to survey or examine the fractured surface and identifying the root causes of samples failure during any loading condition. In SEM first samples are coated with highly conductive material to make composite fracture surface conductive. High electron beam strikes the surface and trace the detail of the surface on a display unit to examine the root causes. TEM micrographs show the dispersion of graphene nano fillers in developed composites. TEM micrographs gives the inside details of graphene dispersion in graphene based nano composites.

5 Applications of Graphene Based Nano Composites

Extra ordinary properties of graphene increase the application spectrum of graphene based nano composites. High electric and thermal properties make graphene based nano fillers as a favorable reinforced material for the development of composites in various applications. It has numerous applications in field of energy, medicine, electronics, household and industrial applications [38]. The incorporation of graphene for the development of composites for various applications shown in Fig. 9. Graphene

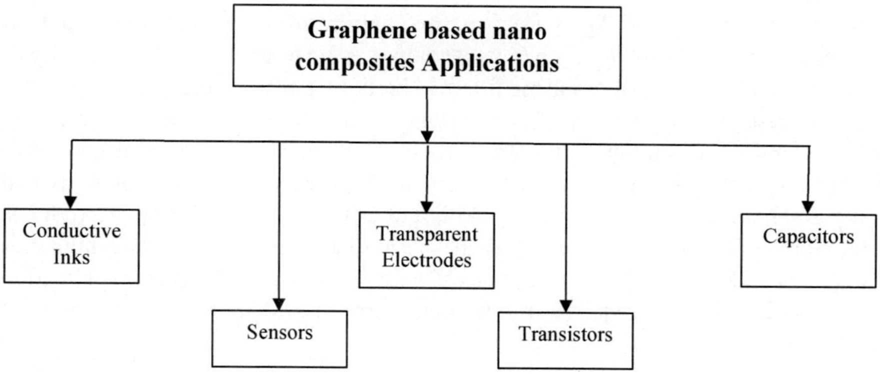

Fig. 9 Application of graphene based nano composites

based nano composites have better thermal and electric properties but have limited mechanical properties. Apart from tensile and flexural properties, graphene-based composites have better impact properties. Presence of nano fillers in composites arrest the crack formation during impact load and absorb maximum energy during impact load before failure. Hence it is better to use graphene based nano composites for impact loading applications instead of other mechanical applications. The reinforcement of graphene as fillers restrict the mobility of molecules at higher temperature, which result the glass transition temperature of its associated composites were increased. This increase in glass transition temperature of graphene based nano composites makes it suitable for dynamic mechanical applications in which material has to withstand high temperature under cyclic loading conditions. As biopolymer-based composites have degradable in nature rapidly after its service life as comparative to synthetic polymer-based composites. Secondly, biopolymer can sustain at lower temperature compare to synthetic polymers. So, it is difficult to used it neat form for engineering applications. Incorporation of graphene in biopolymer makes composites sustain at higher temperature and have better properties as comparative to neat polymer. Thermoplastic polymer can recycle again and again without effects its property. Similarly, graphene-based biopolymer-based composites can be recycled and mold for desired shape application without affecting its properties. The incorporated graphene nano form makes biopolymer-based composites used in nano applications such as in sensors, transistor and in small inbuilt batteries.

6 Limitation and Future Aspects

There are only few limitations are associated with graphene based nano composites. As biopolymers have least glass transition temperature, incorporation of graphene hikes the glass transition temperature but not that much it can be used in extreme temperature application. Apart from thermal application, the mechanical application of graphene-based biopolymer nano composites is limited. The discontinuity or absence of interlinking of graphene fillers in biopolymer-based composites lowered the load capability of developed composites. Although, if graphene reinforcement used in laminates form the better structural or mechanical application of graphene-based biopolymer composites were obtained. As, thermoplastic biopolymers and nano graphene particles are easily mold in any complicated geometry of products. Therefore, a complicated geometry shapes used in aerospace and marine industries can be easily developed by graphene-based composites. Apart from these application some modification are required in reinforced form of graphene to use in structural applications.

7 Conclusion

In this chapter graphene-based biopolymer nano composites fabrication, characterization and application have been discussed and following conclusions are drawn.

1. Different varying techniques are used for manufacturing of graphene based nano composites. The fabrication processes used to fabricate the nano composites are based on its application.
2. Graphene based nano composites are used for nano applications such as used in sensors, capacitors and batteries etc. High thermal and electric conductivity of graphene based nano composites makes a superior choice in varying application.
3. Nano size of nano fillers makes proper dispersion or blending of graphene nano fillers in polymers. The variation of graphene concentration in bio polymer affects the different properties of developed composites.
4. In situ-polymerization is better technique for fabrication of graphene based nano composites compared to other fabrication techniques. SEM micrographs results indicated that in-situ polymerization is the better fabrication technique.
5. Graphene based nano composites characterization was done on the basis of its application where it is used. Mechanical, chemical, physical and thermal characterization is done to exhibit the properties of developed composites. Incorporation of graphene in biopolymer improves the mechanical and thermal properties of its associated composites.
6. Nano size of graphene fillers in biopolymers mold easily in any type of complicated parts. From nano to big complicated mold products it can be used as better reinforcement filers compared to other.
7. Further development of different graphene (GO, GNS and GNP) and the different blending agents in biopolymer makes used in various scientific applications.

References

1. Mohan VB, Lau KT, Hui D, Bhattacharyya D (2018) Graphene-based materials and their composites: a review on production, applications and product limitations. Compos B 142:200–220
2. Geim AK, Novoselov KS (2007) The rise of graphene. Nat Mater 6(3):183–191
3. Castro Neto AH, Guinea F, Peres NMR, Novoselov KS, Geim AK (2009) The electronic properties of graphene. Rev Mod Phys 81(1):109–162
4. Hung P-y, Lau K-t, Fox B, Hameed N, Lee JH, Hui D (2018) Surface modification of carbon fibre using graphene–related materials for multifunctional composites. Compos B Eng 133:240–257
5. Pati MK, Pattojoshi P, Roy GS (2016) Synthesis of graphene-based nanocomposite and investigations of its thermal and electrical properties. J Nanotechnol https://doi.org/10.1155/2016/5135420
6. Becerril HA, Mao J, Liu Z, Stoltenberg RM, Bao Z, Chen Y (2008) Evaluation of solution-processed reduced graphene oxide films as transparent conductors. ACS Nano 2(3):463–470
7. Stankovich S, Dikin DA, Piner RD, Kohlhass KA, Klienhammes A, Jia Y, Wu Y, Nguyen ST, Ruoff RS (2007) Synthesis of graphene-based nanosheets via chemical reduction of exfoliated graphite oxide. Carbon 45(7):1558–1565
8. Novoselov KS, Jiang D, Schedin F, Booth TJ, Khotkevich VV, Morozov SV, Geim AK (2005) Two-dimensional atomic crystal. Geim. Proc Natl Acad Sci USA 102:10451–10453
9. Blake P, Brimicombe PD, Nair RR, Booth TJ, Jiang D, Schedin F, Ponomarenko LA, Morozov SV, Gleeson HF, Hill EW, Geim AK, Novoselov KS (2008) Graphene based liquid crystal display. Nano Lett 8(6):1704–1708
10. Wang M, Yan C, Ma L (2012) Graphene nanocomposites. Web Sci. https://doi.org/10.5772/50840
11. Li X, Xiao Y, Bergeret A, Longerey M, Che J (2014) Preparation of polylactide/graphene composites from liquid-phase exfoliated graphite sheets. Poly Compos. 10.1002/pc.22673
12. Loeffen A, Cree D, Sabzevari M, Wilson Lee (2018) Graphene oxide reinforced bio-epoxy polymers. Conference: Canadian Society for Mechanical Engineering (CSME) International Congress. https://doi.org/10.25071/10315/35411
13. Barett JSF, Abdala AA, Srienc F (2018) Poly(hydroxyalkanoate) elastomers and their graphene nanocomposites. Macromolecules. https://doi.org/10.1021/ma500022x
14. Botta L, Scaffaro R, Sutera F, Mistretta MC (2017) Reprocessing of PLA/graphene nanoplatelets nanocomposites. Polym. https://doi.org/10.3390/polym10010018
15. Gao Y, Picot OT, Bilotti E, Peijs T (2016) Influence of filler size on the properties of poly(lactic acid) (pla)/graphene nanoplatelet (gnp) nanocomposites. Eur Polym J. https://doi.org/10.1016/j.eurpolymj.2016.10.045
16. Wang H, Qiu Z (2012) Crystallization kinetics and morphology of biodegradable poly(l-lactic acid)/graphene oxide nanocomposites: Influences of graphene oxide loading and crystallization temperature. Thermochim Acta 527:40–46
17. Shen Y, Dasari A (2012) Chemical and thermal reduction of graphene oxide and its electrically conductive polylactic acid nanocomposites. Compos Sci Techno 72:1430–1435
18. Mortazavi B, Hassouna F, Laachachi A, Rajabpour A, Ahzi S, Chapron D, Toniazzo V, Ruch D (2013) Experimental and multiscale modeling of thermal conductivity and elastic properties of PLA/expanded graphite polymer nanocomposites. Thermochimica 552: 106–113
19. Han SO, Karevan M, Bhuiyan MdA, Park JH, Kalaitzidou K (2011) Effect of exfoliated graphite nanoplatelets on the mechanical and viscoelastic properties of poly (lactic acid) biocomposites reinforced with kenaf fibers. J Mater Sci 47:3535–3543
20. Ray SC (2015) Application and uses of graphene. 10.1016/B978-0-323-37521-4.00001-7
21. Bae S et al (2010) Roll-to-roll production of 30-inch graphene films for transparent electrodes. Nat Nanotech 5:574–578
22. Rouf TB, Kokini JL (2016) Biodegradable biopolymer–graphene nanocomposites. J Mater Sci. 10.1007/s10853-016-0238-4

23. Zhang HB, Zheng WG, Yan Q et al (2010) Electrically conductive polyethylene terephthalate/graphene nanocomposites prepared by melt compounding. Polym 51:1191–1196
24. Sridhar V, Lee I, Chun HH, Park H (2012) Graphene reinforced biodegradable poly(3-hydroxybutyrate-co-4-hydroxybutyrate) nano-composites. eXPRESS Polym Lett 7(4):320–328
25. Das TK, Prusty S (2013) Graphene-based polymer composites and their applications. Polym Plast Tech Eng 52(4):319–331
26. Wanga WP, Pana CY (2004) Preparation and characterization of polystyrene/graphite composite prepared by cationic grafting polymerization. Polym 45:3987–3995
27. Kalaitzidou K, Fukushima H, Drzal LT (2007) A new compounding method for exfoliated graphite-polypropylene nanocomposites with enhanced flexural properties and lower percolation threshold. Compos Sci Technol 67:2045–2051
28. Kim SK, Kim NH, Lee JH (2006) Effects of the addition of multiwalled carbon nanotubes on the positive temperature coefficient characteristics of carbon-black-filled high density polyethylene nanocomposites. Scripta Mater 55:1119–1122
29. Asmatulu R, Khan WS, Ceylan M (2014) Synthesis and analysis of injection-molded nanocomposites of recycled high-density polyethylene incorporated with graphene nanoflakes. Polym Compos. 10.1002/pc.23063
30. Potts JR, Dreyer DR, Bielawski CW, Ruoff RS (2011) Graphene-Based Polymer Nanocomposites. Polymer 52:5–25
31. Liang J, Huang Y, Zhang L, Wang Y, Ma Y, Guo T et al (2009) Molecular-level dispersion of graphene into poly(vinyl alcohol) and effective reinforcement of their nanocomposites. Adv Funct Mater 19:2297–2302
32. Mu Q, Feng S (2007) Thermal conductivity of graphite/silicone rubber prepared by solution intercalation. Thermochim Acta 462:70–75
33. Shen JW, Chen XM, Huang WY (2002) Structure and electrical properties of grafted polypropylene/graphite nanocomposites prepared by solution intercalation. J Appl Polym 88:1864–1869
34. Abdala A, Iqbal M, Barrett J, Srienc F (2013) Synthesis of poly-(R)-3-hydroxyalkanoic acids (PHA) and their graphene nanocomposites. American Chemical Society. Div Polym Chem 53:305–306
35. Norazlina H, Kamal Y (2015) Graphene modifications in polylactic acid nanocomposites: a review. Polym Bull 72:931–961
36. Jang JY, Kim MS, Jeong HM, Shin CM (2009) Graphite oxide/poly (methyl methacrylate) nanocomposites prepared by a novel method utilizing macroazo initiator. Compos Sci Technol 69:186–191
37. Zhang J, Zhang C, Madbouly SA (2015) In situ polymerization of bio-based thermosetting polyurethane/graphene oxide nanocomposites. J Appl Polym Sci 132:41751
38. Lawal AT (2019) Graphene-based nano composites and their applications. A review. Biosens Bioelectron 141: 111384

Structural Applications of Graphene Based Biopolymer Nanocomposites

Sanjeev Gautam, Bhasha Sharma, and Purnima Jain

Abstract From last few years, researchers are paying more attention to fabricating graphene based biopolymer nanocomposites due to their extensive applications. Graphene is a 2D material that contains long chain of C atom. It is the most versatile material used to reinforce the properties of biopolymer nanocomposites. There are various methods used for the preparation of graphene based biopolymer nanocomposites. Nowadays, graphene is widely used very in the biomedical applications. Graphene and is derivatives such as graphene oxide (GO), reduced graphene oxide (rGO) and functionalized graphene oxide (FGO) are incorporated with different polymers in various applications. They would be more utilized in structural applications due to the hard, thin and tough behavior. They also improve the mechanical strength of biopolymers. GO is a stable or potential material suitable for various applications such as energy storage, aircraft components etc. It has excellent absorbing property than other nanoparticles. Graphene based biopolymers nanocomposites are used in the preparation of structural components of the materials. Graphene based biopolymer nanocomposites have unique property of degradation. Graphene based biopolymers have found its application in different packaging materials. This chapter aims to discuss the different structural applications of graphene based biopolymer nanocomposites. Graphene based biopolymer nanocomposites are used as an absorbent, energy storage material and drug delivery agents.

Keywords Graphene · graphene oxide · Reduced graphene oxide · Functionalized graphene oxide · Biopolymer · Nanocomposites

1 Introduction

Nowadays, many scientists are using biopolymers in various applications instead of plastics. Synthetic polymers are responsible for environmental pollution hence

S. Gautam (✉) · B. Sharma · P. Jain
Department of Chemistry, Advance Centre for Polymer Science, Netaji Subhas University of Technology, Delhi, India
e-mail: sgsanjeevgautam88@gmail.com

B. Sharma and P. Jain (eds.), *Graphene Based Biopolymer Nanocomposites*, Composites Science and Technology,
https://doi.org/10.1007/978-981-15-9180-8_4

various biopolymers are utilized [1]. Biopolymers or biodegradable polymers are easily degraded without polluting the atmosphere when coming in contact with bacteria or fungi [2]. Biopolymers are categorized into two types which are natural and man-made polymer [3]. Atmosphere decomposition is done by three processes such as physical, chemical or biologically [4]. Deterioration was appeared by change in weight and color of the material. Biopolymers entirely break down into water and carbon dioxide molecules during degradation [5]. Biopolymers have extensive properties and are utilized in various applications. Nanocomposites are those materials in which at least one material is in nanosized range (a dimension size range is 1–1 nm) [6]. It has unique physio-chemical properties that cannot be obtained from single phase material [7, 8]. Generally, it differs from conventional composites so that inferior properties can arise from complex interaction between different phases of the material. In addition, nanocomposites have different conformational properties from a macroscopic sample of that particular material [9]. The researcher's shows rapid interest in study of nanocomposite materials. This is due to its super performance with novel properties [10]. The idea behind the using of nanocomposite in building blocks is to design it in the nanometer [11]. Hence, it builds a new material with inferior mechanical and chemical properties. Nature has mastered in the use of nanocomposites.

Human are also learning from surroundings as usual. Earlier, nanocomposites are available in the form of carbohydrates, lipids and various natural fibers [12]. There are also some strong nanocomposites such as bones, wood, shells, etc. [13]. In 1950s, organo-clay has been used to study structural properties, flow behavior or gel components [14]. In the early 1990s, Toyota Central Research Laboratories in Japan reported working of Nylon-6 as a nanocomposite [15]. The polymer matrix is strongly bonded together with nanofillers in continuous phase to form polymer nanocomposite [16]. The matrix plays an important role in determining various properties such as processability, tensile, compressive, shear or heat resistance [17]. Typically, conventional polymers can be classified into different groups based on their processing mechanism. During polymerization mechanism, small amount of nano filler is loaded to improve the thermal and mechanical properties of polymer [18]. The interaction of nanofillers into the polymer matrix may enhance the synergistic effect of each phase in polymer nanocomposite [19]. The nanofillers are highly capable of reinforcing both chemical and structural nanocomposite [20]. The dispersion or addition of nanofillers is a key parameter for significant change in mechanical and chemical properties of polymer nanocomposites [21]. The shear stress of nanocomposite was depending on the interaction between the filler and matrix [22]. The nanocomposite has a high surface to volume ratio that dramatically changes its properties [23]. The result is that the nanocomposite can be improved several times with respect to neat polymer matrix. Some nanocomposites are 1000 times harder than pristine polymer matrix [24] (Fig. 1).

In the medical field, nanocomposite offers a large potential for controlled the degradation and temporary support for a medical implant [25]. The particle size and composition of nanofillers can highly be influenced by degradation mechanism [26]. The incorporation of nanofillers in the polymer matrix is very important to obtain highly dispersed nanocomposite [27]. The morphology shows three

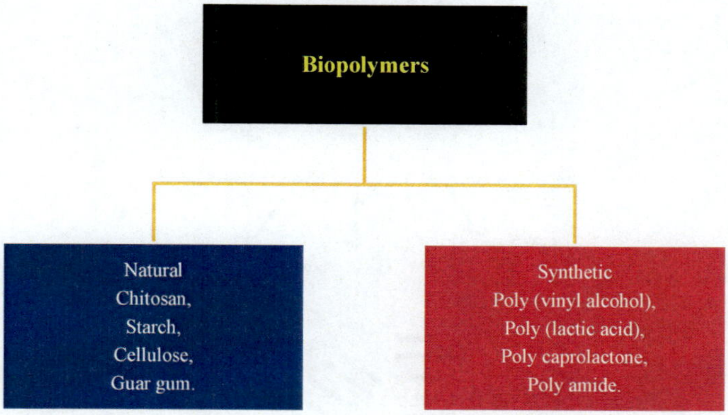

Fig. 1 Classification of Biopolymers based on natural and synthetic derivatives

types of nanocomposites that are based on the degree of separation of nanoparticles (conventional composite (micro composite), intercalated nanocomposite and exfoliated nanocomposite) [28]. The present chapter was focused on structural applications of graphene based biopolymer nanocomposites. Biopolymer nanocomposites are ecofriendly materials that have one material in nanometer range. Graphene is used as a nano-filler and biopolymer as a polymer matrix for the preparation of graphene based biopolymer nanocomposites (Figs. 2 and 3).

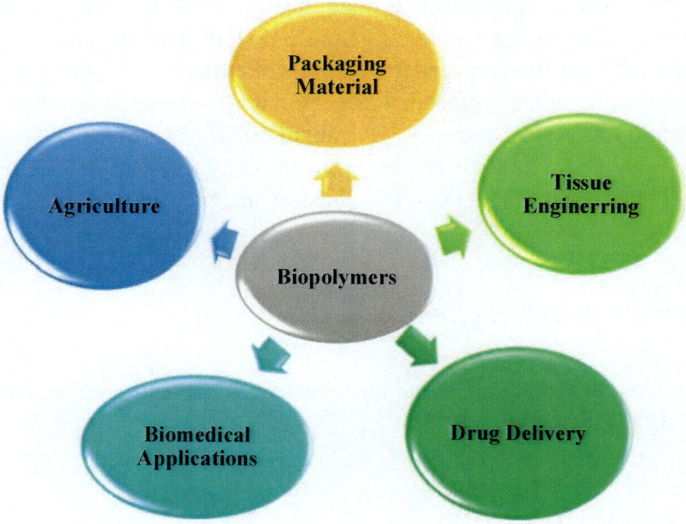

Fig. 2 Versatile applications of biopolymers

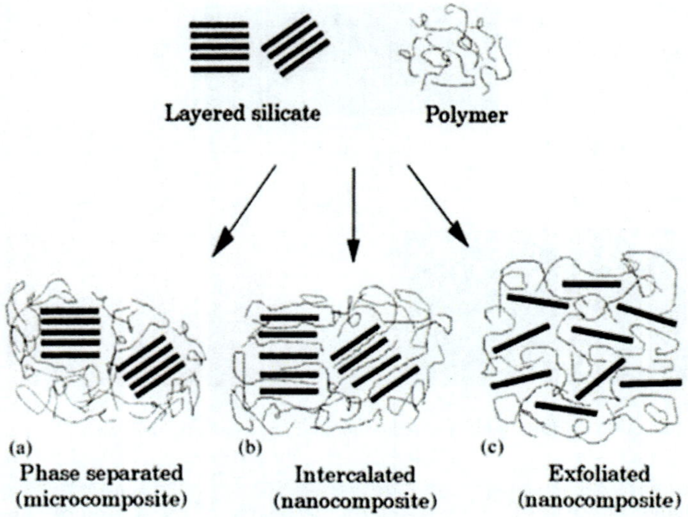

Fig. 3 Structure of polymer nanocomposites

2 Graphene

Graphene is a single layer (monolayer) of carbon atom which is tightly bound in a hexagonal honey comb lattice [29]. It is an allotrope of carbon, in the form of a plane of sp^2-bound atoms, with a molecular bond of 0.145 nm [30]. The graphene layers were stocked on the top of each part of graphite [31]. In graphite, different layers of graphene are held together via vander wall force of attraction [32]. After exfoliation of graphene from graphite, graphene has a special set of properties that set it apart from other allotropes of carbon [33]. It is about 100 times stronger than steel with respect to its thickness [34]. It is transparent or conducts heat or electricity very efficiently. It shows larger and more non-linear diamagnetism than graphite [35].

2.1 Derivatives of Graphene

There are three types of graphene and its derivatives. Graphene and its derivative are as follows (Fig. 4):

2.1.1 Graphene Oxide (GO)

Graphene is expensive and relatively difficult to synthesize [36]. Great efforts are made to make it inexpensive. GO is an oxidized form of graphene [37]. It is a single-layered layered material formed by the oxidation of graphite which is an inexpensive

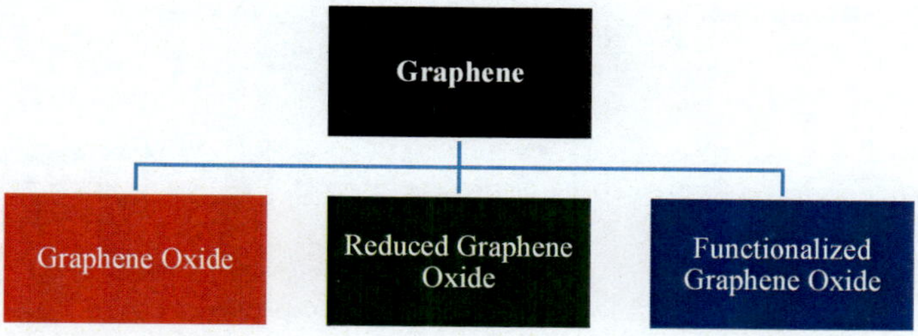

Fig. 4 Graphene and its derivatives

process [38]. It is dispersible in water and other solvents. GO is a bad conductor of heat or electricity [39]. It is commonly available in the powder form.

Synthesis of Graphene Oxide (GO)

GO was prepared by modified hummers method [40]. In detail, 3 g of graphite was mixed with H_2SO_4: H_3PO_4 (60:40) and placed in an ice bath for stirring. 15 g of $KMnO_4$ was added slowly so that the temperature of the mixture remained below 5 °C. The obtained suspension was kept for stirring for 12 h without ice bath. The solution was turned into golden brown color after addition of 15 ml of H_2O_2. The solution was changed to yellow color after continuously stirring of 2 h. The solution was placed for settling down of yellow coloured insoluble particles. The solution was filtered and washed with the help of 10% HCl solution and deionized water The solid particles was kept in vaccum oven for drying at 40 °C for 24 h to obtain fine powder of GO (Fig. 5).

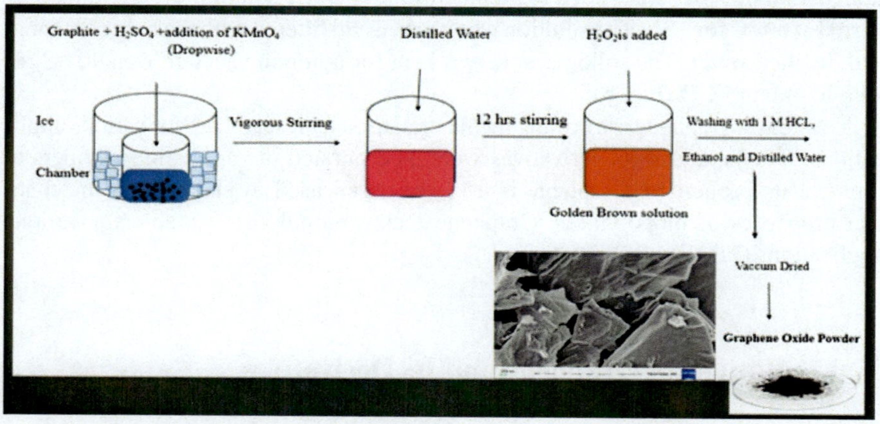

Fig. 5 Preparation of graphene oxide (GO)

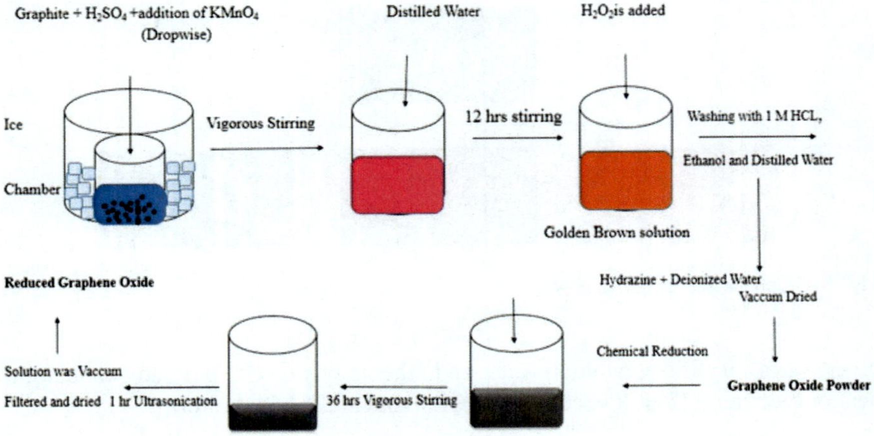

Fig. 6 Preparation of Reduced graphene oxide (rGO)

2.1.2 Reduced Graphene Oxide (rGO)

GO is a form of graphene that contains epoxide and carboxylic linkages as functional group [41]. Its many properties are different from GO. Graphene reduces its oxide group during preparation of GO, which is often called as reduced graphene oxide (rGO). It is also obtained from graphite oxide, that composed of several layers of GO. The series reduction of GO turns into reduced graphene oxide (rGO) [42].

rGO was obtained by chemical reduction of GO with the help of hydrazine hydrate [43]. 1% of GO solution was sonicated for 2 h for complete dispersion. After sonication, 0.4 ml of N_2H_4 was added to prepared solution. The solution was continuously stirred at 60 °C for 12 h. The solution was washed and filtered with the help of ethanol and distilled water. The solid particle was kept for dying in vaccum oven 40° C for 24 h to obtain FGO (Fig. 6).

Nowadays, graphene is a promising material used by researchers for various applications. Graphene and its derivatives were incorporated in various biopolymers to improve its properties. Graphene is a key material used to enhance the mechanical properties of biopolymers. Graphene makes biopolymer suitable for various applications (Fig. 7).

3 Limitations of Graphene and Its Derivatives

The method for production of graphene is very expensive and required some toxic chemicals [44]. Some chemical modification is required. The graphene products are available at high cost product [38]. Therefore its derivatives are prepared easily in the lab. It is done carefully because reaction is explosive [45]. Its derivatives have

Fig. 7 Application of Graphene and its derivatives

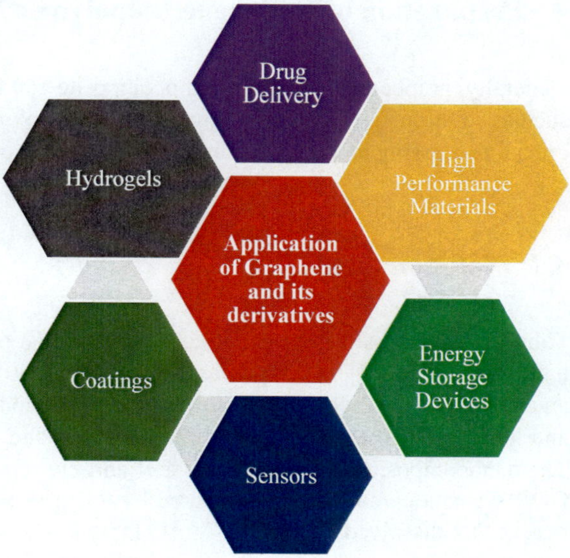

different properties than pristine graphene [46]. The advantages and disadvantages of graphene and its derivative are as follows (Table 1):

Table1 Graphene and its derivatives

S. No	Material	Advantages	Disadvantages	Limitations	References
1	Graphene	✓ Harder and tough ✓ Transfer electron ✓ Perfect barrier	✓ Expensive ✓ Toxic reaction	✓ Required modification	Cygan et al. [47]
2	Graphene oxide (GO)	✓ Easily prepared ✓ Strong	✓ Expensive production ✓ Low conductivity	✓ Explosive reaction	Hou et al. [48]
3	Reduced graphene oxide (RGO)	✓ Hydrophobic ✓ Flexible ✓ Surface 2023 modified	✓ Insoluble ✓ Low density	✓ Chemically reduced	Wang et al. [49] Kim et al. [50]

4 Preparation of Graphene Biopolymer Nanocomposites

Generally, graphene biopolymer nanocomposite can be prepared by three methods such as solution casting, melt mixing and in situ polymerization [51]. These are the crucial processing methods to fabricate polymer nanocomposites (Fig. 8).

4.1 Solution Casting

Solution Casting is the easiest process used for the preparation of graphene based nanocomposites [52]. In this process the polymer matrix dissolves in suitable solvent and nanofiller is dispersed in the solvent via sonication process [53]. Various organic and inorganic solvent such as water, chloroform and acetic acid may also be used. The nanocomposites were obtained after entirely removal of solvent. For example, GO/PVA nanocomposite was prepared by this process [54]. PVA act as a polymer matrix that dissolved in water at 50 °C for 1 h. GO is dispersed in the solvent via sonication. The solution was placed in oven at 50 °C for 24 h after complete mixing to fabricate GO/PVA nanocomposite. This nanocomposite is used in the packaging and biomedical application.

4.2 Melt Mixing

Melt Mixing is an easy and expensive technique used for the preparation of graphene based polymer nanocomposites [55]. The mixing is done via different process such as extrusion, compression or injection molding [56]. Nanofillers are directly dispersed into the melted polymer matrix. It is used for the mass production of polymer nanocomposites. Graphene is used as nanofiller that dispersed into molten polymer matrix at high temperature [57]. For example, GO/PLA nanocomposite was prepared by melt mixing technique [58]. In this process GO and PLA was mixed in a grinder.

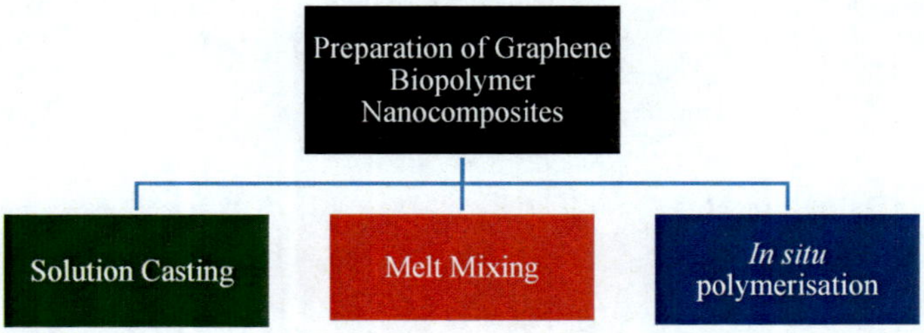

Fig. 8 Preparation techniques of graphene biopolymer nanocomposites

The batch was placed in compression molding at 100 °C to obtained GO/PLA nanocomposite. GO/PLA nanocomposites were used as filament in 3D printing.

4.3 In Situ Polymerization

This process involves completely dispersion of nanofiller into polymer matrix [59]. This process does not depend upon solvents. The nanofillers are adjoined into polymer matrix via covalent or H bonding with the help of initiator [60]. This is used for the preparation of nanocomposites by thermally unstable polymers. For example, GO/PLLA nanocomposite was prepared by in-situ polymerization technique [61]. GO was completely dispersed into poly PLLA with help of initiator at given temperature. GO improves the stability and electrical conductivity of PLLA. Therefore, GO/PLLA nanocomposite was used in electrical applications.

5 Role of Matrix in Graphene Based Nanocomposites

5.1 Graphene Based Biopolymer Nanocomposites

The role of matrix is very important phenomena for the preparation of graphene based biopolymer nanocomposites. Biopolymer such as PVA, PCL, PLA is used as a matrix in graphene biopolymer nanocomposites. The matrix provides strength, structure and stiffness to the nanocomposites. Biopolymers have poor mechanical properties; hence various nanoparticles are incorporated to improve their mechanical properties. Nowadays researchers are paying more attention towards biopolymers. The biopolymers have variety of application in various fields such as construction, aircraft and electrical applications. Graphene and its derivatives are used as nanomaterial to enhance mechanical properties of biopolymers. Various application of graphene based biopolymer nanocomposites are as follows (Table 2):

5.2 Graphene Based Synthetic Polymer Nanocomposites

Graphene based synthetic polymer nanocomposites are those in synthetic polymers (poly olefins, polyurethane etc.) are used as matrix and graphene as reinforcement. It has dynamic mechanical and chemical properties. It is used in packaging, sensing and automobile applications. The major drawback of graphene based synthetic polymer

Table 2 Synthesis and Application of graphene based biopolymer nanocomposites

S. No	Nanofiller	Polymer matrix	Processing technique	Properties	Remarks	References
1	GO	Starch	Solution casting	Excellent thermal and mechanical properties	Fabrication of ecofriendly nanocomposites used in photochemical treatment	Peregrino et al. [62]
2	GO	PVA	In-situ polymerization	Mechanical measurements	Enhancement in tensile strength from 42.3 to 50.8 MPa and Young's modulus from 1477 to 2123 MPa. Used in packaging applications	Li et al. [63]
3	GO	PLA	Solution casting	Morphological properties	Used in preparation of tissues and scaffolds	Zhao et al. [64]
4	GO	PCL	Solution casting	Electrical properties	Electrical conductivity increases from 2.0×10^{-5} to 2.19×10^{-13} S/cm on adding of GO	Nezakati et al. [65]
5	Graphene oxide	Chitosan	Electro spinning method	Drug releasing property	Used as a drug delivery agent in lung cancer treatment	Ardeshirzadeh et al. [66]

nanocomposites is that it is non-decomposable. It creates pollution in the environment. This is the big reason for more use of graphene based biopolymer nanocomposites as compared to these nanocomposites. The properties of various graphene based synthetic nanocomposites are as follows (Table 3):

5.3 *Graphene Nanocomposites with Other Nanofiller*

Various are used as reinforcement with graphene to enhance physical, chemical and mechanical properties of nanocomposites. These nanocomposites have superior application in the field of biomedical and packaging areas. The application depends

Table 3 Synthesis and application of graphene based synthetic nanocomposites

S. No	Nanofiller	Polymer matrix	Processing technique	Properties	Remarks	References
1	RGO	PU	In-situ method	Excellent mechanical and thermal properties	Increase in tensile strength 27.8 Mpa or tensile modulus 36.3 Mpa and used in biomedical applications	Thakur et al. [67]
2	FGO	Poly-aniline	In-situ polymerization	Electro-chemical Measurements	Increment in electrical conductivity 8.66 S/cm. Used in preparation of super capacitor	Kumar et al. [68]
3	GO	Epoxy	Solution casting	Barrier and corrosion properties	Improvement in corrosion properties and used as a coating material for steel material	Pourhashem et al. [69]
4	GO	PVC	Solution casting	Surface energy	Surface energies was measured at ($44 \leq lv \leq 72.8$ mJ/m^2)	Deshmukh et al. [70]
5	GO	LDPE	Melt blending	Electrical charge properties	Enhancement in electrical properties and used in electrical insulation	Mancinelli et al. [71]

on the choice of matrix. The properties and applications of different graphene and other nanofiller nanocomposites are as follows (Table 4):

Table 4 Synthesis and application of graphene and different nanofillers based nanocomposites

S. No	Nanofiller	Grapheneand its derivative	Processing technique	Properties	Remarks	References
1	Ag	rGO	Solution casting	Enhancement in antibacterial and absorbing properties	Biomedical applications	Xu et al. [72]
2	Au	GO	Solution casting	Used as photo thermal therapy in cancer treatment	Sensing and biomedical applications	Turcheniuk et al. [73]
3	ZnO	GO	Solution casting	Improvement in morphological and optical properties	Gas sensing applications	Marlinda et al. [74]
4	TiO_2	GO	Solution casting	Improvement in anticorrosion or corrosion properties	Coating application	Yu et al. [75]
5	Iron Oxides	GO	Solution casting	Increase magnetic and electric properties	Biomedical applications, drug delivery and photo thermal therapy in cancer treatment	Ma et al. [76]
6	CNT	GO	Solution casting	Improvement in surface area and absorption properties	Used as removal of metal ions from the environment	Liu et al. [77]
7	MMT	GO	Electro spinning method	Enhancement in thermal and mechanical properties	Packaging applications	Wang et al. [78]

6 Structural Application of Graphene Based Biopolymer Nanocomposites

6.1 Sensors

The electronic applications are of various types such as sensor. A sensor is an electronic device that can measure the physical changes [79]. Sensors are used in needs of human daily life. This device is used in various fields such as machines, vehicles, medicines, space technology etc. Various non-degradable electronic devices are available in the markets. Currently, the demand of environmentally friendly electronics devices increases rapidly. Kafy et al. reported the preparation of graphene/cellulose based bionanocomposites [80]. Graphene is a 2D honey comb structure that is strongly attached with C atoms [81]. Graphene and its derivative are used as a reinforcement to improve the properties of nanocomposites. Cellulose is a biodegradable polymer that is full of organic resources [82]. It has non-toxic behavior. Cellulose is a thermoplastic biopolymer consists of primarily cellulose ester (cellulose acetate and nitrocellulose) and its derivatives (celluloid). For example, cellulose acetate is a costly material and utilized in packaging applications [83]. Cellulose and its derivative have been used to make sensors. Graphene/cellulose based bionanocomposites are used as biodegradable materials. Graphene/cellulose nanocomposite is an inexpensive material used for the preparation of ecofriendly sensor for solvent [84]. This nanocomposite is used for the production of highly flexible and portable super capacitors [85]. Highly stable ecofriendly energy storage devices are also prepared by these nanocomposites [86]. It is also used in the preparation of sensors used for detecting finger [87]. It improves the electrical conductivity of electronic devices [88].

6.2 Supercapacitor

Supercapacitor is high energy storage device as compared to other capacitors [89]. It has tendency to store high energy at low voltage. It plays as a role of bridge between rechargeable batteries and electrolytic capacitors [90]. Currently, it is used in various fields such as automotive, aircraft etc. for energy storage. Nowadays, various biopolymer nanocomposites are used for the preparation of supercapacitors. Graphene and its derivatives play a key role in enhancing physiochemical properties of biopolymers [91]. It has high surface area, excellent thermal conductivity and charge carrier materials. It is doped in biopolymers to improve their electrical properties. Graphene was incorporated in various biopolymers for the preparation of supercapacitors. Graphene based biopolymer nanocomposites exhibits excellent energy storage material. Therefore, various researchers used graphene based biopolymers nanocomposites for the production of supercapacitors. PVA/GO/ZnO bionanocomposites are used for the preparation of high energy storage capacitors [92]. ZnO/GO nanoparticles are also used for the larger manufacturer of power and energy storage ecofriendly capacitors [93]. GO improves the performance of ecofriendly capacitors [94]. GO improves the physical and chemical properties of supercapacitors [95]. GO incorporated in the nano-fiber for the preparation of electrode used in supercapacitors [96].

6.3 Biomedical Applications

Graphene and its derivatives have found its wide range of application in biomedical field due to its excellent structural and chemical properties [97]. The biomedical applications include scaffolds for tissue engineering and drug delivery agent. Tissue engineering is a mechanism of engineering and material methods that replicates biological cells with the combination of various cells [98]. During tissue engineering, it provides a structure to the tissue in the form of scaffold [99]. The scaffolds are basically used for the replacement of old tissues [100]. Drug delivery system is the process of sending drug into the human body [101]. GO is hard, tough and excellent mechanical properties. Therefore, graphene based biopolymer nanocomposites are used for the preparation of long time stable drug delivery agents and scaffolds [102]. GO induced in PVA and nano fibre for the preparation of ecofriendly composite via electro spinning method [103]. These nanocomposites were utilized in bone tissue engineering and drug delivery system. GO/ZnO nanoparticles doped with starch to form degradable nanocomposites [104]. The antibacterial property of these nanocomposites was improved due to presence of ZnO. The antibacterial behavior enhances synergistic effect of nanocomposites. Therefore, these nanocomposites are utilized in medical applications. GO/gold nanocomposites are used for the preparation of drug delivery system for cancer patients [105].

6.4 Civil Infrastructure

Nowadays, the building and their infrastructures are developing very fast in whole over the world. Therefore, cement is largely using in the construction field. Cement is used as a hardener to connect the two bricks with each other firmly [106]. Nanoparticles are used with cement to improve overall properties of concrete [82]. GO improves the thermal, physical and mechanical properties of cement. Hunain et al. reported that graphene improves overall performance of graphene/cement nanocomposite [107]. It improves tensile strength and interfacial strength of cement. Hence, graphene/cement based nanocomposites are utilized in infrastructure applications. GO is most abundant material used to enhance properties of cement or concrete. GO improves workability, hardness and tensile strength of cement [108]. Hence construction industries are utilizing graphene/cement based material. Graphene improves the absorbing power of electromagnetic waves of cement composites [85]. Cement becomes hard and durable due to presence of graphene [109].

6.5 *Hydrogels*

Hydrogels are combination of hydrophilic polymer chain and it is cross-linked in 3D manner [110]. It doesn't dissolve in water but it has excellent absorbing property. Hydrogels involve various applications such as adhesives, gels, gums, contact lens etc. [111]. Graphene improves the performance of hydrogels. It was reported that absorbing property of environment friendly PVA/CMC hydrogels was improved [112]. This is used as absorbent of dyes in waste water treatment system. GO/chitosan hydrogels are used for the manufacturing of column in water purification system [113]. The stable or ecofriendly hydrogels were fabricated to absorb of anionic dyes [114]. The main application of graphene based biopolymer nanocomposite is that it is used for the removal of pollutants and toxic materials from water [115]. GO/polysaccharides nanocomposites are used in wound dressing applications [116].

7 Conclusion

This chapter discussed about various structural applications of graphene based biopolymer nanocomposites. Graphene-based biopolymer nanocomposites are prepared by three methods. Graphene with various polymers such as PLA, PVA etc. are used in various applications.

- Graphene/Cellulose nanocomposites have excellent degradation behavior. GO improves the electrical performance of graphene cellulose nanocomposites. It is potentially utilized as a preparation of sensing material.
- Ecofriendly capacitors are prepared by graphene based biopolymer nanocomposites. ZnO/GO ecofriendly nanocomposites are responsible for the production of high energy storage capacitors.
- GO shows some antibacterial property with other biopolymers. Hence most of the graphene based biopolymer nanocomposites are used in tissue engineering applications.
- GO with some nano-fibre improves the structural performance of cement nanocomposites. GO improve overall properties of cements such as mechanical, oxidative and physical properties.
- GO has excellent absorbing behavior. It improves the absorbing property of biopolymer nanocomposites. Hence, it is used as hydrogel for the preparation of various products such as column for water purifier, absorbing material for water filtration product.

Acknowledgements The authors would like to thank Department of Chemistry, Netaji Subhas University of Technology for their financial supports.

References

1. Sharma B, Malik P, Jain P (2018) Biopolymer reinforced nanocomposites: a comprehensive review. Mater Today Commun 16:353–363
2. Giri J, Adhikari R (2020) Biodegradable copolyester-based natural fibers–polymer composites: morphological, mechanical, and degradation behavior. In: Advances in sustainable polymers. Springer, Singapore, pp 289–319
3. Sharma B, Malik P, Jain P (2019) To study the effect of processing conditions on structural and mechanical characterization of graphite and graphene oxide-reinforced PVA nanocomposite. Polym Bull 76(8):3841–3855
4. de Bruyn WJ, Clark CD, Senstad M, Toms•N, Harrison AW (2020) Biological degradation of ethanol in Southern California coastal seawater. Mar Chem 218:103703
5. Brebu M (2020) Environmental degradation of plastic composites with natural fillers—a review. Polymers 12(1):166
6. Sharma B, Shekhar S, Gautam S, Jain P (2018) Dynamic shear rheology behavior and long term stability kinetics of reduced graphene oxide filled poly (vinyl alcohol) biofilm. Polym Testing 69:583–592
7. Mullani N, Ali I, Dongale TD, Kim GH, Choi BJ, Basit MA, Park TJ (2020) Improved resistive switching behavior of multiwalled carbon nanotube/TiO_2 nanorods composite film by increased oxygen vacancy reservoir. Mater Sci Semicond Process 108:104907
8. Sharma B, Gautam S, Shekhar S, Sharma R, Rajawat DS, Jain P (2019) Facile synthesis of poly (vinyl alcohol) nanocomposite & its potential application to enhance electrochemical performance. Polym Testing 74:119–126
9. Schmidt D, Shah D, Giannelis EP (2002) New advances in polymer/layered silicate nanocomposites. Curr Opin Solid State Mater Sci 6(3):205–212
10. Wu H, Fahy W, Kim S, Kim H, Zhao N, Pilato L, Kafi A, Bateman S, Koo JH (2020) Recent developments in polymers/polymer nanocomposites for additive manufacturing. Progr Mater Sci:100638
11. Jain P, Shekhar S, Chaudhary V (2019) Synergistic reinforcement of graphene oxide-polyvinyl alcohol bionanocomposite to enhance mechanical performance. Res Rev J Phys 8(3):20–26
12. Preethi R, Leena MM, Moses JA, Anandharamakrishnan C (2020) Biopolymer nanocomposites and its application in food processing. In: Green nanomaterials. Springer, Singapore, pp 283–317
13. Murugesan S, Scheibel T (2020) Copolymer/clay nanocomposites for biomedical applications. Adv Funct Mater 1908101
14. Sreedharan V, Sivapullaiah PV (2014) Swell behavior of organo clay and organo clay-bentonite mixtures. In: Geo-congress 2014: geo-characterization and modeling for sustainability, pp 1940–1950
15. Shelley JS, Mather PT, DeVries KL (2001) Reinforcement and environmental degradation of nylon-6/clay nanocomposites. Polymer 42(13):5849–5858
16. Hegde M, Yang L, Vita F, Fox RJ, van de Watering R, Norder B, Lafont U, Francescangli O, Madsen LA, Picken SJ, Samulski ET, Dingemans., T.J. (2020) Strong graphene oxide nanocomposites from aqueous hybrid liquid crystals. Nat Commun 11(1):1–7
17. Sun BG, Lei Q, Guo Y, Shi HQ, Sun JB, Yang KX, Zhou H, Li YQ, Hu N, Wang H, Fu SY (2019) Enhanced mechanical properties at 400° C of carbon fabric reinforced phthalonitrile composites by high temperature postcure. Compos B Eng 166:681–687
18. Bhasha SS, Jain P, Thin films prepared using sol gel dip coating method. Metal Chalcogenide Nanostruct Charact Synth 86
19. Wang P, Xiong Z, Xiong H, Cai J (2020) Synergistic effects of modified TiO_2/multifunctionalized graphene oxide nanosheets as functional hybrid nanofiller in enhancing the interface compatibility of PLA/starch nanocomposites. J Appl Polym Sci
20. Shekhar S, Sarkar A, Sharma B, Jain P (2019) Electrochemical evaluation of functionalized graphene oxide filled PVA-chitosan biohybrid for supercapacitor applications. J Appl Polym Sci: 48610

21. Tanaka T, Montanari GC, Mulhaupt R (2004) Polymer nanocomposites as dielectrics and electrical insulation-perspectives for processing technologies, material characterization and future applications. IEEE Trans Dielectr Electr Insul 11(5):763–784
22. Manral A, Ahmad F, Sharma B (2019) Advances in curing methods of reinforced polymer composites. Reinforced polymer composites: processing, characterization and post life cycle assessment
23. Aslfattahi N, Saidur R, Arifutzzaman A, Sadri R, Bimbo N, Sabri MFM, Maughan PA, Bouscarrat L, Dawson RJ, Said SM, Goh BT, Sidlik NAC (2020) Experimental investigation of energy storage properties and thermal conductivity of a novel organic phase change material/MXene as a new class of nanocomposites. J Energy Storage 27:101115
24. Wohlhauser S, Kuhnt T, Meesorn W, Montero de Espinosa L, Zoppe JO, Weder C (2020) One-component nanocomposites based on polymer-grafted cellulose nanocrystals. Macromolecules 53(3):821–834
25. Pina S, Oliveira JM, Reis RL (2015) Natural-based nanocomposites for bone tissue engineering and regenerative medicine: A review. Adv Mater 27(7):1143–1169
26. Ramanathan T, Stankovich S, Dikin DA, Liu H, Shen H, Nguyen ST, Brinson LC (2007) Graphitic nanofillers in PMMA nanocomposites—an investigation of particle size and dispersion and their influence on nanocomposite properties. J Polym Sci Part B Polym Phys 45(15):2097–2112
27. Bhasha S, Singh S, Jain P, Malik P (2015) Synthesis and characterization of nanocrystalline zinc oxide thin films via green chemistry. J Nanoanal 2(1):10–16
28. Paul DR, Robeson LM (2008) Polymer nanotechnology: nanocomposites. Polymer 49(15):3187–3204
29. Jung J, DaSilva AM, MacDonald AH, Adam S (2015) Origin of band gaps in graphene on hexagonal boron nitride. Nat Commun 6(1):1–11
30. Jin Y (2004) Atomistic simulations of fracture of 2d graphene systems and the elastic properties of carbon nanotubes
31. Thakur AK, Majumder M, Singh SB (2019) Graphene and its derivatives for secondary battery application. In: Surface engineering of graphene. Springer, Cham, pp. 53–80
32. Li C, Chou TW (2003) Elastic moduli of multi-walled carbon nanotubes and the effect of van der Waals forces. Compos Sci Technol 63(11):1517–1524
33. Terrones M, Botello-Méndez AR, Campos-Delgado J, López-Urías F, Vega-Cantú YI, Rodríguez-Macías FJ, Elías AL, Munoz-Sandoval E, Cano-Márquez AG, Charlier JC, Terrones H (2010) Graphene and graphite nanoribbons: morphology, properties, synthesis, defects and applications. Nano Today 5(4):351–372
34. Peng XY, Liu XX, Diamond D, Lau KT (2011) Synthesis of electrochemically-reduced graphene oxide film with controllable size and thickness and its use in supercapacitor. Carbon 49(11):3488–3496
35. Liaros N, Tucek J, Dimos K, Bakandritsos A, Andrikopoulos KS, Gournis D, Zboril R, Couris S (2016) The effect of the degree of oxidation on broadband nonlinear absorption and ferromagnetic ordering in graphene oxide. Nanoscale 8(5), 2908–2917
36. Zhao G, Wen T, Chen C, Wang X (2012) Synthesis of graphene-based nanomaterials and their application in energy-related and environmental-related areas. RSC Adv 2(25):9286–9303
37. Long Y, Zhang C, Wang X, Gao J, Wang W, Liu Y (2011) Oxidation of SO 2 to SO 3 catalyzed by graphene oxide foams. J Mater Chem 21(36):13934–13941
38. Nishina Y, Siegfried E (2020) Chemical and electrochemical synthesis of graphene oxide–a generalized view. Nanoscale
39. Zhu Y, Wan T, Guan P, Wang Y, Wu T, Han Z, Tang G, Chu D (2020) Improving thermal and electrical stability of silver nanowire network electrodes through integrating graphene oxide intermediate layers. J Colloid Interface Sci 566:375–382
40. Sharma B, Shekhar S, Malik P, Jain P (2018) Study of mechanism involved in synthesis of graphene oxide and reduced graphene oxide from graphene nanoplatelets. Mater Res Expr 5(6):065012

41. Zhen XJ, Huang YF, Yang SS, Feng ZZ, Wang Y, Li CH, Miao YJ, Yin H (2020) The effect of 500 keV proton irradiation on reduced graphene oxide paper. Mater Lett 260:126880
42. Huang HH, Joshi RK, De Silva KKH, Badam R, Yoshimura M (2019) Fabrication of reduced graphene oxide membranes for water desalination. J Membr Sci 572:12–19
43. Sharma B, Shekhar S, Gautam S, Sarkar A, Jain P (2018) Nanomechanical analysis of chemically reduced graphene oxide reinforced poly (vinyl alcohol) nanocomposite thin films. Polym Test 70:458–466
44. Deka MJ, Das P, Boruah PK, Barthakur P, Gogoi A, Das MR (2020) Plasmonic nanoparticles decorated graphene sheets for detection of water pollutants. In: Sensors in water pollutants monitoring: role of material. Springer, Singapore, pp 79–106
45. Muniyalakshmi M, Sethuraman K, Silambarasan D (2020) Synthesis and characterization of graphene oxide nanosheets. Mater Today Proc 21:408–410
46. Han C, Xu YJ (2020) The surface chemistry of graphene-based materials: functionalization, properties, and applications. In: Interface science and technology, vol 31). Elsevier, pp 453–474
47. Cygan T, Petrus M, Wozniak J, Cygan S, Teklińska D, Kostecki M, Jaworska L, Olszyna A (2020) Mechanical properties and tribological performance of alumina matrix composites reinforced with graphene-family materials. Ceram Int 46(6):7170–7177
48. Hou S, Su S, Kasner ML, Shah P, Patel K, Madarang CJ (2010) Formation of highly stable dispersions of silane-functionalized reduced graphene oxide. Chem Phys Lett 501(1–3):68–74
49. Wang K, Pang J, Li L, Zhou S, Li Y, Zhang T (2018) Synthesis of hydrophobic carbon nanotubes/reduced graphene oxide composite films by flash light irradiation. Front Chem Sci Eng 12(3):376–382
50. Kim J, Jeon JH, Kim HJ, Lim H, Oh IK (2014) Durable and water-floatable ionic polymer actuator with hydrophobic and asymmetrically laser-scribed reduced graphene oxide paper electrodes. ACS Nano 8(3):2986–2997
51. Rouf TB, Kokini JL (2016) Biodegradable biopolymer–graphene nanocomposites. J Mater Sci 51(22):9915–9945
52. Hegde M, Yang L, Vita F, Fox RJ, van de Watering R, Norder B, Lafont U, Francescangeli U, Madsen LA, Picken SJ, Samulski ET, Dingemans TJ (2020) Strong graphene oxide nanocomposites from aqueous hybrid liquid crystals. Nat Commun 11(1):1–7
53. Huang X, Li R, Zeng L, Li X, Xi Z, Wang K, Li Y (2020) A multifunctional carbon nanotube reinforced nanocomposite modified via soy protein isolate: a study on dispersion, electrical and mechanical properties. Carbon 161:350–358
54. Cobos M, De-La-Pinta I, Quindós G, Fernández MJ, Fernández MD (2020) Synthesis, physical, mechanical and antibacterial properties of nanocomposites based on poly (vinyl alcohol)/graphene oxide-silver nanoparticles. Polymers 12(3):723
55. Graziano A, Garcia C, Jaffer S, Tjong J, Sain M (2020) Novel functional graphene and its thermodynamic interfacial localization in biphasic polyolefin systems for advanced lightweight applications. Compos Sci Technol 188:107958
56. Marathe UN, Bijwe J (2020) High performance polymer composites-influence of processing technique on the fiber length and performance properties. Wear 446:203189
57. Majumdar P, Chakrabarti A (2020) Diverse applications of graphene-based polymer nanocomposites. In: Diverse applications of organic-inorganic nanocomposites: emerging research and opportunities. IGI Global, pp 47–82
58. Wang Y, Lei M, Wei Q, Wang Y, Zhang J, Guo Y, Saroia J (2020) 3D printing biocompatible l-Arg/GNPs/PLA nanocomposites with enhanced mechanical property and thermal stability. J Mater Sci 55(12):5064–5078
59. Divakaran N, Zhang X, Kale MB, Senthil T, Mubarak S, Dhamodharan D, Wu L, Wang J (2020) Fabrication of surface modified graphene oxide/unsaturated polyester nanocomposites via in-situ polymerization: comprehensive property enhancement. Appl Surf Sci 502:144164
60. Uzunpinar C (2011) Carbon nanofiller reinforced epoxy nanocomposites. Doctoral dissertation, Auburn University

61. Layek RK, Nandi AK (2013) A review on synthesis and properties of polymer functionalized graphene. Polymer 54(19):5087–5103
62. Peregrino PP, Sales MJ, da Silva MF, Soler MA, da Silva LF, Moreira SG, Paterno LG (2014) Thermal and electrical properties of starch–graphene oxide nanocomposites improved by photochemical treatment. Carbohyd Polym 106:305–311
63. Ma J, Li Y, Yin X, Xu Y, Yue J, Bao J, Zhou T (2016) Poly (vinyl alcohol)/graphene oxide nanocomposites prepared by in situ polymerization with enhanced mechanical properties and water vapor barrier properties. RSC Adv 6(55):49448–49458
64. Zhao L, Liu X, Zhang R, He H, Jin T, Zhang J (2015) Unique morphology in polylactide/graphene oxide nanocomposites. J Macromolec Sci Part B 54(1):45–57
65. Nezakati T, Tan A, Seifalian AM (2014) Enhancing the electrical conductivity of a hybrid POSS–PCL/graphene nanocomposite polymer. J Colloid Interface Sci 435:145–155
66. Ardeshirzadeh B, Anaraki NA, Irani M, Rad LR, Shamshiri S (2015) Controlled release of doxorubicin from electrospun PEO/chitosan/graphene oxide nanocomposite nanofibrous scaffolds. Mater Sci Eng C 48:384–390
67. Thakur S, Karak N (2014) Multi-stimuli responsive smart elastomeric hyperbranched polyurethane/reduced graphene oxide nanocomposites. J Mater Chem 2(36):14867–14875
68. Kumar NA, Choi HJ, Shin YR, Chang DW, Dai L, Baek JB (2012) Polyaniline-grafted reduced graphene oxide for efficient electrochemical supercapacitors. ACS Nano 6(2):1715–1723
69. Pourhashem S, Vaezi MR, Rashidi A, Bagherzadeh MR (2017) Exploring corrosion protection properties of solvent based epoxy-graphene oxide nanocomposite coatings on mild steel. Corros Sci 115:78–92
70. Deshmukh K, Khatake SM, Joshi GM (2013) Surface properties of graphene oxide reinforced polyvinyl chloride nanocomposites. J Polym Res 20(11):286
71. Mancinelli P, Fabiani D, Schirone F, Vanga-Bouanga C, Fréchette MF (2015) Space charge behavior of LDPE graphene-oxide nanocomposites. In: 2015 IEEE conference on electrical insulation and dielectric phenomena (CEIDP). IEEE, pp 547–550
72. Xu WP, Zhang LC, Li JP, Lu Y, Li HH, Ma YN, Wang WD, Yu SH (2011) Facile synthesis of silver@ graphene oxide nanocomposites and their enhanced antibacterial properties. J Mater Chem 21(12):4593–4597
73. Turcheniuk K, Boukherroub R, Szunerits S (2015) Gold–graphene nanocomposites for sensing and biomedical applications. J Mater Chem B 3(21):4301–4324
74. Marlinda AR, Huang NM, Muhamad MR, An'Amt MN, Chang BYS, Yusoff N, Harrison I, Lim HN, Chia CH, Kumar SV (2012) Highly efficient preparation of ZnO nanorods decorated reduced graphene oxide nanocomposites. Mater Lett 80:9–12
75. Yu Z, Di H, Ma Y, He Y, Liang L, Lv L, Ran X, Pan Y, Luo Z (2015) Preparation of graphene oxide modified by titanium dioxide to enhance the anti-corrosion performance of epoxy coatings. Surf Coat Technol 276:471–478
76. Ma X, Tao H, Yang K, Feng L, Cheng L, Shi X, Li Y, Guo L, Liu Z (2012) A functionalized graphene oxide-iron oxide nanocomposite for magnetically targeted drug delivery, photothermal therapy, and magnetic resonance imaging. Nano Res 5(3):199–212
77. Liu T, Gao B, Fang J, Wang B, Cao X (2016) Biochar-supported carbon nanotube and graphene oxide nanocomposites for Pb (II) and Cd (II) removal. Rsc Adv 6(29):24314–24319
78. Wang Q, Li G, Zhang J, Huang F, Lu K, Wei Q (2014) PAN nanofibers reinforced with MMT/GO hybrid nanofillers. J Nanomater
79. Kruss S, Hilmer AJ, Zhang J, Reuel NF, Mu B, Strano MS (2013) Carbon nanotubes as optical biomedical sensors. Adv Drug Deliv Rev 65(15):1933–1950
80. Kafy A, Yadav M, Kumar K, Kumar K, Mun S, Gao X, Kim J (2014) Synthesis and characterization of graphene/cellulose nanocomposite. In: Nanosensors, biosensors, and info-tech sensors and systems 2014, vol 9060. International Society for Optics and Photonics, p 906008
81. Ronci F, Colonna S, Flammini R, De Crescenzi M, Scarselli M, Salvato M, Berbezier I, Jardali F, Lechner C, Pochet P, Vach H, Castrucci P (2020) High graphene permeability for room temperature silicon deposition: he role of defects. Carbon 158:631–641

82. Decroix C, Chalamet Y, Sudre G, Caroll V (2020) Thermo-mechanical properties and blend behaviour of cellulose acetate/lactates and acid systems: natural-based plasticizers. Carbohydr Polym 116072
83. Azammi AN, Ilyas RA, Sapuan SM, Ibrahim R, Atikah MSN, Asrofi M, Atiqah A (2020) Characterization studies of biopolymeric matrix and cellulose fibres based composites related to functionalized fibre-matrix interface. In: Interfaces in particle and fibre reinforced composites. Woodhead Publishing, pp 29–93
84. Liu L (2006) Bioplastics in food packaging: Innovative technologies for biodegradable packaging. San Jose State Univ Pack Eng 13:1348–1368
85. Kafy A, Sadasivuni KK, Akther A, Min SK, Kim J (2015) Cellulose/graphene nanocomposite as multifunctional electronic and solvent sensor material. Mater Lett 159:20–23
86. Weng Z, Su Y, Wang DW, Li F, Du J, Cheng HM (2011) Graphene–cellulose paper flexible supercapacitors. Adv Energy Mater 1(5):917–922
87. Kafy A, Sadasivuni KK, Kim HC, Akther A, Kim J (2015) Designing flexible energy and memory storage materials using cellulose modified graphene oxide nanocomposites. Phys Chem Chem Phys 17(8):5923–5931
88. Sadasivuni KK, Kafy A, Zhai L, Ko HU, Mun S, Kim J (2015) Transparent and flexible cellulose nanocrystal/reduced graphene oxide film for proximity sensing. Small 11(8):994–1002
89. Zhang C, Xu S, Cai D, Cao J, Wang L, Han W (2020) Planar supercapacitor with high areal capacitance based on Ti3C2/Polypyrrole composite film. Electrochim Acta 330:135277
90. Shekhar S, Sharma R, Sharma S, Sharma B, Sarkar A, Jain P (2020) An exploration of electrocatalytic analysis and antibacterial efficacy of electrically conductive poly (D-Glucosamine)/graphene oxide bionanohybrid. Carbohydr Polym 116242
91. Wan CF, Cui Y, Gai WX, Cheng ZL, Chung TS (2020) Nanostructured membranes for enhanced forward osmosis and pressure-retarded osmosis. In: Sustainable nanoscale engineering. Elsevier, pp 373–394
92. Deshmukh K, Ahamed MB, Deshmukh RR, Pasha SK, Chidambaram K, Sadasivuni KK, Ponnamma D, AlMaadeed MAA (2016) Eco-friendly synthesis of graphene oxide reinforced hydroxypropyl methylcellulose/polyvinyl alcohol blend nanocomposites filled with zinc oxide nanoparticles for high-k capacitor applications. Polym-Plast Technol Eng 55(12):1240–1253
93. Bhirud A, Sathaye S, Waichal R, Park CJ, Kale B (2015) In situ preparation of N-ZnO/graphene nanocomposites: excellent candidate as a photocatalyst for enhanced solar hydrogen generation and high performance supercapacitor electrode. J Mater Chem 3(33):17050–17063
94. El-Gendy DM, Afifi IM, Allam NK (2019) Eco-friendly, one-step synthesis of cobalt sulfide-decorated functionalized graphene for high-performance supercapacitors. J Energy Storage 24:100760
95. Sethi M, Bantawal H, Shenoy US, Bhat DK (2019) Eco-friendly synthesis of porous graphene and its utilization as high performance supercapacitor electrode material. J Alloy Compd 799:256–266
96. Sahoo S, Dhibar S, Hatui G, Bhattacharya P, Das CK (2013) Graphene/polypyrrole nanofiber nanocomposite as electrode material for electrochemical supercapacitor. Polymer 54(3):1033–1042
97. Eslahi N, Mahmoodi A, Mahmoudi N, Zandi N, Simchi A (2020) Processing and properties of nanofibrous bacterial cellulose-containing polymer composites: a review of recent advances for biomedical applications. Polym Rev 60(1):144–170
98. Chandy T (2020) Biocompatibility of materials and its relevance to drug delivery and tissue engineering. In: Biointegration of medical implant materials. Woodhead Publishing, pp 297–331
99. Chen Y, Guo C, Manousiouthakis E, Wang X, Cairns DM, Roh TT, Du C, Kaplan DL (2020) Bi-layered tubular microfiber scaffolds as functional templates for engineering human intestinal smooth muscle tissue. Adv Funct Mater 2000543

100. Christy PN, Basha SK, Kumari VS, Bashir AKH, Maaza M, Kaviyarasu K, Arasu MV, Al-Dhabi NA, Ignacimuthu S (2020) Biopolymeric nanocomposite scaffolds for bone tissue engineering applications–a review. J Drug Deliv Sci Technol 55:101452
101. Pacheco C, Sousa F, Sarmento B (2020) Chitosan-based nanomedicine for brain delivery: where are we heading? React Funct Polym 146:104430
102. Liu J, Cui L, Losic D (2013) Graphene and graphene oxide as new nanocarriers for drug delivery applications. Acta Biomater 9(12):9243–9257
103. Qi YY, Tai ZX, Sun DF, Chen JT, Ma HB, Yan XB, Liu B, Xue QJ (2013) Fabrication and characterization of poly (vinyl alcohol)/graphene oxide nanofibrous biocomposite scaffolds. J Appl Polym Sci 127(3):1885–1894
104. Sandhya PK, Jose J, Sreekala MS, Padmanabhan M, Kalarikkal N, Thomas S (2018) Reduced graphene oxide and ZnO decorated graphene for biomedical applications. Ceram Int 44(13):15092–15098
105. Wang C, Li J, Amatore C, Chen Y, Jiang H, Wang XM (2011) Gold nanoclusters and graphene nanocomposites for drug delivery and imaging of cancer cells. Angew Chem Int Ed 50(49):11644–11648
106. CAFÉ, C. (2020). 1.1 Related documents. Construction of new safari Café building at Cape May County Park & Zoo, 220356
107. Silvestre J, Silvestre N, De Brito J (2016) Review on concrete nanotechnology. Eur J Environ Civil Eng 20(4):455–485
108. Alkhateb H, Al-Ostaz A, Cheng AHD, Li X (2013) Materials genome for graphene-cement nanocomposites. J Nanomech Micromech 3(3):67–77
109. Chuah S, Pan Z, Sanjayan JG, Wang CM, Duan WH (2014) Nano reinforced cement and concrete composites and new perspective from graphene oxide. Constr Build Mater 73:113–124
110. Sun S, Ding S, Han B, Dong S, Yu X, Zhou D, Ou J (2017) Multi-layer graphene-engineered cementitious composites with multifunctionality/intelligence. Compos B Eng 129:221–232
111. Han B, Zheng Q, Sun S, Dong S, Zhang L, Yu X, Ou J (2017) Enhancing mechanisms of multi-layer graphenes to cementitious composites. Compos Appl Sci Manuf 101:143–150
112. Huebsch N, Kearney CJ, Zhao X, Kim J, Cezar CA, Suo Z, Mooney DJ (2014) Ultrasound-triggered disruption and self-healing of reversibly cross-linked hydrogels for drug delivery and enhanced chemotherapy. Proc Natl Acad Sci 111(27):9762–9767
113. Ahmadi F, Oveisi Z, Samani SM, Amoozgar Z (2015) Chitosan based hydrogels: characteristics and pharmaceutical applications. Res Pharm Sci 10(1):1
114. Dai H, Huang Y, Huang H (2018) Eco-friendly polyvinyl alcohol/carboxymethyl cellulose hydrogels reinforced with graphene oxide and bentonite for enhanced adsorption of methylene blue. Carbohyd Polym 185:1–11
115. Chen Y, Chen L, Bai H, Li L (2013) Graphene oxide–chitosan composite hydrogels as broad-spectrum adsorbents for water purification. J Mater Chem 1(6):1992–2001
116. Omidi S, Kakanejadifard A (2018) Eco-friendly synthesis of graphene–chitosan composite hydrogel as efficient adsorbent for Congo red. RSC Adv 8(22):12179–12189
117. Wang H, Li J, Ding N, Zeng X, Tang X, Sun Y, Lei T, Lin L (2020) Eco-friendly polymer nanocomposite hydrogel enhanced by cellulose nanocrystal and graphitic-like carbon nitride nanosheet. Chem Eng J 124021
118. Deng H, Yu Z, Chen S, Fei L, Sha Q, Zhou N, Chen Z, Xu C (2020) Facile and eco-friendly fabrication of polysaccharides-based nanocomposite hydrogel for photothermal treatment of wound infection. Carbohyd Polym 230:115565

Graphene Functionalized PLA Nanocomposites and Their Biomedical Applications

Ifrah Kiran, Naveed Akhtar Shad, M. Munir Sajid, Yasir Jamil, Yasir Javed, M. Irfan Hussain, and Kanwal Akhtar

Abstract Nanocomposites have been emerged a major class of materials mainly due to providing multifunctional properties in a single system. Graphene is a 2-dimensional carbon-based material and has applications in many fields due to their extraordinary properties. Polylactic acid (PLA) is a biopolymer with huge scope in the biomedical field and especially in bio-engineering but has limitations due to low crystallization and week thermal and mechanical properties. Recently, Graphene-PLA based nanocomposites are investigated to overcome constraints faced by PLA polymer materials, especially for biomedical engineering. These nanocomposites have improved crystallization of PLA polymers and also enhanced mechanical strength of the polymer required for preparing different artificial organs such as tissues, bones etc. In this book chapter, a brief discussion is made on PLA and graphene, then synthesis methods for graphene composites and their functionalization on the graphene surface. Finally, biomedical applications of Graphene functionalized PLA nanocomposites have been discussed.

Keywords PLA · Graphene · Nanocomposites · Biomedical applications · Tissue engineering

---

I. Kiran · Y. Jamil · Y. Javed (✉) · K. Akhtar
Department of Physics, Magnetic Materials Laboratory, University of Agriculture, Faisalabad, Pakistan
e-mail: myasi60@hotmail.com

N. A. Shad
Department of Physics, Government College University, Faisalabad, Pakistan

M. I. Hussain
Institute of Materials Science and Advanced Ceramics, University of Science and Technology Beijing, Xueyuan Road 30th, Haidian District, Beijing, China

M. M. Sajid
School of Physics, Henan Normal University, Xinxiang 453007, China

B. Sharma and P. Jain (eds.), *Graphene Based Biopolymer Nanocomposites*, Composites Science and Technology,
https://doi.org/10.1007/978-981-15-9180-8_5

1 Introduction

Polylactic acid (PLA) based polymers are prepared and attained attention through renewable agriculture resources as a result of polymerization of lactides that are cyclic diester of lactic acid. For many applied purposes, lactic acid acts as PLA monomer. Recent advances involve the manufacturing of monomers from renewable feedstocks in cost-effective means have placed PLA based polymers on the forefront of many evolving biodegradable resources in materials science [1, 2]. Lactic acid is a chiral based molecule with D and L isomers, "polylactic acid" belongs to the polymer group such as poly-D,L-lactic acid, poly-L-lactic and poly-D-lactic. L-isomer act as biological metabolite and consists upon major fractions of PLA derived through renewable resources, meanwhile, most of the lactic acid obtained from organic sources occur in this form. Based on the optically active D, L- and L enantiomers, PLA crystallizes in the three basic forms (α, β and γ) [3, 4]. Intrinsic structural, physiochemical and thermal properties of PLA based polymers make them important candidate in biotechnological applications to manipulate and tune with other nanomaterials. Review proposed by Stold and Sodergart presented many polymerization schemes related to the lactic acid chemistry [5]. In the presence of acidic catalysts, dimerization of lactic acid requires water removal. Dimer formation with greater yield and selectivity needs special use of catalysts having weakly basic in nature. The use of zinc and tin oxide, organo-titanates and -stannates for this purpose is very common [6].

PLA based polymers have been proved as guided tissue, surgical implant materials, drug delivery systems, bone regeneration platform and porous scaffolds for new tissues growth. PLA based composites gave tremendous uses in biomedical applications due to their thermo-mechanical properties and biodegradability [7]. Many medical devices, including drug releasing microscale nanoparticles, degradable suture and porous scaffolds have been synthesized through different types of PLA. PLA copolymers have been also considered as attractive biodegradable material in nanomedicine. Biodegradation of poly (lactic-co-glycolic acid) (PLGA) happens as hydrolysis of metabolite monomers, lactic acid and glycolic acid [8]. Since these monomers are endogenous in nature and metabolize effortlessly by the body through Krebs cycle, nominal systemic toxicity is linked with PLGA as a drug releasing vehicle and biomedical material. But major limitations associated with them include mechanical limitations, chemical reactivity, biodegradation rates and hydrolytic degradation. PLA is brittle polymer with higher stress level and poor toughness [9].

Graphene, on the other hand, is considered amongst most promising materials being examined today, not due to the academic curiosity, but their role in potential applications. In the hexagonal lattice, graphene is a single carbon packed layer, where distance between carbon–carbon atoms is 0.142 nm in 2D crystalline materials, which are stable at ambient environments. Graphene possessed a variety of fascinating properties with enhanced mechanical properties (Young's modulus ~ 1 TPa) and higher electron mobility (250,000 cm^2/Vs) at room conditions with extraordinary values of thermal conductivity (5000 W m^{-1} K^{-1}) [10, 11]. Graphene sheets showed excellent mechanical, electronic and thermal properties with their use in variety of applications including batteries, hydrogen storage systems, sensors, supercapacitors and their role as reinforcement fillers of nanocomposites. The scotch-tape method, an original top-down approach has synthesized high quality samples, but this technique is neither high-yield nor high throughput, chances of finding the discrete good quality graphene sheets is very low. To exfoliate the single carbon sheet, the van der Waals attraction amid first and second layers must intimidate which also cause a slight disturbance in the subsequent sheets [12, 13]. Many synthesis protocols with alternatives to mechanical exfoliation have been established to produce graphene on the large-scale including pyrolysis, chemical vapor deposition, electrically assisted synthesis, scotch-tape method, arc discharge, solvothermal, molecular beam epitaxy, carbon nanotube cutting, direct sonication and exfoliation methods. All of these methods offer pros and cons, dealing with scalability and cost. The reduction in graphene derivatives into graphene oxide (GO) provides the primary strategy that can produce graphene sheets in bulk amounts that are highly processable, although, not defect free. The availability of graphene in bulk quantities as powders and colloidal dispersion is a particularly attractive factor, which permit the possible synthesis of carbon-based nanomaterials [14, 15].

Through reduction of GO-oxygenated graphene sheets shielded with carboxyl, epoxy and hydroxyl groups can produce large amounts of graphene, which provides access to graphene-based nanocomposites. At GO surface, the presence of oxygen functionalities is interesting according to the chemical point of view as they provide many chemically reactive sites for enhancing carbon surface chemistry [16]. Graphene functionalized PLA nanocomposites promisingly enhance PLA properties as thermo-mechanical properties of these polymers connected with biomedical performance when encounter living systems. Hence, these properties required optimization of design criteria including strength, crystallinity, modulus, biocompatibility and morphology and in return, these properties affect in vivo degradation, tissue regeneration and cell response [17]. Moreover, recent advances in this field indicate the thermal effects on crystallinity on PLA based polymers improved significantly with graphene contents. Graphene and its derivatives influenced thermo-mechanical properties linked with semi-crystalline plastics (biopolymers), prudently synthesized and engineered nanocomposites of PLAs functionalized with graphene can remediate thermal, degradation and structural related limitations of such polymers making them strong candidate for many medical applications [18]. In this chapter, the synthesis methods for graphene based nano-composites, functionalization methods for graphene and promising applications in biomedical fields have been discussed.

2 Synthesis Protocols for Graphene Nanocomposites

2.1 Covalent Reactions

The successful transformation synthesis of graphene nanocomposite is indispensable for new catalytic applications [19]. Numerous chemical strategies have been adopted to introduce pristine graphene; generally, exfoliation confined graphite, thermal reduction or chemical quintessence of GO, graphite intercalative expansion, epitaxial growth using CVD method [20–29]. A pioneer work reported from Tour and Co-Worker with segregation of monolayer contained graphene through graphite dispersed inside the ortho-dichlorobenzene [30]. Similarly, Bourlinose et al. [31] succeeded to obtain graphite dispersion within different organic solvents, for example, per-fluorinated confined aromatic molecules, chloroacetate and pyridine. The dispersible impregnation includes monolayer and confined graphene layers, as well as few-layered of ultrathin graphite sheets that have limited defects. The dispersion ability during organic solvents supports the functionalization of graphene via different features. From a functionalization perspective, graphene sheets with aromatic groups have been prepared. The prime objective is to promote the dispersibility of graphene that exist in standard organic solvents. Dispersion of graphene sheets inside organic solvents provides a prerequisite step towards the development of nanocomposites with other materials. The functional organic groups also provide noteworthy effect, for example, chromophores offer an appropriate combined outcome that meet the better properties of graphene, including conductivity. Clearly, when aromatic functional molecules are covalently bonded on graphene surface, its extended organic behavior which promotes effective control and boost electronic features [32]. The covalent functionalization reaction of organic materials with graphene involves two regular pathways: (a) the covalent bond formation between dienophiles or free radicals of C=C bonds with graphene (b) by virtue of covalent formation during graphene confined oxygen and functional organic groups (Fig. 1) [33]. Based on both the theoretical calculations and experimental investigations with carbon nanotubes and fullerene, an escalating organic species reaction were observed with sp^2 carbon confined graphene, dienophiles and free radicals of organic material [19].

2.2 Solvothermal/Hydrothermal Method

Solvothermal and Hydrothermal synthesis methods are considered the top priority for producing composites particularly under elevated temperature/pressure in an autoclave container [34]. Typically, in hydrothermal process, aqueous solution is used, while with a solvothermal process, non-aqueous solution is preferred. For precise synthesis, the hydrothermal method is extensively employed owing to different

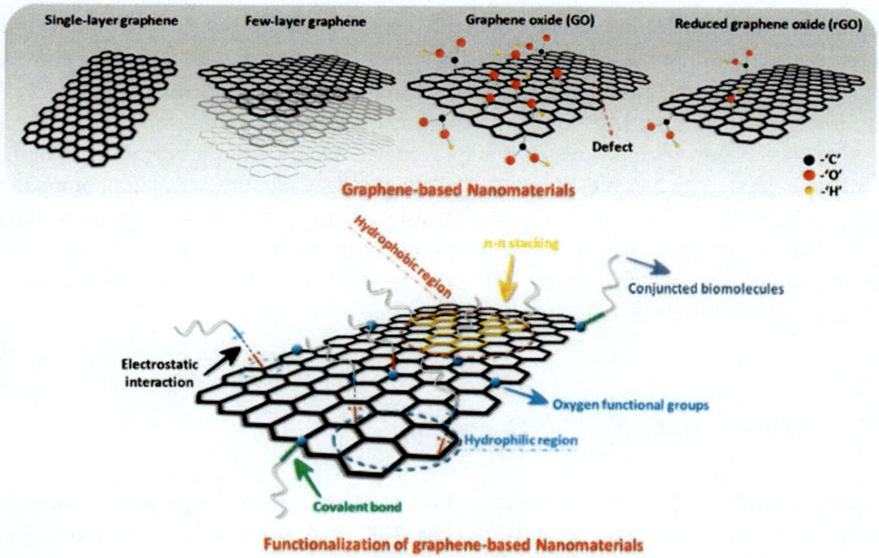

Fig. 1 Illustrations of different graphene-based nanomaterials structures and representation covalent/non-covalent functionalization [33]

advantages such as simple, facile, ecofriendly and economical. Nanocomposite materials such as GO-TiO_2 nanocomposite [35], Graphene-CeO_2 nanocomposite [36, 37]. N-doped graphene-Fe_2O_3 and GO-CuO nanocomposites [38] were obtained by this protocol. Similarly, solvothermal synthesis is used for the production carbon composites, for example, graphene at gram-scale [39], diamond [40] and CNT [41] materials. First time Bao's group [42] synthesized N-doped graphene contained 1–6 layers by solvothermal method. One further step by Yang et al. [43] who deposited silver NPs on graphene with a facile way to synthesized nanocomposites of graphene-Ag with exceptional electro-ductility for deionized water, hydrazine and ethylene glycol. Another experiment was reported with nanocomposites of graphene-Ag [44] reveals electroconductibility on near 2.94 s cm^{-1}. Using hydrothermal route Lee et al. [45] synthesized graphene wrapped TiO_2 NPs which displayed excellent photocatalytic activity under visible light.

2.3 *In Situ Electroless Chemical Deposition*

In situ process based electrochemical deposition technique is a popular control synthesis route, in which a solid metal film is deposited holding the ions solution of electro-conducting surface [46]. Considering the rapid synthesis progress, this scalable, inexpensive and facile method offers a viable potential advantage, such as high purity deposition of material and persistent control of composition. For example, monocrystalline Zinc oxide deposited nanorods are an effective donor for conductive rGO films, while, integrated organic–inorganic nanocomposites showed drastically increased efficiency for solar cell applications [47]. Similarly, nano-wall hybrids contained graphene-MnO_2 were fabricated through a one-step electrochemical approach and obtained materials displayed benchmark efficiency in electrochemical supercapacitors and biosensors [48]. Gao et al. [49] synthesized nanocomposite of graphene-PtNi via simple single-step assisted ultrasonication, which exhibits excellent performance to check non-enzymatic glucose in urine of human samples. Using same approach, Xie et al. [50] prepared rGO-Cu films which showed high electroactivity for electrical conversion materials.

2.4 Mixing and Physical Deposition

Solution mixing approach is broadly employed to synthesize novel graphene confined metal and metal oxide nanocomposites (Fig. 2) [51]. Apart from mixing, deposition is another liquid-phase mechanism that can be achieved at ambient temperature; rapid segregation as well as atomic dispersion provide support for composites along with uniform dispersion reinforcement [52]. Zeng et al. [53] reported aqueous GO with Al powder and used ultrasonic process for preparation, Al-GO nanocomposites exhibited tensile strength about 255 MPa. Prolonged and better mechanical features of graphene metal or metal oxide regularly employed to demonstrate the renewable energy and climate related problems. Williams et al. [54] reported an efficient

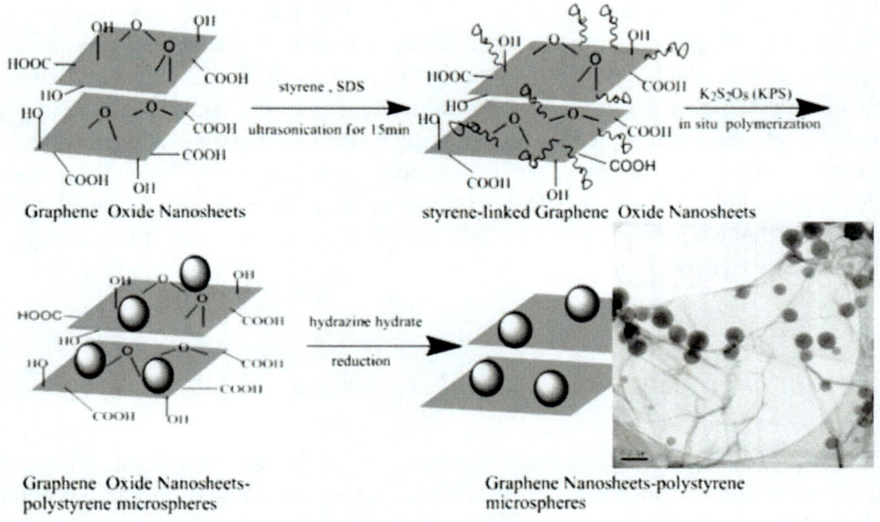

Fig. 2 The synthesis of graphene/polystyrene composites prepared by in situ emulsion polymerization. Reprinted with permission from [51]

pathway to prepare graphene-TiO_2 for UV-supported photocatalytic mitigation with GO sheets. Similar strategy was adopted by Akhavan et al. [55] formed GO-TiO2 thin films and demonstrated antibacterial efficiency with thin films towards *Esherichia coli* bacteria.

2.5 Ball-Milling Approach

Ball milling is a top-down approach which widely applied for mechanical alloying powder mixture. During the fabrication process, required sample is placed within a concealed container and governed with energy intensive collision among balls with specific diameters [56]. Indeed, this method is current interest due to facile control, high energy grinding medium, particularly for improvement in mechanical properties of materials. Using this mechanism, graphite is allowed to rupture and then produced graphene sheets that may further mixed with other materials to generate nanocomposite (Fig. 3) [57]. As synthesized graphite powder is usually ball-milled in an Argon environment to produce layered graphene [58]. Alternatively, most of the metal phthalocyanines are attached on the graphene surface using the ball milling method under Argon atmosphere [59]. For instance, He et al., synthesized graphene-Al_2O_3 powders using ball milling route and revealed that the addition of graphene

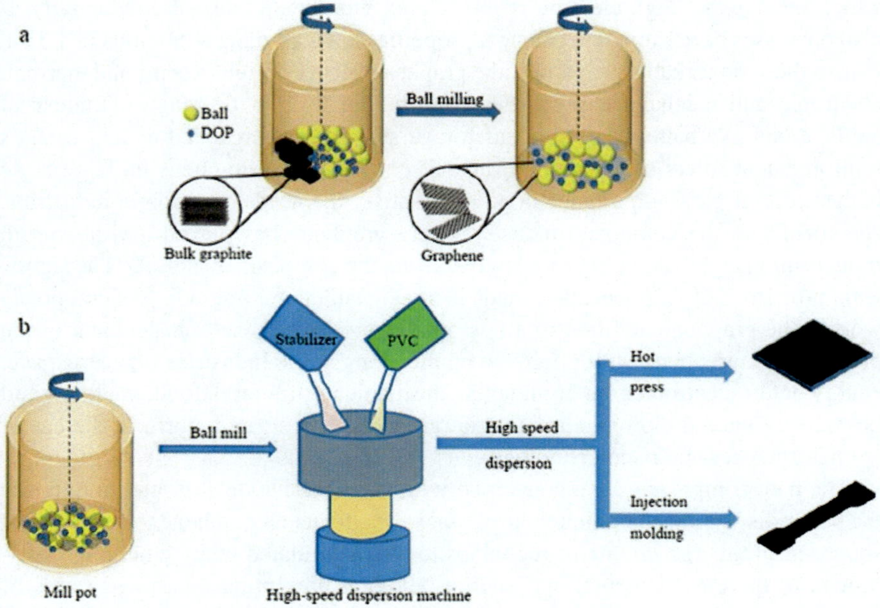

Fig. 3 **a** Exfoliation of graphite with dioctyl phthalate using mechanical the ball milling method, **b** Schematic for preparation of graphene-poly vinyl chloride composites. Reprinted with permission from [57]

can hamper grain growth and fine Al_2O_3 are generated [60]. Notably, benchmark studies manifested the hardness and strength was drastically decreased on graphene-Al composite due to the formation of aluminum carbide [61]. Similar attempt from Bastwors et al. [62] achieved a 1 wt% graphene support aluminum (A16061) by ball milling and indicated their flexural strength up to 47% and compared with reference Al6061 prepared by typical-process. Mechanical and tribological properties contained by graphene sheets supported Ni_3Al matrix, synthesized by ball-milling approach [63]. To increase the capability of multilayer graphene and lower frication coefficient along with enhance wear resistance from the nanocomposite, Recently, Yue et al. [64] fabricated GNS-Cu nanocomposite using the ball-milling route and observed tunable GNS content into a prolonged fracture and maximum tensile strength.

3 Functionalization Methods for Graphene Nanocomposites

Graphene is a material of consisting of a single carbon layer arranged in the form of a hexagonal lattice in which the carbon to carbon atom distance is 0.142 nm and proved to be purely a two-dimensional material. The graphene is a stable material at room temperature and pressure. It has stimulating properties at room temperature which involves its high electron mobility and remarkable thermal conductivity. It also possesses exceptional mechanical properties having young's modulus of 1 TPa. Due to these remarkable properties, the graphene sheets exhibit exceptional thermal, electronic and mechanical properties contributing for a wide range of industrial applications like batteries, hydrogen storage systems, supercapacitors and used as reinforcement fillers of nanocomposites. The synthesis of graphene on large scale is therefore of great importance in order to satisfy the needs of different industries, specifically in the composite industry where graphene has played a vital role in transforming a global market for manufacturing the composite materials. The significant properties of graphene have made it "magic bullet" for upgrading of composite world. The graphene addition to a host matrix produces great enhancement in the properties of graphene with improved applications in the industries like aerospace, energy fields, electronics, environmental, mechanical, structural, food, medicine and beverage. Hence the promising material graphene has created storm in the era of applied sciences and nanotechnology.

The nanocomposites have concerned with an excessive deal of attention due to the better dispersion and unique morphology. The focus on graphene to be a general source of producing unique nanocomposites has stimulated many emerging possibilities in the environmental and energy fields by presenting controlled series of functional graphene building blocks. The functionalized graphene nanocomposites are of great importance for showing their advanced applicability in different areas of

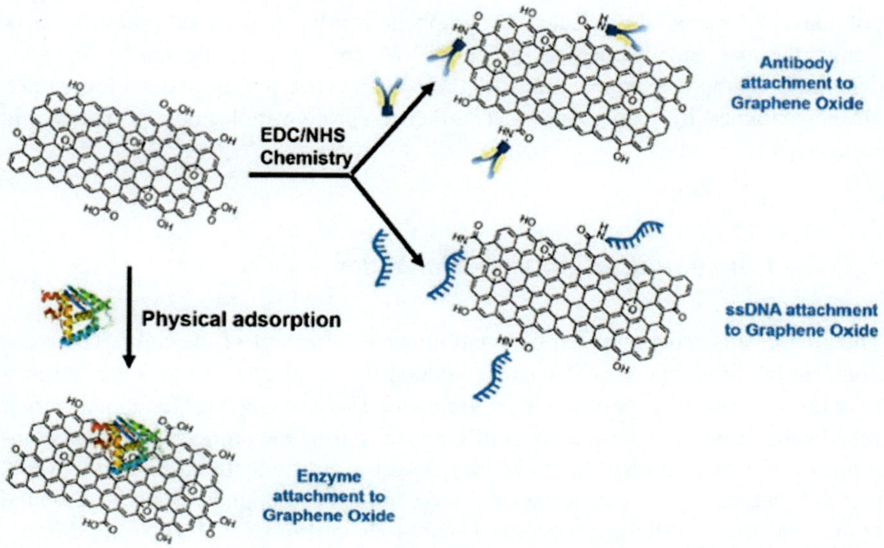

Fig. 4 Functionalization methods of different bioreceptors on the surface of graphene [65]

energy and environmental aspects. They have shown promising applications in environmental sensing purposes and the monitoring devotions to remediation. They are widely implemented in sensing the inorganic ions, bacteria detection and removal, organic species degradation, organisms and biomolecules, environmental gas sensors and also utilized in removing the hazardous species from environment. There are a variety of functionalization methods for preparing functionalized graphene nanocomposites (Fig. 4) [65]. The functionalization methods involve the hydrothermal and photochemical reaction, solvo-thermal growth, electrochemical and electrophoresis deposition, physical deposition and mixing, covalent and non-covalent methods. The controlled functionalized graphene nanocomposites are also prepared with nanoscale objects, molecules and also with the polymers which upon inserting to graphene lead the functionalized graphene nanocomposites to certain energy and environmental applications. Every functionalization method exhibits number of advantages and also faces some of shortcomings. Hence, the production of graphene nanocomposites can be enhanced in quality measures by adopting best suitable manufacturing process.

3.1 *Functionalization with Molecules and Nanoscale Objects*

There are basically three main functionalization methods involving functionalization with molecules, nanoscale objects and polymers. The integration of molecules, polymers and nanoscale objects with graphene results in modified photonic, electronic, thermal and mechanical properties which the graphene nanocomposites contain. The graphene nanocomposites therefore possess multifunctionalities which are not obtained by pure graphene or other integrating molecules, polymers and nano-objects.

3.2 (i) Functionalization with Molecules

The graphene is functionalized by small molecules in most of the ways. The functionalization of graphene is also carried out with biological molecules which influence the bio-related properties of pure graphene. The biomolecular functionalization specifically affects the bio-reactivity of graphene during the production of nanocomposites and nanohybrids. The molecules used for attachment can be DNA, RNA, proteins, peptides, pyridine, surfactants and complex compounds which increase the optical, electronic, solution processing capabilities and bio-related properties. Molecular functionalization on graphene enhances the solubility of graphene in various solvents because the chemically generated graphene shows poor solubility and is not suitable to be used in solution processing techniques on a large scale. Prussian Blue (PB), a dark blue pigment fabricated by ferrous ferrocyanide salts containing porous structure and having chemical formula $Fe_7(CN)_{18}$, and Styrene–butadiene–styrene tri-block copolymer (SBS) are used to functionalize graphene and attain high solubility of graphene which results in the possible conductive or biological analysis of solution-based fabrications. The proteins and DNA were employed in functionalization of graphene to improve solubility because these biomolecules have charges which are used to stabilize graphene in water. One more advantage of functionalization with molecules is that it increases the optical performance of graphene to a great extent. The Porphyrin-functionalized graphene showed higher optical limiting effects which were highly improved as compared with the standard optical limiting material (C_{60}). The molecular functionalization also improves the electronic properties of graphene. The tetracyanoquinodimethane (TCNQ) grafting on the surface of graphene generated a strong p-doping in graphene and produced a great improvement in its intrinsic electronic properties. The NO_2 molecules were also introduced on the surface of graphene, which also generated a p-type doping in graphene and affect charge transport. The carrier density of the graphene material depends on annealing temperature.

3.3 (ii) Functionalization with Nanoscale Objects

The graphene grafting can also be done with nanoscale objects such as nanoparticles, nanorods, nanowires and nanosheets and can obtain different graphene heterostructures and composites. These nanoobjects possess different intrinsic characteristics. On this basis, the nanoobjects offer different functionalities which are not present in pure graphene. The graphene nanocomposites functionalized by nanoobjects present unique, electronic and optical properties which include advantages of both graphene and nanoobjects leading the nanoobjects graphene nanocomposites towards various electronic as well as optoelectronic applications. The semiconductor nanomaterials contain excellent optical properties and play an important role in improving the low absorption drawback of pure graphene. For instance, the semiconductor CdS quantum dots or nanoparticles on combining with graphene significantly enhanced the photoelectrical responses and photo-absorption. The ZnO nanomaterials possess wide bandgap and highly ultraviolet UV-active materials. Hence, The ZnO nanorods/nanowires on making the graphene-ZnO heterostructures increased the UV response of graphene. The TiO_2 nanoparticles are also combined with graphene to form graphene nanocomposites, consequently, TiO_2 nanoparticles offered better photoelectrical and photocatalytic performances by generating close contact between graphene and TiO_2 nanoparticles. The graphene nanocomposites with nanomaterials demonstrated improved electrical and electrocatalytic properties in the hydrogen evolution process. The $Ni(OH)_2$ nanosheet-graphene nanocomposites exhibited the great increase in the electrochemical capacitance. The Co_3O_4-graphene nanocomposites revealed strong oxygen-reduction activity as compared to C-Pt catalysts, whereas, pure form of Co_3O_4 nanoparticles show less oxygen-reduction activity. The metallic nanoparticles also play vital role in forming metal graphene nanocomposites. The metal nanoparticle functionalized graphene nanocomposites for instance; Pt nanoparticles graphene nanocomposites gave highly exceptional electrocatalytic performance in the methanol oxidization process.

3.4 Inorganic Functionalization of Graphene Nanocomposites

Graphene is an attractive material for immobilization of inorganic nanoparticles. In the recent decade, the suitable combination of the metallic nanoparticles with the graphene sheets results into uniquely distinctive magnetic, catalytic and optoelectronic materials. The dispersion of metallic nanoparticles on surface of graphene sheets makes available the graphene nanocomposites applicably suitable in various applications involving energy and hydrogen storage systems, catalysts and chemical sensor applications. The advancement in the inorganic functionalized graphene nanocomposites comprise of specific properties of inorganic nanoparticles as well involving the magnetic, catalytic, electrical and optical properties. Most of the inorganic nanoparticles used for this purpose are the noble metals like gold, silver, palladium and platinum whereas other metals like copper, iron, tin and cobalt are also being considered to be used in this process. The gold has unique identity among the metallic materials due to its exclusive surface and optical properties and is promisingly used in the optical, catalytic and nanobiotechnology applications. The silver nanoparticles are also widely utilized in inorganic functionalization of graphene nanocomposites. The silver nanoparticles retain optical, biological and electrical properties, also due to antibacterial properties more attractive for biomedical applications. The silver nanoparticles can be grafted over the graphene sheets to result the silver functionalized graphene nanocomposites. The distinct association of these silver nanoparticles on the graphene sheets steps forward towards the exploration of GO matrix. The silver nanoparticles can disperse on the GO sheet with the help of simultaneous reduction of Ag^+ by the involvement of hydrazine hydrate which behaves as a modest reducing agent. The Ag + and GO can pass through reduction and Ag^+ can simply nucleate on the surface of graphene.

3.5 Organic Functionalization of Graphene Nanocomposites

The functionalized graphene polymer nanocomposites are usually fabricated with homogeneous dispersion using solvent-associated method. The functionalized graphene solution in an organic solvent gives rise to better dispersion of functionalized graphene in the considered polymers matrix. The interfacial structure of the manufacturing nanocomposites can be desirably controlled according to the physical and chemical interactions occurring between graphene and functionalized polymeric matrices due to the functionalities on functionalized graphene. The key principle of the process is to choose the best suitable polymer with the solvents which are being employed for exfoliating functionalized graphene. The different polymers are used along with the functionalized graphene through simple fabrication method such as polyurethane, polymers, poly (methyl methacrylate), and polycarprolactone (PCL). The GO nanosheets exfoliated in water support the in situ emulsion polymerization of the respective styrene monomers. The solvent-associated exfoliation technique for functionalized graphene is not found suitable for practical purposes. There is a solvent-free technique named thermal expansion used to split out GO into single GO sheets. The thermal expansion of GO involves its exfoliation in polycarbonate by the process of melting compound and thus created a conductive network within polymeric matrix. There is a technique which uses modified twin-extruder in order to grind the polypropylene pellets possessing the unmodified graphite. Another unique synthetic method depends on coating of the polymer powders such as nylon, polypropylene and polyethylene powders in the water or other organic solvents along with exfoliated functionalized graphene. The powders coated with functionalized graphene are then referred for processing through injection modeling and twin-extruder. The mechanical properties of the functionalized graphene nanoparticles i.e. ultimate strength, elastic modulus and young's modulus of the product nanocomposites are improved due to the concentration of functionalized graphene. The establishment of strong chemical bonding of various polymers having different functionalities with functionalized graphene is the best effective approach to provide strong interface. This technique provides highly stiff graphene nanocomposites. The semi-crystalline polymers basically face strong nucleating ability of which generate strong crystalline interface. Hence, it is concluded that the enhancement of crystallinity directly comes from a strong functionalized graphene nucleated crystalline interface. The graphene is also found favorable to increase the thermal stability of polymers. Hence, the functionalized graphene nanocomposites are going to be the attractive area of future research interest.

3.6 Graphene Functionalized PLA Nanocomposites

Polylactic acid is a part of the group of polyesters, usually comprised of hydroxyl acids. The PLA is basically a polymer which involves the modification of the stereochemical structures by polymerization of well-organized mixture of L and D isomers to attain the amorphous or semi-crystalline polymers with high molecular weight. The PLA lactic acid is found to be a chiral molecule existing in L and D isomers. The L isomers contain the hydroxy group attached with the left side of asymmetric carbon atom situated at certain distance from carbonyl whereas D-isomers possess the hydroxy group on right side of asymmetric carbon. The PLA properties can be improved by adding fillers, plasticides or biopolymers. The PLA polymers are manufactured and attained from sustainable agricultural resources by process of polymerization of lactide, basically cyclic di-ester of lactic acid. The lactic acid can be taken as monomer for PLA to be used in the practical applications. The PLA contains different molecular weights but in the industrial applications only higher molecular weight is being employed. The ultimate properties of PLA are basically dependent on the optical impurities present between LA enantiomers inside the PLA chains. The PLA is generally unstable thermally and shows spontaneous decrease in molecular weight in effect of main chain scissions. The degradation processes during the thermal reactions involve various reactions such as de-polymerization, hydrolysis, and intra or inter molecular transesterification to oligomeric esters and monomers. The thermal degradation of PLA starts from temperature lower than PLA melting point and rate quickly rises above melting point. The PLA possess young's modulus of approximately 3 GPa and have impact strength of about 2.5 kJ/m^2. The semi-crystalline Poly (L-lactic acid) possess mechanical properties which are more attractive as compared to other commodity polymers. The mechanical, barrier and physical properties of respective material, PLA are directly affected by the crystallinity and solid-state morphology of the material.

The PLA functionalized graphene nanocomposites have now become one of the most attractive materials due to their superb combined properties attained from both the materials. For the fabrication of respective nanocomposites, the chemical modification of pristine graphene is certainly required for the proper dispersion of graphene within PLA matrix. The PLA-graphene nanocomposites also show intense suitability for the biomedical applications depending on the electrospinning techniques used for their manufacturing. The presence of graphene also influences the crystallinity of PLA polymers. The graphene nanosheets significantly affect both cold and melt crystallization aspects of PLA polymers and it is observed that the addition of graphene produces decrease in cold crystallization rate. On the other hand, the addition of graphene results into heterogeneous nucleating effect in case of melt crystallization and the overall crystallization process is accelerated. It is better to use isothermal quenching for the proper crystallization of the PLA and composites from molten state. The three-dimensional growth of PLA is not affected by the addition of graphene sheets. The graphene nanosheets resist the formation of large crystalline domains and therefore give a degraded crystallite structure. The PLA-graphene nanocomposites possess markedly improved properties and these properties depend on loading content of graphene and its dispersion. The graphene seems to strengthen the PLA resin but at a certain concentration the graphene disturbs the polymer chains interactions which reduce the mechanical properties. The hardness of the functionalized graphene with PLA also enhances with the increase in the graphene loading.

The GO has superiority due to its highly dispersive nature in the polymer matrix, producing constant dispersibility in organic or aqueous solutions with electrostatic repulsion, due to their unique oxygenated surface functionalities and potential interactions of these functionalities in GO sheets which are fabricated with polymers. The PLA composites upon filled with ethylene glycol can also be investigated. There was a clear reduction in the average molecular weight by increasing in the ethylene glycol loading with an amount of 2–6% and this reduction is produced due to the impurities present like metallic ions or residual products, acidic species which motivated the PLA degradation during the process of melt blending. Overall the comparison of the graphene PLA nanocomposites with PLA attributed with reinforcing effect of various fillers, the graphene functionalized PLA nanocomposites exhibit great enhancement in the mechanical properties.

4 Biomedical Applications

4.1 Bone Substitutes and Repairing

With the growing industrialization and number of vehicles, traffic accidents are very commonly happening in big cities which are creating need of exploring new materials for bone replacements and as bone repairing materials. There are higher chances that bones can be damaged or lost during the accident or over the time owing to pathogenic changes, consequently, reconstruction of bones is an important area of interest in biomedical field which ultimately boost generation of materials that can sustain bone related problems [66]. Defect repairment of bones is very frequent in orthopedic surgery, whereas, fabricated bones are the future in medical field, therefore, it is imperative to identify suitable materials to produce artificial bones [67]. The appropriate bone substitute must be biocompatible, osteoinductive and osteoconductive characteristics. Previously, polymer-based pins, plates and screws have been utilized in orthopedic and craniofacial surgeries. PLA is an auspicious material to fabricate scaffolds and repair bones [68]. Due to its biodegradable nature, no second operation is needed to remove implants and consequently, can alleviate patients's pain. In addition, the utilization of PLA refrain stress block and minimize chances of operation failure. Still, PLA polymer do not have reasonably strong bone-bonding force and lack regeneration efficacy as compare to PLA based nanocomposites [69]. Gong et al. [70] prepared PLA/hydroxyapatite/GO nanocomposites and checked thermal stability and mechanical properties. GO addition into PLA enhanced thermal properties significantly. The mechanical properties such as tensile strength and hardness increased extraordinary by the addition of GO into PLA polymer matrix along with hydroxyapatite. Wu et al. [71] studied the effect of graphene nanosheets on crystallization of PLA and observed that nanosheets act as inert fillers. Marques et al. [72] prepared PLLA/hydroxyapatite/GO composites and study their mechanical properties important for biomedical applications. Nanocomposites with 30% hydroxyapatite and 1% GO by weight percent exhibited highest mechanical hardness and modulus values. This specifies better load transfer between fillers and polymer. These nanocomposites have potential application in bone screws. Bao et al. [73] produced graphene/PLA nanocomposites by melt blending method. Addition of graphene significantly improved crystallization of the polymer and mechanical properties. These properties depend on dispersion and content level of graphene in the matrix. They emphasized that graphene addition reinforced PLA chains up to certain concentration, however, further increase in concentration disturbed interactions of polymer chains, consequently, declined mechanical properties. They observed increase in mechanical strength (35%) up to 0.08 wt% of graphene contents and reduced on further increase in graphene concentration.

Graphene and its different forms were incorporated in PLA to form scaffolds. This can tune and improve diverse thermomechanical properties appropriate relevant to biomedical engineering. Nieto et al. [74] prepared biocompatible scaffolds with higher mechanical strength using graphene foam and PLA-poly-ε-caprolactone solution. This results in improved strength and ductility in scaffold based on graphene foam and poly-ε-caprolactone. It was witnessed that human mesenchymal stem cells survived and proliferated through scaffold based on graphene foam and graphene-poly-ε-caprolactone. Cell growth was more appropriate in graphene-poly-ε-caprolactone as compare to only graphene foam. Chen et al. [75] prepared nanocomposites based on polyurethane/PLA/GO in 3D form using solvent mixing and fused deposition techniques. Compression modulus and tensile adhesion improved 167% and 75.5% respectively with the presence of GO contents (Fig. 5). These nanocomposites have shown improve thermal stability (90 °C) and better crystallinity. 3D printed scaffold with different GO concentrations have shown excellent cell viability

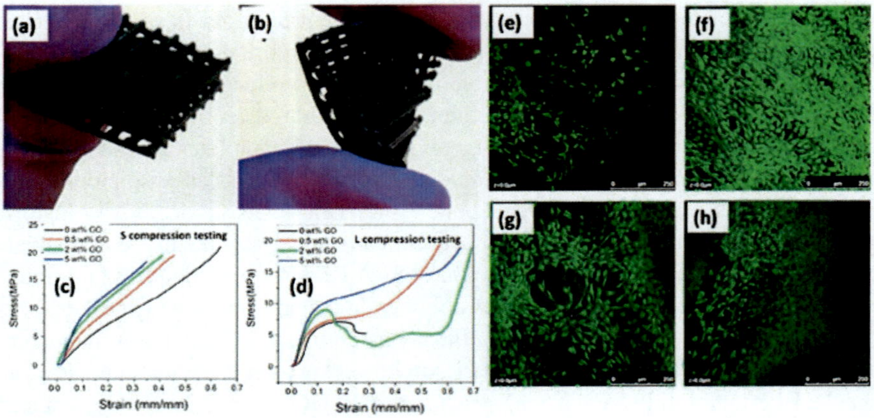

Fig. 5 **a** 3D printed microlattice based on polyurethane/PLA/GO nanocomposites, **b** bending of the microlattice, **c** S, **d** L compression curves with different GO loadings, **e–h** cell viability results of nanocomposites with different GO contents, 0, 0.5, 2 and 5 wt% respectively. Green color reparents live cells. Reprinted with permission from [75]. Copyright (2017) American Chemical Society

and exhibited that minimal GO addition has no significant toxicity whereas small quantity of GO is advantageous for cell proliferation (Fig. 5).

4.2 *Tissue Engineering*

The tissue engineering is primarily a process which highlights the formation and regeneration of the organs and tissues which has become an emerging method in the field of healthcare. It has provided numerous facilities to many humans based on its unique long-term dedication to structure, growth mechanism and function of the biological tissues which is contrary to the background history of the cellular biology and history of area of bioengineering development [76]. The scaffolds are the main carriers in cell adhesion and growth of cells and play significant part in tissue engineering and have been employed in many branches like blood vessels, bone regeneration and neural system [77]. The designing of the scaffolds involves not only the suitable construction that advantageous to the cells growing but also includes the remarkable bio-functionality because local environment can easily influence the behavior of the cells. The PLA materials are also extensively employed as scaffolds due to their admirable biocompatibility. The random poly(L-lactic acid) nano/micro-fibrous scaffolds comprising of different efficacy have been compared for the neural tissue engineering [78]. The results obtained revealed that the neural stem cells and their neurite growth possessed tendency to be extended in the parallel direction to the PLLA fibers for the exclusive aligned scaffolds. The results also revealed that there was no relation between the fiber arrangement and the cell differentiation rate. The nanofibers showed improved differentiation performance as compare to the microfibers which indicated the promising potential of the aligned nanofibrous PLLA scaffold for the process of neural tissue engineering. There are many biomaterials including collagen and GO which are presented to polymerize with PLA or PLGA for obtaining the remarkable optimized scaffold structures. In an experiment [79], the type I collagen was blended with the d,L-PLA (PDLLA) and the examination of the mixture showed that the scaffolds in which type I collagen engaged 40% proportion that PDLLA/collagen scaffolds exhibited highest stability, osteogenic differentiation and cell proliferation after the cultivation of five weeks. The different PDL60/Col and PDL60/Gel scaffolds were prepared and compared and both the PDL scaffolds exhibited same size, possessing diameter raging between 500–1000 nm. The PDL60/Col scaffolds contained greater number of cells attached with them as compared to the cells attached to PDL60/Gel. The shape of the cells was found to be different. The cells present on the PDL60/Gel were of spherical shape whereas the cells attached on the PDL60/Col scaffolds were flat. The shapes of the cells specify that the PDL60/Col scaffolds possess improved cell adhesion property as compared to the PDL60/Gel.

In an experiment [80], the hybrid fiber matrices GO-PLGA-Col were successfully fabricated which were composed of PLGA and the collagens saturated with GO by applying the electrospinning technique. The analysis of the prepared hybrid fiber matrices exhibited a well-proportioned GO distribution over the GO-PGLA-Col matrices. The small addition of the Col and GO produced an extensive enhancement in the hydrophilicity of the matrices. The results also revealed that the prepared hybrid matrices also helped in inducing spontaneous myogenesis. These properties made formulation suitable candidate for the process of skeletal tissue engineering. The electrospinning has also shown wonderful properties including its high mass production capability and better surface area of the scaffolds, due to which it has become an effective procedure to fabricate the scaffolds. There are different number of methods used for this purpose including phase separation method, fiber bonding, gas formation, particulate leaching/solvent casting and emulsion freeze-drying. In an experiment [81], the PLGA scaffolds were successfully synthesized with demineralized bone particle (DBP) through the technique of salt leaching/solvent casting method. The smooth muscle cells revealed upregulated gene expression and improved cell growth due to involvement of DBP in PLGA scaffolds. In another investigation [76], a chain of coatings with β-tricalcium phosphate scaffolds was prepared with the help of dip and dry coating and the prepared scaffolds were set to further analyze the topography and angularity of the scaffolds. The pore size of the scaffolds and their interconnectivity are bit difficult to be controlled and designed. The organic solvents involved also showed risk of damaging the cells or tissues.

There is also an emerging technique developed known as 3D printing established based on computer-aided design (CAD) which has ability of solving this problem and has produced a great revolution in the field of manufacturing industry. Shim et al. [82] presented the preparation of resorbable semi-dome-shaped polycaprolactone (PCL)/PLGA/β-TCP membrane by using 3D printing system. The product contained PLGA showed quick degradation rate and high elasticity of PCL. The in vivo preclinical experiment, in vitro mechanical and cytology test were conducted and the comparison test with the titanium membrane was also done. The properties involve the ability of forming the new bone around implants and osseointegration. The coordination of this biomaterial with the 3D printing technique gave an evolving choice for the field of tissue engineering. Generally, the huge efforts have been done to enhance the cell proliferation and the cell adhesion on these scaffolds surface in the tissue engineering field [83]. The different types of collagen have been supplemented and many other additives have been introduced to enhance the surface reactions of PLA and PLGA [84]. The tissue regeneration technologies are presented to be mature in this field whereas the feasibility of PLA based nanocomposites for being employed in neural system and the blood vessel is required to be further established through plenteous experiments which may lead to a new developing future approach of PLA based nanocomposites.

4.3 Drug Delivery System

Efficacy of traditional drugs is usually hampered by irregular therapeutic drug levels in different organs which results least effect for a disease. This demand repeated drug in take to keep definite drug level which can cause unnecessary side effects and can create burden mentally and financially. Therefore, drug delivery is crucial link in drug therapy because this helps in maintaining drug blood concentration by transferring the medicine in the microenvironment precisely without decomposition/degradation of the drugs [83, 84]. For drug delivery, different types of nanoparticles i.e. inorganic, organic and biodegradable polymers are extensively investigated by researchers during several years, however, due to certain limitations, many nanoparticles systems are still under pre-clinical trials. The drug delivery agents based on biodegradable polymers can be classified into two groups i.e. nano-capsules and nanospheres [85]. The drug molecules are usually embedded into the polymer or attached on the surface. PLA nanoparticles are used to synthesize biodegradable nanoparticles using different methods such as solvent displacement [86], solvent evaporation, salting out [87] and solvent diffusion. Consequently, PLA based drug carriers have shown excellent efficacy for the incorporation of psychotic drugs, hormones and proteins including reduce the drug intake amount and render liver and kidney damage. The loose polymer structure of the PLA can release the extra drug amount which can somehow compensate the total reduced drug dose and outcomes a sustained delivery system.

PLA with other copolymers have also been investigated as drug delivery agent. Sanchez et al. [88] observed 60% release of the gentamicin in one week with PLA/amorphous tricalcium phosphate nanoparticles. Poly-lactic glycolic acid (PLGA), a derivative of PLA polymers has been focused more for drug delivery systems. Govender et al. [89] loaded PLGA with procaine hydrochloride using nano-precipitation and investigated release mechanisms with the change in pH value of the aqueous phase. It was observed that by changing pH value, procaine hydrochloride substituted with procaine dihydrate. However, there are still many shortcomings to use PLA nanocomposites such as preparation and storage of drug-polymer formulation is challenging, protein stability is in question in an acidic environment during the degradation of the PLA [90]. Continuous degradation can accumulate acidic monomers inside the polymer matrix and reduce release of further drug and can denature proteins easily [91].

5 Conclusions

PLA polymer is considered as the most appropriate biopolymer for biomedical engineering; however, pure form of PLA has certain disadvantages such as inferior thermal and mechanical properties. To improve these properties, composites of PLA with GO are considered as promising candidate. PLA-graphene interface engineering has improved thermal and mechanical properties significantly. Biomedical applications of composites have shown promising response for artificial bone formation and tissue engineering. However, systematic studies are needed to see the effect of graphene sheets' length and number of layers, as these are important parameters that can affect biomedical related properties of the nanocomposites. Another important point is kept quantity of graphene to threshold limit to keep overall formulation biocompatible.

References

1. Nampoothiri KM, Nair NR, John RP (2010) An overview of the recent developments in polylactide (PLA) research. Biores Technol 101(22):8493–8501
2. Dedenaro G, Costa S, Rugiero I, Pedrini P, Tamburini E (2016) Valorization of agri-food waste via fermentation: production of L-lactic acid as a building block for the synthesis of biopolymers. Appl Sci 6(12):379
3. Ajioka M, Enomoto K, Suzuki K, Yamaguchi A (1995) Basic properties of polylactic acid produced by the direct condensation polymerization of lactic acid. Bull Chem Soc Jpn 68(8):2125–2131
4. Sha L, Chen Z, Chen Z, Zhang A, Yang Z (2016) Polylactic acid based nanocomposites: promising safe and biodegradable materials in biomedical field. Int J Polymer Sci 2016.
5. Södergård A, Stolt M (2002) Properties of lactic acid based polymers and their correlation with composition. Prog Polym Sci 27(6):1123–1163
6. Lee HH, Sang Shin U, Lee JH, Kim HW (2011) Biomedical nanocomposites of poly (lactic acid) and calcium phosphate hybridized with modified carbon nanotubes for hard tissue implants. J Biomed Mater Res B Appl Biomater 98(2):246–254
7. Datta R, Henry M (2006) Lactic acid: recent advances in products, processes and technologies—a review. J Chem Technol Biotechnol Int Res Process Environ Clean Technol 81(7):1119–1129
8. Lasprilla AJ, Martinez GA, Lunelli BH, Jardini AL, Maciel Filho R (2012) Poly-lactic acid synthesis for application in biomedical devices—a review. Biotechnol Adv 30(1):321–328
9. Jamshidian M, Tehrany EA, Imran M, Jacquot M, Desobry S (2010) Poly-Lactic Acid: production, applications, nanocomposites, and release studies. Compr Rev Food Sci Food Saf 9(5):552–571

10. Marques P, Gonçalves G, Cruz S, Almeida N, Singh M, Grácio J, Sousa A (2011) Functionalized graphene nanocomposites. Adv Nanocompos Technol 11:247–272
11. Soldano C, Mahmood A, Dujardin E (2010) Production, properties and potential of graphene. Carbon 48(8):2127–2150
12. Ushio S, Yoshii A, Tamai N, Ohtani N, Kaneko T (2011) Wide-range temperature dependence of epitaxial graphene growth on 4H-SiC (0 0 0–1): A study of ridge structures formation dynamics associated with temperature. J Cryst Growth 318(1):590–594
13. Janowska I, Ersen O, Jacob T, Vennégues P, Bégin D, Ledoux M-J, Pham-Huu C (2009) Catalytic unzipping of carbon nanotubes to few-layer graphene sheets under microwaves irradiation. Appl Catal A 371(1–2):22–30
14. Eswaraiah V, Jyothirmayee Aravind S, Balasubramaniam K, Ramaprabhu S (2013) Graphene-functionalized carbon nanotubes for conducting polymer nanocomposites and their improved strain sensing properties. Macromol Chem Phys 214(21):2439–2444
15. Feng X, Wang X, Cai W, Qiu S, Hu Y, Liew KM (2016) Studies on synthesis of electrochemically exfoliated functionalized graphene and polylactic acid/ferric phytate functionalized graphene nanocomposites as new fire hazard suppression materials. ACS Appl Mater Interfaces 8(38):25552–25562
16. Athanasiou KA, Niederauer GG, Agrawal CM (1996) Sterilization, toxicity, biocompatibility and clinical applications of polylactic acid/polyglycolic acid copolymers. Biomaterials 17(2):93–102
17. Valantin M-A, Aubron-Olivier C, Ghosn J, Laglenne E, Pauchard M, Schoen H, Bousquet R, Katz P, Costagliola D, Katlama C (2003) Polylactic acid implants (New-Fill)® to correct facial lipoatrophy in HIV-infected patients: results of the open-label study VEGA. Aids 17(17):2471–2477
18. Narayanan G, Vernekar VN, Kuyinu EL, Laurencin CT (2016) Poly (lactic acid)-based biomaterials for orthopaedic regenerative engineering. Adv Drug Deliv Rev 107:247–276
19. Georgakilas V, Otyepka M, Bourlinos AB, Chandra V, Kim N, Kemp KC, Hobza P, Zboril R, Kim KS (2012) Functionalization of graphene: covalent and non-covalent approaches, derivatives and applications. Chem Rev 112(11):6156–6214
20. Li D, Müller MB, Gilje S, Kaner RB, Wallace GG (2008) Processable aqueous dispersions of graphene nanosheets. Nat Nanotechnol 3(2):101
21. Dreyer DR, Park S, Bielawski CW, Ruoff RS (2010) The chemistry of graphene oxide. Chem Soc Rev 39(1):228–240
22. Stankovich S, Piner RD, Chen X, Wu N, Nguyen ST, Ruoff RS (2006) Stable aqueous dispersions of graphitic nanoplatelets via the reduction of exfoliated graphite oxide in the presence of poly (sodium 4-styrenesulfonate). J Mater Chem 16(2):155–158
23. Bourlinos AB, Gournis D, Petridis D, Szabó T, Szeri A, Dékány I (2003) Graphite oxide: chemical reduction to graphite and surface modification with primary aliphatic amines and amino acids. Langmuir 19(15):6050–6055
24. McAllister MJ, Li J-L, Adamson DH, Schniepp HC, Abdala AA, Liu J, Herrera-Alonso M, Milius DL, Car R, Prud'homme RK (2007) Single sheet functionalized graphene by oxidation and thermal expansion of graphite. Chem Mater 19(18):4396–4404
25. Schniepp HC, Li J-L, McAllister MJ, Sai H, Herrera-Alonso M, Adamson DH, Prud'homme RK, Car R, Saville DA, Aksay IA (2006) Functionalized single graphene sheets derived from splitting graphite oxide. J Phys Chem B 110(17):8535–8539
26. Huh S, Park J, Kim YS, Kim KS, Hong BH, Nam J-M (2011) UV/ozone-oxidized large-scale graphene platform with large chemical enhancement in surface-enhanced Raman scattering. ACS Nano 5(12):9799–9806
27. Li X, Zhang G, Bai X, Sun X, Wang X, Wang E, Dai H (2008) Highly conducting graphene sheets and Langmuir-Blodgett films. Nat Nanotechnol 3(9):538–542
28. Bao C, Guo Y, Song L, Kan Y, Qian X, Hu Y (2011) In situ preparation of functionalized graphene oxide/epoxy nanocomposites with effective reinforcements. J Mater Chem 21(35):13290–13298

29. Grayfer ED, Makotchenko VG, Nazarov AS, Kim S-J, Fedorov VE (2011) Graphene: chemical approaches to the synthesis and modification. Russ Chem Rev 80(8):751
30. Hamilton CE, Lomeda JR, Sun Z, Tour JM, Barron AR (2009) High-yield organic dispersions of unfunctionalized graphene. Nano Lett 9(10):3460–3462
31. Bourlinos AB, Georgakilas V, Zboril R, Steriotis TA, Stubos AK (2009) Liquid-phase exfoliation of graphite towards solubilized graphenes. Small 5(16):1841–1845
32. Niyogi S, Bekyarova E, Itkis ME, Zhang H, Shepperd K, Hicks J, Sprinkle M, Berger C, Lau CN, Deheer WA (2010) Spectroscopy of covalently functionalized graphene. Nano Lett 10(10):4061–4066
33. Bei HP, Yang Y, Zhang Q, Tian Y, Luo X, Yang M, Zhao X (2019) Graphene-based nanocomposites for neural tissue engineering. Molecules 24(4):658
34. Feng S-H, Li G-H (2017) Hydrothermal and solvothermal syntheses. In: Modern inorganic synthetic chemistry. Elsevier, Amsterdam, pp 73–104
35. Liang Y, Wang H, Casalongue HS, Chen Z, Dai H (2010) TiO 2 nanocrystals grown on graphene as advanced photocatalytic hybrid materials. Nano Res 3(10):701–705
36. Srivastava M, Das AK, Khanra P, Kim NH, Lee JH (2013) A facile one-step hydrothermal synthesis of graphene/CeO_2 nanocomposite and its catalytic properties. In: Advanced materials research, 2013. Trans Tech Publications, pp 242–245
37. Pu N-W, Chen C-Y, Qiu H-X, Liu Y-M, Song C-H, Lin M-H, Ger M-D (2018) Hydrothermal synthesis of N-doped graphene/Fe_2O_3 nanocomposite for supercapacitors. Int J Electrochem Sci 13:6812–6823
38. Song J, Xu L, Zhou C, Xing R, Dai Q, Liu D, Song H (2013) Synthesis of graphene oxide based CuO nanoparticles composite electrode for highly enhanced nonenzymatic glucose detection. ACS Appl Mater Interfaces 5(24):12928–12934
39. Choucair M, Thordarson P, Stride JA (2009) Gram-scale production of graphene based on solvothermal synthesis and sonication. Nat Nanotechnol 4(1):30
40. Li Y, Qian Y, Liao H, Ding Y, Yang L, Xu C, Li F, Zhou G (1998) A reduction-pyrolysis-catalysis synthesis of diamond. Science 281(5374):246–247
41. Jiang Y, Wu Y, Zhang S, Xu C, Yu W, Xie Y, Qian Y (2000) A catalytic-assembly solvothermal route to multiwall carbon nanotubes at a moderate temperature. J Am Chem Soc 122(49):12383–12384
42. Deng D, Pan X, Yu L, Cui Y, Jiang Y, Qi J, Li W-X, Fu Q, Ma X, Xue Q (2011) Toward N-doped graphene via solvothermal synthesis. Chem Mater 23(5):1188–1193
43. Yang J, Zang C, Sun L, Zhao N, Cheng X (2011) Synthesis of graphene/Ag nanocomposite with good dispersibility and electroconductibility via solvothermal method. Mater Chem Phys 129(1–2):270–274
44. Lin-jun H, Yan-xin W, Jian-guo T, Hui-min W, Hai-bin W, Jian-xiu Q, Yao W, Ji-xian L, Jing-quan L (2012) Synthesis of graphene/metal nanocomposite film with good dispersibility via solvothermal method. Int J Electrochem Sci 7:11068–11075
45. Hu C, Chen F, Lu T, Lian C, Zheng S, Zhang R (2014) Aqueous production of TiO_2–graphene nanocomposites by a combination of electrostatic attraction and hydrothermal process. Mater Lett 121:209–211
46. Abdel-Karim R (2016) Electrochemical synthesis of nanocomposites. Electrodeposition of Composite Materials, p 1
47. Yin Z, Wu S, Zhou X, Huang X, Zhang Q, Boey F, Zhang H (2010) Electrochemical deposition of ZnO nanorods on transparent reduced graphene oxide electrodes for hybrid solar cells. Small 6(2):307–312
48. Chengzhou Z, Shaojun G, Youxing F, Lei H, Erkang W, Shaojun D (2011) One-step electrochemical approach to the synthesis of Graphene. MnO 2:648–657
49. Gao H, Xiao F, Ching CB, Duan H (2011) One-step electrochemical synthesis of PtNi nanoparticle-graphene nanocomposites for nonenzymatic amperometric glucose detection. ACS Appl Mater Interfaces 3(8):3049–3057
50. Xie G, Forslund M, Pan J (2014) Direct electrochemical synthesis of reduced graphene oxide (rGO)/copper composite films and their electrical/electroactive properties. ACS Appl Mater Interfaces 6(10):7444–7455

51. Hu H, Wang X, Wang J, Wan L, Liu F, Zheng H, Chen R, Xu C (2010) Preparation and properties of graphene nanosheets–polystyrene nanocomposites via in situ emulsion polymerization. Chem Phys Lett 484(4):247–253
52. Mohan S, Abraham J, Oluwafemi OS, Kalarikkal N, Thomas S (2016) Rheology and processing of inorganic nanomaterials and quantum dots/polymer nanocomposites. Rheol Process Polymer Nanocompos 355–382
53. Zeng X, Teng J, Yu J-g, Tan A-s, Fu D-f, Zhang H (2018) Fabrication of homogeneously dispersed graphene/Al composites by solution mixing and powder metallurgy. Int J Minerals Metallur Mate 25(1):102–109
54. Williarris G, Seger B, Kamat P (2008) TiO2-graphene nanocomposites. UV-Assist Photocatal Reduct Graphene Oxide ACS Nano 2:1487
55. Akhavan O, Ghaderi E (2009) Photocatalytic reduction of graphene oxide nanosheets on TiO_2 thin film for photoinactivation of bacteria in solar light irradiation. J Phys Chem C 113(47):20214–20220
56. Delogu F, Gorrasi G, Sorrentino A (2017) Fabrication of polymer nanocomposites via ball milling: present status and future perspectives. Prog Mater Sci 86:75–126
57. Wei Z-B, Zhao Y, Wang C, Kuga S, Huang Y, Wu M (2018) Antistatic PVC-graphene composite through plasticizer-mediated exfoliation of graphite. Chin J Polym Sci 36(12):1361–1367
58. Deng D, Yu L, Pan X, Wang S, Chen X, Hu P, Sun L, Bao X (2011) Size effect of graphene on electrocatalytic activation of oxygen. Chem Commun 47(36):10016–10018
59. Deng D, Chen X, Yu L, Wu X, Liu Q, Liu Y, Yang H, Tian H, Hu Y, Du P (2015) A single iron site confined in a graphene matrix for the catalytic oxidation of benzene at room temperature. Sci Adv 1(11):e1500462
60. He T, Li J, Wang L, Zhu J, Jiang W (2009) Preparation and consolidation of alumina/graphene composite powders. Mater Trans 50(4):749–751
61. Bartolucci SF, Paras J, Rafiee MA, Rafiee J, Lee S, Kapoor D, Koratkar N (2011) Graphene–aluminum nanocomposites. Mater Sci Eng A 528(27):7933–7937
62. Bastwros M, Kim G-Y, Zhu C, Zhang K, Wang S, Tang X, Wang X (2014) Effect of ball milling on graphene reinforced Al6061 composite fabricated by semi-solid sintering. Compos B Eng 60:111–118
63. Zhai W, Shi X, Yao J, Ibrahim AMM, Xu Z, Zhu Q, Xiao Y, Chen L, Zhang Q (2015) Investigation of mechanical and tribological behaviors of multilayer graphene reinforced Ni3Al matrix composites. Compos B Eng 70:149–155
64. Yue H, Yao L, Gao X, Zhang S, Guo E, Zhang H, Lin X, Wang B (2017) Effect of ball-milling and graphene contents on the mechanical properties and fracture mechanisms of graphene nanosheets reinforced copper matrix composites. J Alloy Compd 691:755–762
65. Peña-Bahamonde J, Nguyen HN, Fanourakis SK, Rodrigues DF (2018) Recent advances in graphene-based biosensor technology with applications in life sciences. J Nanobiotechnol 16(1):75
66. Dimitriou R, Jones E, McGonagle D, Giannoudis PV (2011) Bone regeneration: current concepts and future directions. BMC Med 9(1):66
67. Martou G, Antonyshyn M (2011) Advances in surgical approaches to the upper facial skeleton. Curr Opin Otolaryngol Head Neck Surg 19(4):242–247
68. Schwach G, Vert M (1999) In vitro and in vivo degradation of lactic acid-based interference screws used in cruciate ligament reconstruction. Int J Biol Macromol 25(1–3):283–291
69. Santos Jr AR (2010) Bioresorbable polymers for tissue engineering. In: Eberli D (ed) Tissue engineering, pp 225–246
70. Gong M, Zhao Q, Dai L, Li Y, Jiang T (2017) Fabrication of polylactic acid/hydroxyapatite/graphene oxide composite and their thermal stability, hydrophobic and mechanical properties. J Asian Ceramic Soc 5(2):160–168
71. Wu D, Cheng Y, Feng S, Yao Z, Zhang M (2013) Crystallization behavior of polylactide/graphene composites. Ind Eng Chem Res 52(20):6731–6739
72. Marques PA, Gonçalves G, Singh MK, Grácio J (2012) Graphene oxide and hydroxyapatite as fillers of polylactic acid nanocomposites: preparation and characterization. J Nanosci Nanotechnol 12(8):6686–6692

73. Bao C, Song L, Xing W, Yuan B, Wilkie CA, Huang J, Guo Y, Hu Y (2012) Preparation of graphene by pressurized oxidation and multiplex reduction and its polymer nanocomposites by masterbatch-based melt blending. J Mater Chem 22(13):6088–6096
74. Nieto A, Dua R, Zhang C, Boesl B, Ramaswamy S, Agarwal A (2015) Three dimensional graphene foam/polymer hybrid as a high strength biocompatible scaffold. Adv Func Mater 25(25):3916–3924
75. Chen Q, Mangadlao JD, Wallat J, De Leon A, Pokorski JK, Advincula RC (2017) 3D printing biocompatible polyurethane/poly (lactic acid)/graphene oxide nanocomposites: anisotropic properties. ACS Appl Mater Interfaces 9(4):4015–4023
76. Ardjomandi N, Henrich A, Huth J, Klein C, Schweizer E, Scheideler L, Rupp F, Reinert S, Alexander D (2015) Coating of ß-tricalcium phosphate scaffolds—a comparison between graphene oxide and poly-lactic-co-glycolic acid. Biomed Mater 10(4):045018
77. Haugen H, Will J, Fuchs W, Wintermantel E (2006) A novel processing method for injection-molded polyether–urethane scaffolds. Part 1: Processing. J Biomed Mater Res Part B Appl Biomater Off J Soc Biomater Jpn Soc Biomater Austral Soc Biomat Korean Soc Biomater 77 (1):65–72
78. Yang F, Murugan R, Wang S, Ramakrishna S (2005) Electrospinning of nano/micro scale poly (L-lactic acid) aligned fibers and their potential in neural tissue engineering. Biomaterials 26(15):2603–2610
79. Qiao X, Russell SJ, Yang X, Tronci G, Wood DJ (2015) Compositional and in vitro evaluation of nonwoven type I collagen/poly-dl-lactic acid scaffolds for bone regeneration. J Funct Biomater 6(3):667–686
80. Shin YC, Lee JH, Jin L, Kim MJ, Kim Y-J, Hyun JK, Jung T-G, Hong SW, Han D-W (2015) Stimulated myoblast differentiation on graphene oxide-impregnated PLGA-collagen hybrid fibre matrices. J Nanobiotechnol 13(1):21
81. Ziabari M, Mottaghitalab V, Haghi AK (2010) A new approach for optimization of electrospun nanofiber formation process. Korean J Chem Eng 27(1):340–354
82. Shim J-H, Won J-Y, Sung S-J, Lim D-H, Yun W-S, Jeon Y-C, Huh J-B (2015) Comparative efficacies of a 3D-printed PCL/PLGA/β-TCP membrane and a titanium membrane for guided bone regeneration in beagle dogs. Polymers 7(10):2061–2077
83. Davachi SM, Kaffashi B, Zamanian A, Torabinejad B, Ziaeirad Z (2016) Investigating composite systems based on poly l-lactide and poly l-lactide/triclosan nanoparticles for tissue engineering and medical applications. Mater Sci Eng C 58:294–309
84. Shafiq M, Jung Y, Kim SH (2015) Stem cell recruitment, angiogenesis, and tissue regeneration in substance P-conjugated poly (l-lactide-co-ε-caprolactone) nonwoven meshes. J Biomed Mater Res Part A 103(8):2673–2688
85. Masood F (2016) Polymeric nanoparticles for targeted drug delivery system for cancer therapy. Mater Sci Eng C 60:569–578
86. Fessi H, Puisieux F, Devissaguet JP, Ammoury N, Benita S (1989) Nanocapsule formation by interfacial polymer deposition following solvent displacement. Int J Pharm 55(1):R1–R4
87. Reis CP, Neufeld RJ, Ribeiro AJ, Veiga F (2006) Nanoencapsulation I. Methods for preparation of drug-loaded polymeric nanoparticles. Nanomed Nanotechnol Biol Med 2(1):8–21
88. Sanchez E, Baro M, Soriano I, Perera A, Evora C (2001) In vivo–in vitro study of biodegradable and osteointegrable gentamicin bone implants. Eur J Pharm Biopharm 52(2):151–158
89. Govender T, Stolnik S, Garnett MC, Illum L, Davis SS (1999) PLGA nanoparticles prepared by nanoprecipitation: drug loading and release studies of a water soluble drug. J Control Rel 57(2):171–185
90. Mundargi RC, Babu VR, Rangaswamy V, Patel P, Aminabhavi TM (2008) Nano/micro technologies for delivering macromolecular therapeutics using poly (D, L-lactide-co-glycolide) and its derivatives. J Control Rel 125(3):193–209
91. Bittner B, Ronneberger B, Zange R, Volland C, Anderson J, Kissel T (1998) Bovine serum albumin loaded poly (lactide-co-glycolide) microsphe: the influence of polymer purity on particle characteristics. J Microencapsul 15(4):495–514

Graphene Reinforced PVA Nanocomposites and Their Applications

Hafeez Anwar, Muhammad Haseeb, Mariyam Khalid, and Kamila Yunas

Abstract Graphene reinforced poly (vinyl alcohol) (PVA) nanocomposites are fast emerging materials due to their fascinating properties such as their morphology, excellent mechanical, optical and thermal properties. In this present chapter, different synthesis approaches and physical properties of these nanocomposites will be discussed in detail. In addition to this, various potential applications of these nanocomposites will be discussed. These applications include food packaging, biomedical, sensors, energy storage devices, safety glasses etc.

Keywords Graphene · Poly (vinyl alcohol) · Nanocomposites · Food packaging · Energy storage devices

1 Introduction

The field of nano science has flourished following the inventions of 2D structure of carbon atoms. In the 1500s, 3D graphitic form of carbon is confirmed. In the 1980s and 1990s, additional carbon allotropes such as 0D (fullerene) and 1D (carbon nanotubes) are discovered (Fig. 1). There remained an argument on the presence of 2D allotropes of carbon until Andre Geim and Konstantin Novoselov successfully separated a single layer of graphite (graphene) on a sticky tape by scotch tape method [1]. Graphite is a carbon-based material in 3D and prepared by millions of layers of graphene. In graphite structure, oxygenated functional groups are present due to increase of layers separation and also these layers make the material hydrophilic. Using sonication, this property allows graphite oxide to be exfoliated in water, finally create a single or few layer of graphene known as graphene oxide (GO). The major difference between GO and graphite oxide is only the number of layers. A small number of flakes and single layer flakes can be observed in the GO dispersion. Graphene oxide has been used on a large scale for the manufacturing of graphitic

H. Anwar (✉) · M. Haseeb · M. Khalid · K. Yunas
Advanced Nanomaterials and Devices Laboratory, Department of Physics, University of Agriculture, Faisalabad 38040, Pakistan
e-mail: hafeez.anwar@gmail.com

B. Sharma and P. Jain (eds.), *Graphene Based Biopolymer Nanocomposites*, Composites Science and Technology,
https://doi.org/10.1007/978-981-15-9180-8_6

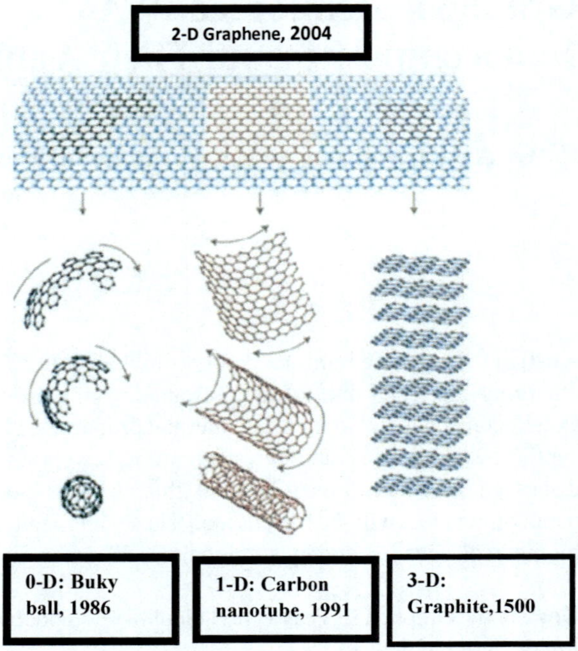

Fig. 1. 2D Hexagonal nanosheets of graphene as a building block of other polymorphs

films and flake. Furthermore, hydrophilicity of GO permits it to uniformly deposit onto the substrates as thin films, having applications in electronics. It is also necessary that the GO changed back into a conductive graphite material in bulk or in thin films and by the chemical reduction to chemically converted graphene (CCG). On the other hand, using these conditions the graphitic structure is not fully recovered and significant defects are established [2]. Graphene is single layer of carbon atoms, tightly bond in a hexagonal honeycomb lattice and in the form of a plane of sp^2—bonded atoms with a molecular bond length of 0.142 nm. Such graphene structure is a unique component of crystalline graphite where layers of graphene are stacked on top of each other form graphite with interplanar spacing ~3.4 nm [3]. Graphene has many potential applications such as sensors, energy conversion and energy storage devices. Due to its attractive electronic, mechanical, and electrochemical properties graphene is a hot topic now-a-days [4].

Graphene oxide is prepared through chemical oxidation of natural graphite although there are small numbers of reports on another electrochemical oxidation [5]. Brodie is the first scientist who prepared the graphite oxide by the addition of potassium chlorate in fuming nitric acid. At least after 40 years, Staudenmaier enhanced this method by changing the two third of nitric acid (HNO_3) with concentrated sulfuric acid (H_2SO_4) and adding the chlorate slowly. Depending on these previous works, Hummer and Offeman introduced another method to oxidize the graphite into graphite oxide within a few hours. Hummer's method is extensively adopted to prepare GO but Hummer's method still faces several problems involving poisons gas production like NO_2, N_2O_4 and low yield etc. In the previous 20 years, different modifications in hummer's method have been proposed to overcome these problems. The main steps can be follows as: firstly, removing the $NaNO_3$ from hummer's method; secondly in the absence of $NaNO_3$addition of per oxidation step before $KMnO_4$ oxidation; thirdly greater the amounts of $KMnO_4$ as an alternate of $NaNO_3$; fourthly, change the Potassium permanganate ($KMnO_4$) with Potassium ferrate (K_2FeO_4). For example, Kovtyukhova et al., used potassium per sulfate ($K_2S_2O_8$) and Phosphorus pentoxide (P_2O_5) to proto oxidize graphite before the Hummer's method. This work showed that the produced GO was highly oxidized but the whole process was time consuming [6]. Marcano et al. performed the improved Hummers method in which amounts of $KMnO_4$ and concentrated H_2SO_4 were in used instead of $NaNO_3$ lead to the higher yield [2]. Peng et al., performed experiment in which K_2FeO_4 based oxidation approach by replacing of $KMnO_4$ was used and attained a monolayer of GO at the room temperature [7]. Tables 1 and 2 showed various parameters for synthesis of GO and rGO respectively available in literature.

2 Poly (Vinyl Alcohol) (PVA)

Poly (vinyl alcohol) (PVA) is a water-soluble polymer which can be prepared by aqueous methods. PVA contains large number of hydroxyl groups and has poor water vapor properties [23]. PVA played a significant role in drug delivery, fuel cells and shape memory applications. It is also used in papermaking, textiles and a variety of coatings [24]. Water resistance property of PVA materials is the most significant for package applications in many fields. PVA is a water-soluble synthetic polymer. It has the idealized formula $[CH_2CH(OH)]_n$. Structure of PVA is given in Fig. 2.

3 Nanocomposites

The term nanocomposites was first used by Roy, Komarneni and their colleagues during 1982–1983 [25]. Word 'composite' means made up from two or more different parts. Nanocomposites are multiphase solid materials in which one of the phases has one, two and three dimensions of in the range of 1–100 nm. In general, nanocomposites mean nano sized particles (semiconductors, metals and dielectric materials) which are embedded in different matrix materials like glass, polymer and ceramics. Nanocomposites consist of two phases; one is primary and other one is secondary phase. Primary phase, having a continuous character is known as 'matrix'. Matrix is usually more ductile and less hard phase. It holds the dispersion phase and shares a load with it. The secondary phase which is embedded in the matrix is called dispersed phase. Dispersed phase is usually stronger than matrix therefore; it is called 'reinforcement' or 'reinforcing material'. These different phases are shown in Fig. 3.

Table 1 Various parameters for synthesis of GO

Methods	Temperature	Carbon source	Oxidizing agent	Reaction duration	Features	Interlayer spacing (nm)	References
Marcano	40–50 °C	Graphite powder	H_2SO_4, H_3PO_4 and $KMnO_4$	12 h	Bi-component acids, high yield	0.95	[8]
Hummer's	>35 °C	Graphite powder	$KMnO_4$, $NaNO_3$ and H_2SO_4	<2 h	water-free, Less than 2 h processing		[9]
Hummer's and improved Hummer's with purification	~40 °C	Graphite powder	$NaNO_3$, H_2SO_4, $KMnO_4$	7.5 days	Validate $NaNO_3$ unnecessary	0.81	[8]
Staudenmaie	Room temperature	Graphite powder	$KClO_3$, HNO_3, H_2SO_4	96 h	Improved efficiency		[10]
Chen	<20 °C	Graphite powder	$KMnO_4$, H_2SO_4, H_3PO_4	<1 h	High-yield		[11]
Su	Room temperature	sonicated graphite powder, <3000 μm	$KMnO_4$, H_2SO_4,	4 h	Large-size GO	0.831	[12]
Rosillo-Lopez	Room temperature	defective arc-discharge carbon	HNO_3	20 h	Nano sized GO		[13]
Brodie	60 °C	Graphite powder	$KClO_3$, HNO_3	3–4 days	Earliest method		[14]
Ranjan	<RT-35–95	Graphite powder	H_2SO_4, H_3PO_4 and $KMnO_4$	24 h	Cool the exothermal reaction to keep safe		[15]
Modified Hummer's	>35	Graphite powder	H_2SO_4, $NaNO_3$, $KMnO_4$	13 h		0.77	[16]
Shen	110	graphite colloidal	Benzoyl peroxide	10 min	Non-acid and fast		[17]

Table 2 Various parameters for synthesis of rGO

Carbon source	Distilled Water	Reducing agent	Reaction duration	Temperature (°C)	Refs.
Graphene oxide	20 ml	Hydrazine	30 s (microwave radiation)	90	[18]
Graphene oxide		Hydrazine monohydrate	24 h	80	[19]
Graphene oxide	10 ml	Rose water	5 h	95	[20]
Graphene oxide		Ammonium bicarbonate	60 s (microwave oven)	80	[21]
Graphene oxide	400 ml	Ascorbic acid	30 min	60	[22]

Fig. 2 Structure of PVA

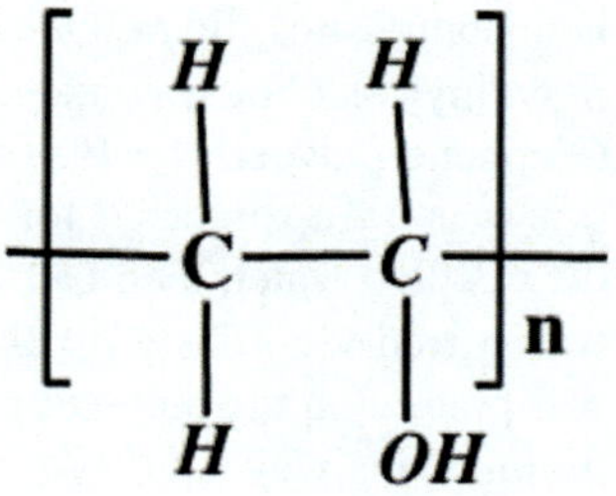

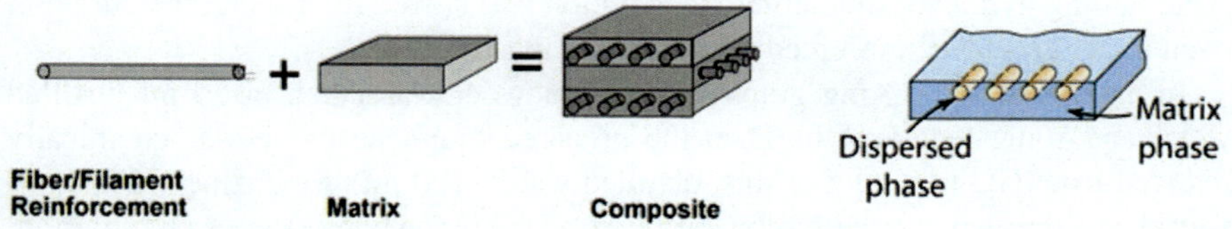

Fig. 3 Phases of nanocomposites

There are different from traditional composites in the sense of smaller size of particles in the matrix materials and mechanically nanocomposites exceptionally high surface to volume ratio of reinforcing, which enhance the electrical, chemical, physical and optical properties of composites nanocomposites are broadly classified in two terms one is polymer-based nanocomposites and other one is non-polymer nanocomposites.

4 Preparation Methods of Graphene Reinforced PVA

There are lot of useful techniques which will be adopted for the synthesis of graphene reinforced PVA nanocomposites. Few of useful and efficient techniques for the preparation of graphene oxide reinforced PVA nanocomposites are discussed below.

Graphene reinforced PVA nanocomposites were prepared by, stirring the polyvinyl alcohol (2 g) in distilled water (30 ml) at 95 °C to produce aqueous solution. Dispersion of graphene oxide was prepared by sonicating (0.3 wt% of PVA) with distilled water (20 ml). When graphene oxide was sonicated with distilled water, it formed a yellowish colloidal solution. Then these two solutions were mixed by stirring for 2 h at 60 °C. As PVA solution is thicker than graphene oxide collides, therefore solutions were vigorously stirred for 2 h. After stirring, homogeneous mixture of GO and PVA was dried in vacuum dryer for next 48 h at 40 °C to form GO-PVA nanocomposites. To reduce the nanocomposites to rGO-PVA nanocomposites, 0.2 ml of hydrazine hydrate was added to these GO-PVA nanocomposites and refluxed it for 24 h at 80 °C, and then brown polymer gradually turned into dark black. After 24 h, dried the 10 ml reduced (dark black polymer) nanocomposite with the help of oven at 40 °C [23]. For the synthesis of graphene reinforced PVA nanocomposites, first GO/PVA nanocomposites were synthesized. For this, 2 g of polyvinyl alcohol was added in distilled water (30 ml) and stirring it at 100 °C. Graphene powder (4 wt% of PVA) was dissolved in distilled water and the solution was ultra-sonicated for 20 min. When PVA and GO completely dissolved in the distilled water, then GO solution was added to the PVA solution. Final solution was stirred vigorously for 0.5 h to make homogenous solution. Then the solution was poured on the polytetrafluoroethylene plate and placed in the oven at 60 °C for drying. This way, GO/PVA nanocomposites were synthesized, and then hydrazine (0.5 ml) and ammonia (0.5 ml) were added to prepare GO/PVA nanocomposites. After adding hydrazine and ammonia, solution was stirred for 1 h and then dried in oven to get graphene reinforced PVA nanocomposites [24].

In another work, 50 mg graphene oxide powder was taken in 50 ml distilled water and sonicated for 1 h. Then the prepared graphene oxide was chemically reduced from GO to rGO. For this, distilled water (100 ml), hydrazine (75 wt% of water) and ammonia (28 wt% of water) were added in the homogenous GO solution. Then mixture was refluxed in the presence of oil bath for 5 h at 100 °C. After 5 h, color of the solution was changed from brown to dark black. This solution was centrifuged at 300 rpm for 15 min. In order to synthesize graphene reinforced PVA nanocomposites, 1 g of PVA was dissolved in 10 ml of distilled water and stirred at 90 °C. When PVA was completely dissolved in distilled water then cooled it down to room temperature. The prepared rGO dispersion (0.5 wt%) was gradually added to the PVA suspension and sonicated for 1 h at room temperature. After 1 h it was placed in vacuum for drying at 60 °C [26]. To prepare graphene reinforce PVA nanocomposites, first GO/PVA nanocomposites were prepared and then reduced. For GO/PVA nanocomposites, 1 mg graphene oxide powder was dissolved in 10 ml distilled water and sonicated to produce homogenous solution. Then PVA was added to the homogenous brown colored graphene oxide solution and stirred for next 6 h at 85 °C. After 6 h, suspension was then cooled down to room temperature and hydrazine (8 μ L) and ammonia (35 μ L) were added into the suspension and stirred for 0.5 h. Solution was heated in the presence of oil bath at 95 °C for 1 h. Color of the suspension gradually changed from brown to dark black. Solution was put onto a clean glass plate. The glass plate was placed into an oven for drying at 60 °C for 24 h [23].

In this work, first reduced graphene oxide was synthesized. In this method, graphene oxide (0.6 g) was taken in distilled water (100 ml) and then hydrazine hydrate (2 ml) was added in the solution. The solution was kept on stirring for 36 h. Then this solution was sonicated for 1 h to get better dispersion. After sonication, sample was washed several times to maintain pH and dried in oven at 60 °C. As synthesized reduced graphene oxide (1 wt%) was taken in distilled water (100 ml) and sonicated for 2 h. Then PVA (4 g) was added to the solution and vigorously stirred for 8 h to get homogeneous dispersion of reduced graphene oxide and polyvinyl alcohol. After this, solution was poured into a petri dish and placed in an oven at 60 °C for 48 h [27]. Yuan used modified Hummer's method to synthesize rGO. The solution of synthesized rGO (1 wt%) was taken in distilled water (100 ml) and sonicated for 10 to 15 min. On the other hand, 10 g polyvinyl alcohol was taken in 200 ml distilled water and heated at 90 °C until PVA completely dissolved. When homogenous solution of PVA was prepared then graphene solution was added to this solution while stirring. Then solution was further stirred at 90 °C for 2 h. After 2 h, dispersion was poured into an aluminium pan and kept it in an oven at 60 °C for drying [28].

Guo et al., synthesized graphene oxide by using modified Hummer's method and then this synthesized graphene oxide was chemically reduced with the help of hydrazine hydrate. Then homogenous dispersion of polyvinyl alcohol (1 g) in distilled water (30 ml) was prepared by stirring at 90 °C. For reduced graphene oxide dispersion, 0.1 mg rGO was dissolved in 10 ml distilled water. When each of dispersion was homogenously prepared then rGO dispersion was gradually added to the polyvinyl alcohol dispersion while stirring. Final solution was stirred for 1 h and poured into a culture dish. To dry the dispersion, culture dish was placed in an oven at 50 °C [29]. Sellam et al., synthesized graphene reinforced PVA nanocomposites. First of all, PVA was dissolved in distilled water under continuous stirring at 90 °C for 4 h. While, graphene oxide was also synthesized by using modified Hummers method. For GO dispersion, 1 g of prepared graphene oxide was dissolved in 10 ml distilled water under sonication and suspension was left aside for 12 h to precipitate thicker aggregates. After 12 h, graphene oxide suspension was diluted by adding ethanol, ratio of water/ethanol was 40:60. After adding ethanol, suspension of graphene oxide was gradually added to the PVA suspension while stirring. When homogeneous suspension was prepared then solution was casted in a petri dish at room temperature. Solvent was evaporated at ambient condition. The obtained nanocomposites were then ready to characterize [30].

Zhou et al., prepared graphene oxide suspension by dissolving graphene oxide in distilled water (20 ml) and sonicated for 15 min at room temperature. In another beaker, polyvinyl alcohol (1 g) was dissolved into 10 ml distilled water by stirring at 98 °C for an hour. When both suspensions were prepared, GO suspension was gradually added to PVA solution while stirring at 98 °C for 30 min. No graphene oxide precipitation was observed in 30 min hour and then homogenous GO/PVA nanocomposites were poured into an aluminium dish. As prepared GO/PVA nanocomposites was chemically reduced. 2 g of NaOH and 0.5 g of $Na_2S_2O_4$ were dissolved into distilled water (50 ml) by stirring for 15 mints to make reducing agent solution. In this suspension, NaOH was acted as a catalyst and stabilizer. Then, prepared GO/PVA nanocomposites were gradually added to the reducing agent solution while stirring for 1 h at 60 °C. After 1 h, GO/PVA nanocomposites were completely reduced to PVA/rGO nanocomposites and its color was changed from brown to dark black. Prepared PVA/rGO nanocomposites were then placed in vacuum to remove any moisture [31]. In another work, first graphene oxide was obtained by Hummer's method. Then dispersion of prepared GO was made by dissolving it in the distilled water. The dispersion was continuously stirred until GO was homogenously mixed into distilled water. On the other hand, 1 g of PVA was dissolved into10 ml distilled water and stirred at 98 °C for 1 h. After 1 h, GO suspension was gradually added to the PVA solution while stirring. The final solution was stirred at 98 °C for 30 min. After stirring, no precipitation was observed in next few hours which showed that no aggregation of graphene oxide occurred. At the end, PVA/GO homogeneous solution was poured into an aluminium dish and aluminium dish was placed in vacuum at 60 °C for 15 h. Then obtained PVA/GO nanocomposites were thermally reduced. For reduction, sample was placed in the furnace which was filled with nitrogen gas (N_2) then temperature of the furnace was gradually increased at the rate of 10 °C per min until it reached at reduction temperature [32].

Goumri et al., prepared graphene oxide by modified Hummer's method and then as prepared graphene oxide was chemically reduced by adding 10% hydrazine hydrates in it. After adding hydrazine hydrate, the sample was continuously stirred for 48 h at 40 °C. After 48 h, solution was ultrasonicated for 1 h at 40 °C and dried in an oven. Now for the nanocomposites, 4 g of polyvinyl alcohol was added in 28 ml of distilled water and stirred at 80 °C for 5 h. To avoid evaporation, sealed containers were used. On the other hand, suspension of reduced graphene oxide was also prepared in distilled water. 7 ml of polyvinyl alcohol solution was gradually added to the suspension of reduced graphene oxide under uniform stirring for 1 h at room temperature. For making homogeneous dispersion, suspension was further stirred for 48 h. Then, solution was poured on glass substrate and placed in an oven at 40 °C for drying [33]. Muralidharan et al. prepared graphene oxide by Hummer's method. Then as prepared graphene oxide was dissolved in distilled water (10 ml) and ultrasonicated for 2 h to make a homogenous brownish dispersion [34]. PVA was also dissolved in distilled water. Now graphene oxide solution was gradually added to the PVA suspension while stirring. The final GO/PVA nanocomposites were sonicated for 15 min to obtain homogenous suspension. To chemically reduce the obtained GO/PVA nanocomposites, 35% of hydrazine monohydrate was added to the GO/PVA suspension. Solution was then heated at 80 °C for while stirring. Hydrazine was acted as a strong reducing agent which removed the oxygen containing functional groups from the graphene oxide within the polymer. After 3 h stirring, color of suspension was turned to dark black from brown. rGO/PVA nanocomposites were casted into film and was placed in vacuum for drying purpose at 60 °C. In this way rGO/PVA nanocomposites films were prepared.

5 Applications of Graphene Reinforced PVA Nanocomposites

Graphene reinforced PVA nanocomposites exhibit superior performance in multiple applications of daily lives such as transistors, electronics and memory devices, flexible packaging, hydrogen storage applications, printing devices, structural constituent for transportation and aerospace applications. Some of those applications are discussed in the following section.

5.1 Sensors

Graphene-PVA based biosensors have the superiority of enhanced sensitivity with high selectivity, stability, rapid response time, and a minimum limit of detection (LOD). A combination of conducting polymers and nano-fillers are responsible for fabricating better and improved sensors. Graphene reinforced PVA nanocomposite materials are used for various sensor applications like temperature sensors, pH sensors, pressure and strain sensors, biomolecules sensors, etc. as these composites possess 2D conjugated structures exhibiting larger surface area and higher conductivity. Furthermore Graphene-PVA composite films shows improved electrochemical stability enhanced electrolytic activity, impervious to molecules of gasses and rapid charge transfer between various elements hence greasing the wheel for numerous sensor applications. These nanocomposites are proven to be good sensors for air pollutants (SO_x, CO_x and NO_x) in environments. Thangamani et al. reported solution casting method for fabrication of White graphene reinforced PVA/WPPy/hBNNP films. These nanocomposite films are used as chemiresistive sensors for flammable liquid petroleum gas (LPG) leakage detection even at room temperature. Graphene reinforced PVA/WPPy/hBNNP films with 6 wt% filler loading exhibited enhanced sensitivity for LPG over ethanol, benzene, acetone and chloroform vapors [35].

Hwang et al., described the fabrication of Poly (dopamine)-treated Graphene Oxide reinforced PVA, and Its application as Humidity Sensors with neat PVA, rGO-PVA films (0.5 wt%) and GO-PVA films (0.5 wt%) [36]. Scanning electron micrographs of these nanocomposites are given in Fig. 4.

Wang et al., investigated (GO)/PVA nanocomposite films coated in fiber Mach–Zehnder interferometer (MZI) as Fiber optic humidity sensor applications as shown in Fig. 5. The two waist-enlarged tapers are used for MZI, having distance of 20 mm between adjacent tapers. Comparative study of Multiple GO/PVA composite films coated MZI revealed that films fabricated by ratio of PVA (0.3 g) along with GO dispersion (10 ml) exhibited improved performance of sensing capabilities. Molecular structure model explained the operation mechanism of water molecules and GO/PVA composite films. The improved stability with relative humidity range from 25 to 80%, improved a linear correlation coefficient of 99.1% and sensitivity level of 0.193 dB/%RH was observed [37]. Conclusion of temperature effect analysis on

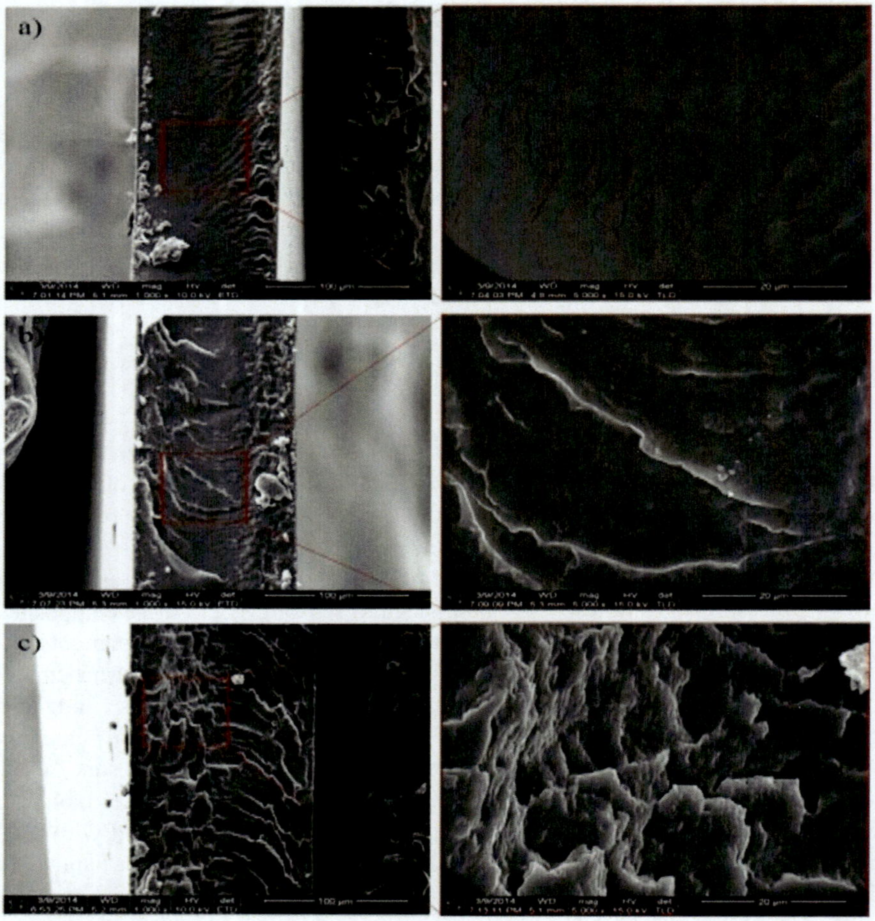

Fig. 4 SEM images of fracture surface of specimens after tensile test; **a** neat PVA, **b** GO-PVA (0.5 wt%), **c** 0.5:1 **d** GO-PVA (0.5 wt%)

the remote fiber optic sensor proposed humidity sensors exhibited a very good range of repeatability as shown in Fig. 6.

5.2 *Energy Storage Devices*

Owing to the effective and easy fabrication of PVA-reinforced graphene nanocomposites has been gaining attention as energy conversion and storage devices exhibiting low recombination rate, increased carrier mobility, longer stability and superior material properties, hence, making them useful for batteries, fuel cells,

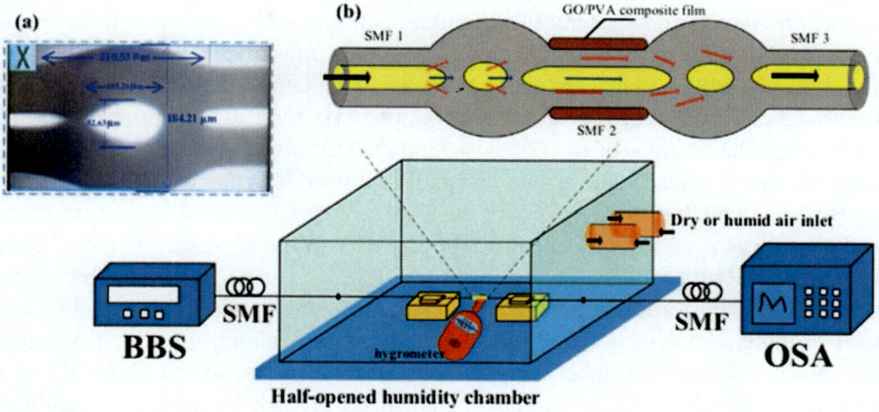

Fig. 5 Schematic diagram of the experimental setup, **a** structure of the waist-enlarged taper, and **b** the partially enlarged drawing of the sensing head

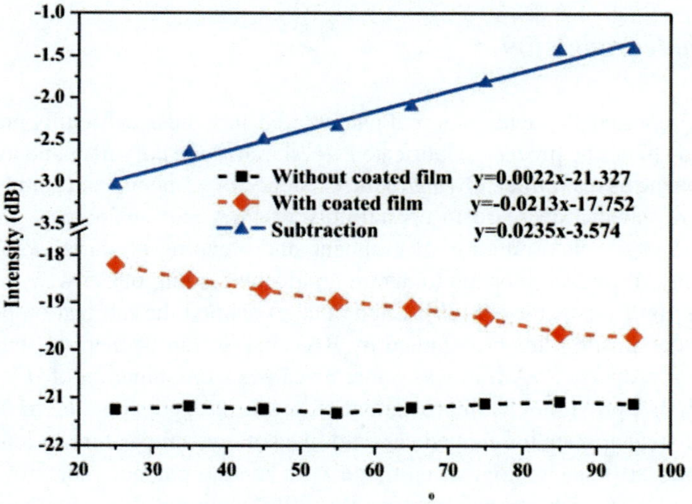

Fig. 6 Temperature effect on the proposed fiber optic humidity sensor

solar cells, supercapacitors, biomedical applications, thermoelectric devices, electric memory chips and energy transporting devices. Deshmukh et al. described fabrication of PVA/PEG/GO nanocomposites with GO (3 wt% loading) having dielectric constant of 644.39 that is 60 times more than that of pure PVA and 20 times more than that of PVA/PEG blend film. Hence, effectively developing flexible and free standing PVA/PEG/ GO films as affirmative material having low dielectric loss but high dielectric constant of energy storage devices like bio-friendly embedded capacitors have been fabricated. Deshmukh et al. reported fabrication of GO reinforced Hydroxyl-propyl Methylcellulose/PVA nanocomposites having ZnO nanoparticles, GO/HPMC/PVA/ZnO nanocomposites with enhanced dielectric constant and reduced dielectric loss at a constant frequency of 50 Hz and 150 °C temperature. With nanocomposite loadings of GO (2.5 wt%) and ZnO (25 wt%) making it a remarkable contender for enhanced automation demands of embedded capacitors [38].

Hummer's method was utilized for the formation of GO and colloidal processing technique was used to subsequently reinforce GO in blend matrix of PVA/PSSA to achieve GO/PVA/PSSA nanocomposites. PSSA is used in conductive polymers as a dopant for the polymerization exhibiting efficient coating capability and film forming, a polyelectrolyte, nontoxic polymer containing great transparency, enhanced dielectric strength. Owing to the effective interaction of GO, PVA and PSSA. GO/PVA/PSSA nanocomposites are used in multiple electronic applications like embedded capacitors with enhanced dielectric constant. Atomic force microscopy of these nanocomposites is shown in Figs. 6 and 7.

5.3 Biodegradability

Jose et al. reported characterization of nanocomposites and eco-friendly preparation via solution blending process to fabricate PVA/starch blend polymer nanocomposites with graphene as nano filler. Owing to the existence of graphene and starch, thermal analyses revealed a decrease in crystallinity of PVA and an increase in thermal stability. Also, on incorporation of graphene and blending of starch, PVA surface morphology shifted from brittle to ductile graphene loading of 0.5 wt%, maximum tensile strength was achieved [39]. Chen et al. explained the fabrication process of PVA/GO composite films crosslinked by BA, observation of thermal stability and mechanical property PVA film was done, revealing a combination of 0.2 wt% GO and 5 wt% BA providing an improvement of tensile strength about 65 to 88.5 MPa, so, its practical application in medical field like soft eye drops, contact lenses, slow release, and artificial biological cartilage with unique characteristics of biocompatibility, biodegradability and nontoxicity [40]. Scanning electron microscopy of nanocomposites with various Gr loadings are shown in Fig. 8.

Yang et al. described Poly (vinyl alcohol)/graphene oxide (PVA/Gr) nanocomposite films. PVA is highly nontoxic, hydrophilic and water soluble having many industrial applications such as flexible paper coating, textile sizing and packaging films. Layered structure of the PVA film with 2 wt% loading of GO. PVA/GO nanocomposites have improved tensile stress and modulus 92.2% and 167% respectively which is greater than pure PVA. Owing to the homogeneous dispersion of GO in PVA matrix, it is stated that GO reinforcement during polymer nanocomposites fabrication as nano-filler and it can be utilized to enhance mechanical and thermal properties of these nanocomposites [41]. He et al. explained Graphene oxide/amylose (GO/amylose) composite films with varied consignments of graphene oxide (GO) with, glycerol and polyvinyl alcohol (PVA) via solution casting method resulted in

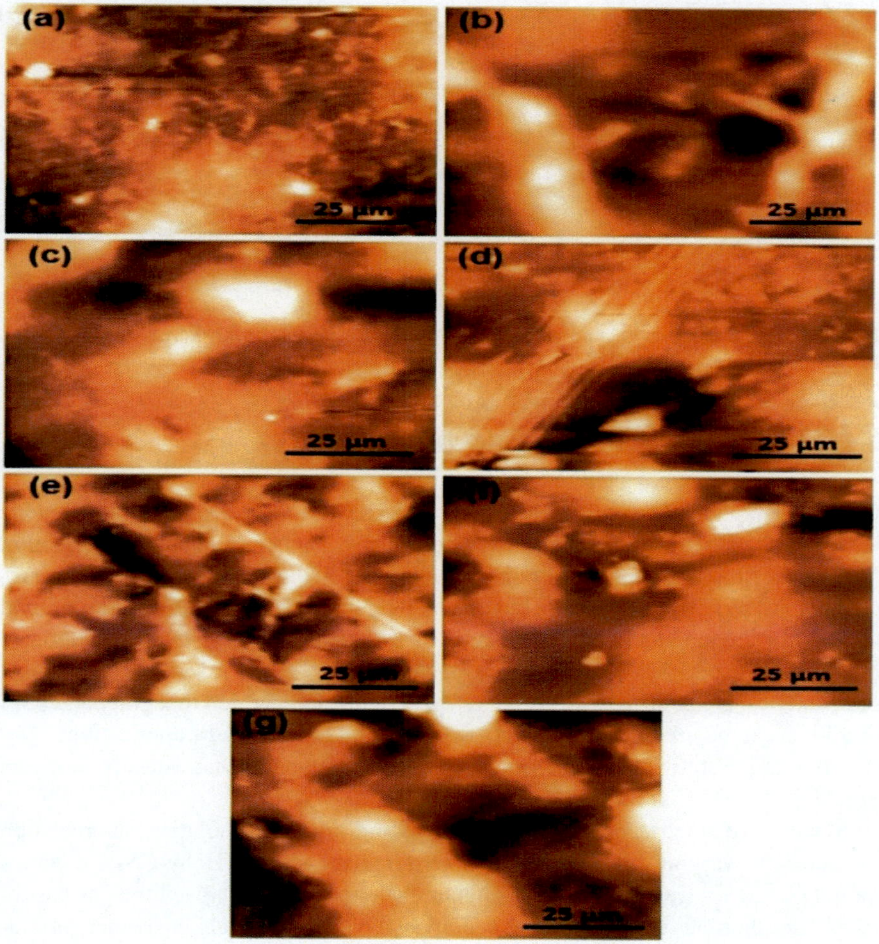

Fig. 7 AFM topographic images of PSSA/PVA blend and PSSA/PVA/GO composite films. **a** PSSA/PVA blend, **b** 0.5 wt% GO, **c** 1 wt% GO, **d** 1.5 wt%, films **a** PSSA/PVA blend, **b** addition to a shoulder peak at 262 nm for all composite samples which indicate the significant interaction between the functional groups of PSSA and PVA 0.5 wt% GO, **c** 1 wt% GO, **d** 1.5 wt% GO, **e** 2 wt% GO, **f** 2.5 wt% GO, **g** 3 wt% GO

immensely enhanced tensile strength of the amylose films. Glycerol served as plasticizer enhancing the elongation at break but reducing tensile strength. On addition of PVA with the nanocomposite, films turned flexible and tough. As GO content in the GO/amylose films enhanced Moisture absorbability of the films to gradually decline. GO/amylose films maintain perfect stability in alkaline as well as in acidic aqueous media, just like oils. Good transmittance, stability and mechanical properties in acid/base media is shown by GO/amylose nanocomposite film with 0.5% GO, 20% glycerol and 4% PVA [42]. Scanning electron microscopy and transmission electron microscopy images of GO/amylose films are shown in Fig. 9.

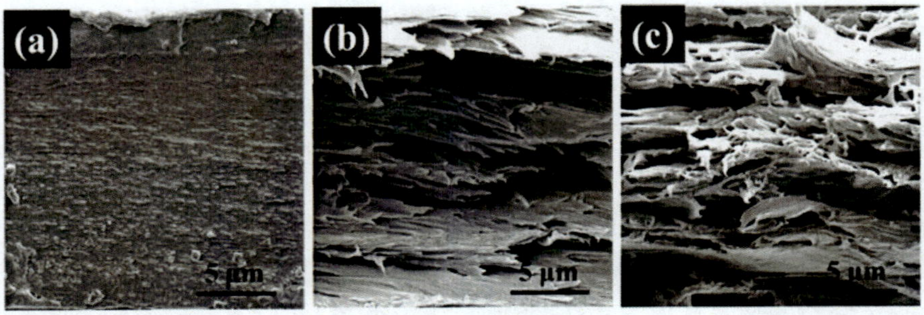

Fig. 8 SEM images of the fracture surface of PVA/GO films, **a** PVA, **b** with 0.5 wt% GO and **c** with 1.0 wt% GO

5.4 Medical and Food Packaging Applications

Cao et al. explained the utilization of casting method for the synthesis of biocide based graphene reinforced Poly (vinyl alcohol) i.e. Graphene/PVA antibacterial thin film. These films had improved antibacterial actions against gram positive and gram-negative bacteria with polymeric biocide loading of 1, 5, and 10wt%. Antimicrobial activity for the loading level of 1, 5, and 10 wt% against gram negative bacteria *E. coli* was 92.0%, 95.8%, and 97.1%, respectively and 92.3, 99.6, and 99.7% against the gram positive *S. aureus* was observed. These results revealed that graphene/PVA thin films can be used in the hygienic, drinking and food package applications [43]. A comparative study of antimicrobial activity of these nanocomposites is shown in Fig. 10.

Shang, explained fabrication of PVA hybrid films via solution intercalation method with varying filler (functionalized graphene sheets (HDA-GSs)) contents ranging 0–10 wt%. Owing to the two main factors, analysis showed the gas barrier properties: thermal stabilities of these composites and polymer chain-segment immobility and detour ratio. Improved barrier properties, reduced solvent uptake, chemical resistance, and flame insulator of polymer hybrids get eased from hindered ways through these nanocomposites [44] Chao et al. reported synthesis of PVA/Gr nanocomposite films by the reduction of GO in graphene nano sheets. Graphene oxide (GO) into PVA aqueous solution, resulted an 34% in elongation at break and with only 0.5 wt% Gr content, tensile strength enhanced up to 212%. Hence it showed a remarkable reduction in water absorption of the Gr/PVA composites ranging 105.2% to 48.8%, resulting in the improved barrier properties. Results from contact angle measurements (CAM) of nanocomposites turned out to be hydrophobic ($\theta > 90°$) in comparison with other Pure PVA hydrophilic composites ($\theta < 90°$). At low loading of graphene, sudden and efficient improvements in mechanical and thermal stability properties were observed. Hence, resulting in the enhanced barrier properties for pure PVA with water [43]. Effects of various graphene contents on nanocomposites are shown in Fig. 11.

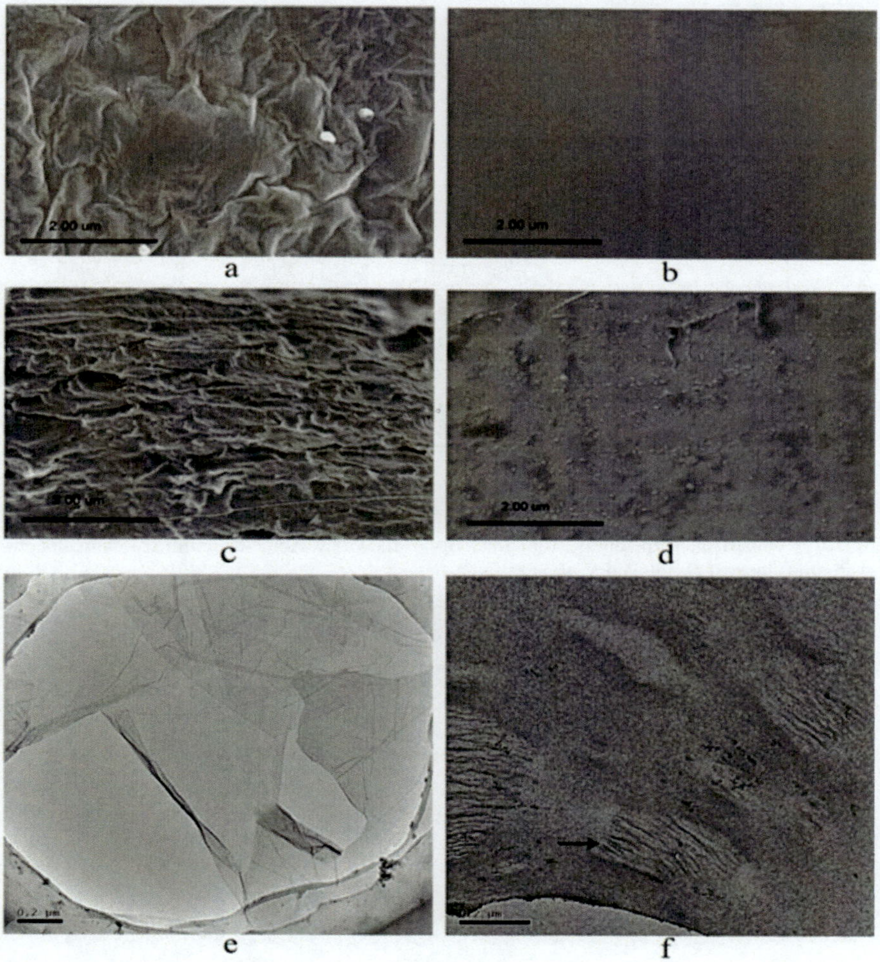

Fig. 9 SEM photos of a GO/amylose film (1.0% GO, Series 1), surface (**a**) and cross-section (**c**); and a pure amylose film, surface (**b**) and cross-section (**d**); and TEM photos of GO (**e**), and the GO/amylose composites (1.0% GO, Series 1) (**f**)

Liu et al. explained the PVA fabrication of layered nanocomposite films in their organic phase embed between parallel GO platelets. Using borate as an agent, crosslinking of the interfacial region, GO/PVA nanocomposite films exhibited improved by the tensile strength to 360 MPa at the cost of ductility, improving toughness and tensile strength as shown in Fig. 12. Hence, these results helped in designing improved nanocomposite films showing enhanced toughness and strength hence making it a compatible and beneficial contender to be used in biomimetic, such as in replacement of nacre and bone [45].

Jankovic et al. explained the synthesis of PVA/Gr in varied configurations of film, hydrogel and colloid dispersion to determine the small quantities variations

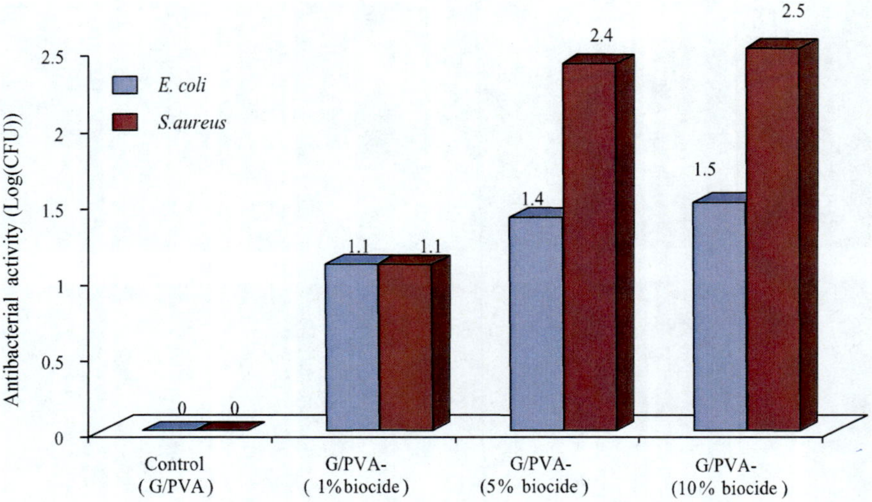

Fig. 10 Comparison of the antimicrobial activity of the G/PV Abiocide antimicrobial nanocomposite films against *E. coli* and *S. aureus*

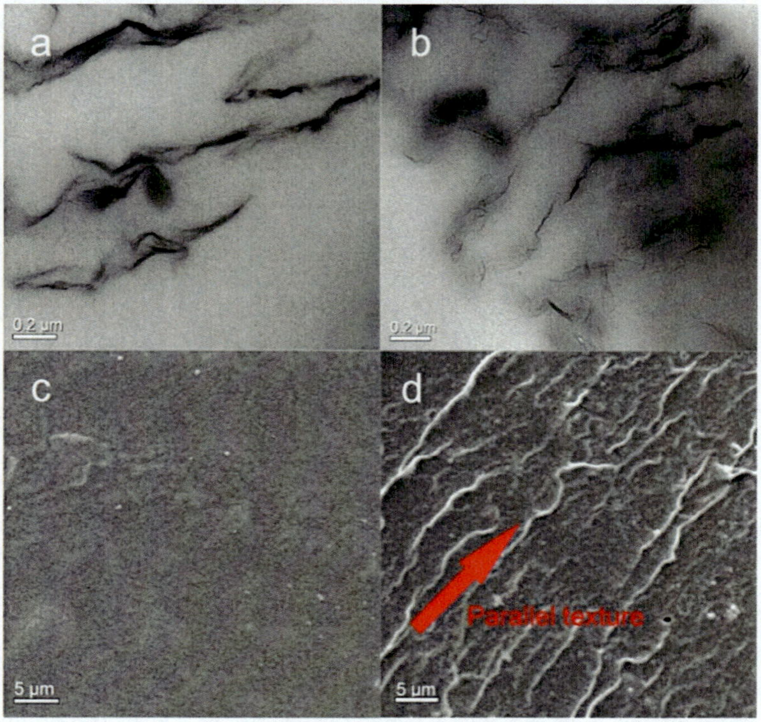

Fig. 11 **a**, **b** TEM images of ultra-thin sections of PVAc/GO nanocomposites with 0.04 wt%. **c**, **d** SEM images of fractured surfaces PVA and PVA 0.04% GO

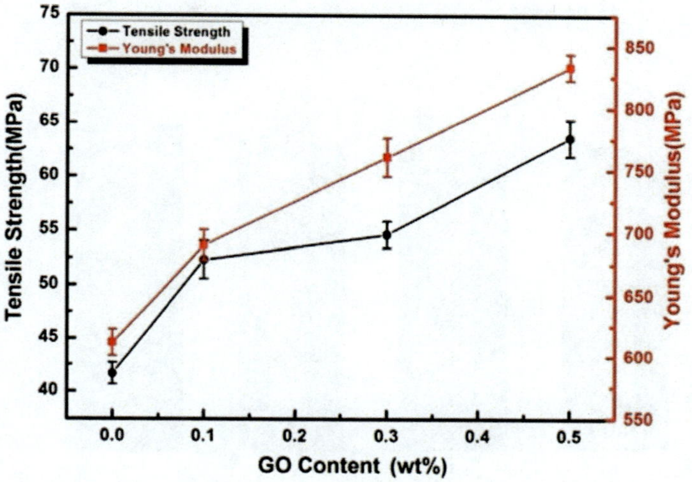

Fig. 12 Mechanical properties of pure PVA and PVA-g-GO/PVA nanocomposite films GO

of graphene on thermal, biological and mechanical properties of PVA. Comparative analysis of PVA/graphene nanocomposite with pure PVA exhibited greater thermal stability and improved mechanical properties, i.e., improved Young's modulus to 20.7% and tensile strength to 56.0% as shown in Fig. 13. Therefore, PVA/Gr hydrogel showed great antibacterial activity against *S. aureus* and was named non-cytotoxic against healthy peripheral blood mono nuclear cells (PBMC) in the MTT assay.

Quantitative study of Antibacterial activity against agents of two common widespread bacterial infections (*S. aureus* and *E. coli*) was done by observing variations in suspension using viable bacterial cells numbers. Therefore, the results presented explained that nano sized PVA/G nanocomposite has proven to be suitable choice for wound dressings and biomedical implants of soft tissue [46]. Hajeeassa et al. stated the fabrication of biologically active polymer nanocomposites based on PVA/Gr/CNT via efficient solution mixing procedure with ultrasonic radiation. Mixed Gr/CNT loadings of 2, 5, 10, 15, and 20 wt% were subjected to host PVA polymer matrix. Gr/CNT mixture was prepared in ratio 50:50 wt%. In zone of inhibition, improved antibacterial activity against both bacterial agents were observed as shown in Fig. 13, thus making them effective to be used in hygiene products, longer shelf life and food packaging [47]. Antibacterial activity of these nanocomposites is shown in Figs. 14 and 15.

Cao et al. described detailed analysis of antibacterial activity of PVA against *E. coli* with multiple loadings of Gr in PVA 1% to 10% resulted in improvement by 92–97.1% and 92.3—99.7% against *S. aureus.* At a relative low load of biocide based Gr nanocomposite thin film showed effective antibacterial activity against both *gram positive* and *gram-negative* bacteria. These nanocomposites can be used in drinking and food packaging and hygiene applications [43].

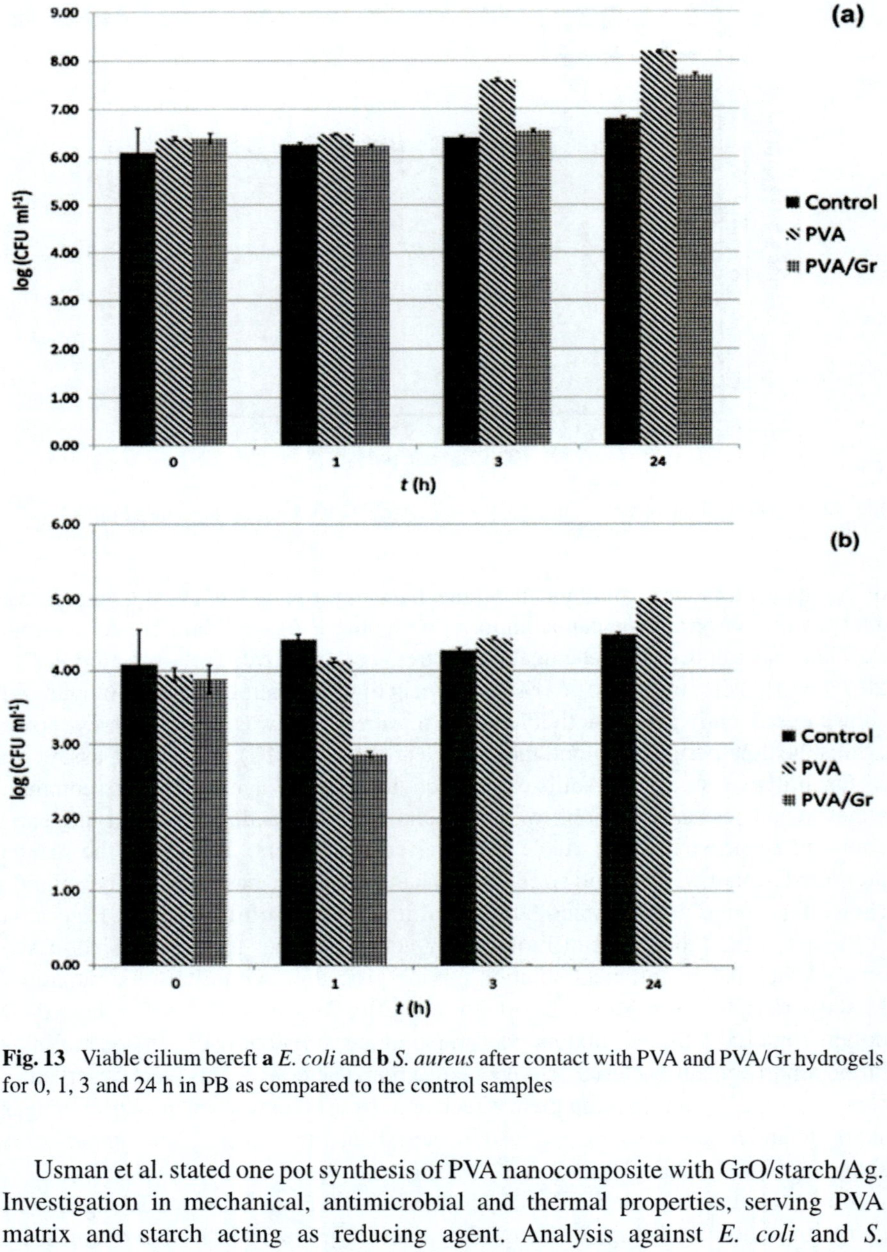

Fig. 13 Viable cilium bereft **a** *E. coli* and **b** *S. aureus* after contact with PVA and PVA/Gr hydrogels for 0, 1, 3 and 24 h in PB as compared to the control samples

Usman et al. stated one pot synthesis of PVA nanocomposite with GrO/starch/Ag. Investigation in mechanical, antimicrobial and thermal properties, serving PVA matrix and starch acting as reducing agent. Analysis against *E. coli* and *S. aureus* bacteria resulted enhancement of antimicrobial activity through grafting of GO/starch/Ag to PVA matrix. Antimicrobial activity of nanocomposites against *E. coli* and *S. aureus* are shown in Fig. 16.

For PVA/GO/Ag/starch nanocomposites, Modulus and tensile strength were increased by about 145%, and 26.81% respectively. The thermal degradation

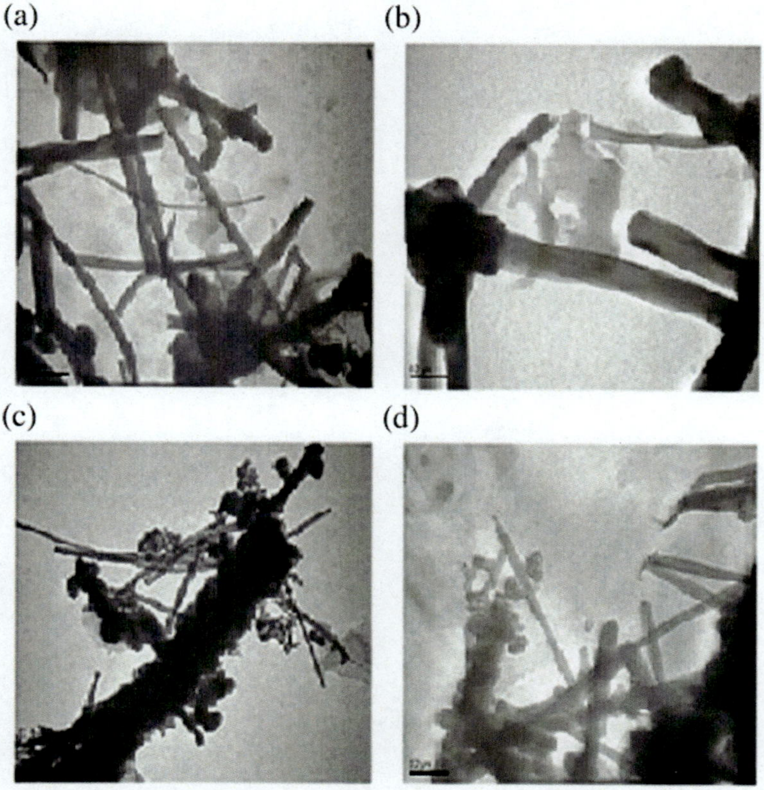

Fig. 14 TEM images of PVA/MGr-CNTs$_b$ nanocinoisutea (**a**, **b**) and PVA/MGr-CNTs nanocomposite (**c**, **d**). TEM: transmission electron microscopy; MGr: mixed graphene; CNT: carbon nanotube

compared to pure PVA was also enhanced. Hence, PVA/GO/starch/Ag nanocomposites films has proven to be effective approach to be utilized as food packaging materials [48]. Shuai et al. described fabrication of nanocomposite scaffolds containing sponge like structure via selective laser sintering (SLS). Results exhibited the immense progress in mechanical properties, i.e. 152, 69 and 60% improvement of Young's modulus, tensile strength and compressive strength obtained at 2.5 wt%, loading of GO, as shown in Fig. 16. Enhanced load transfer owing to the strong hydrogen bonding interactions resulting in homogeneous dispersion of GO in PVA matrix [49] (Fig. 17).

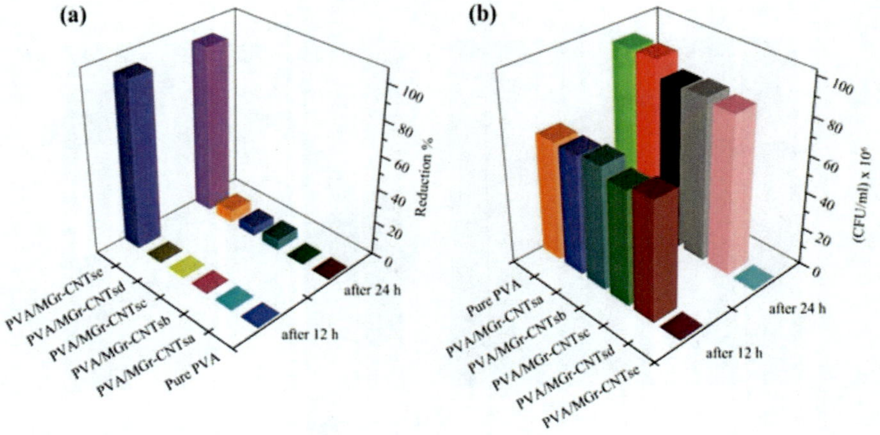

Fig. 15 **a** Reduction percentage of *E. coli* after contact with PVA and PVA/MGr-CNTs$_{a\text{–}e}$ nanocomposites for 12 and 24 h. **b** Viable cell number of *E. coli* after contact with PVA and PVA/MGr-CNTs$_{a\text{–}e}$ nanocomposites for 12 and 24 h. PVA: polyvinyl alcohol; MGr

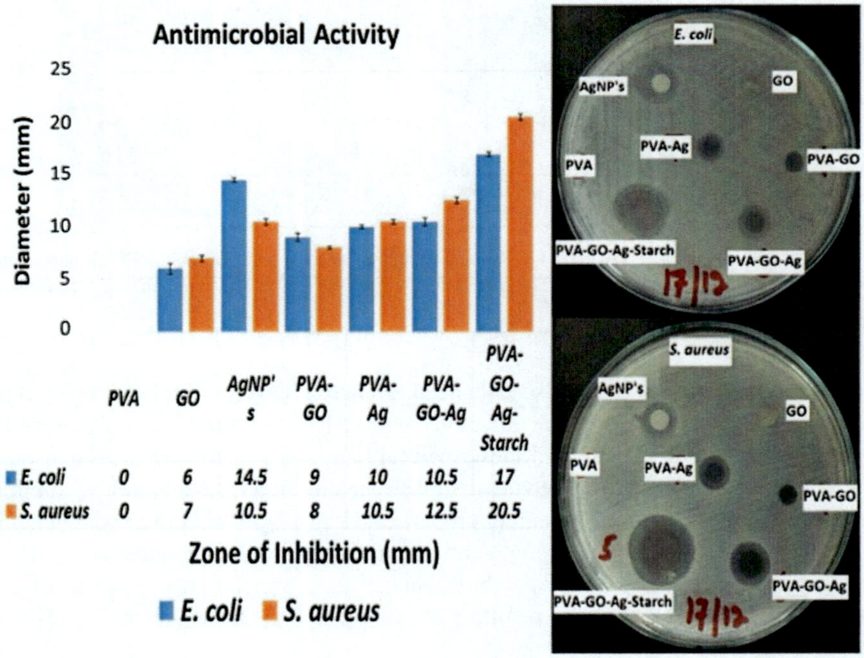

Fig. 16 Statistical representation of anti-microbial activity of polymer nanocomposites films against *E. coli* and *S. aureus*

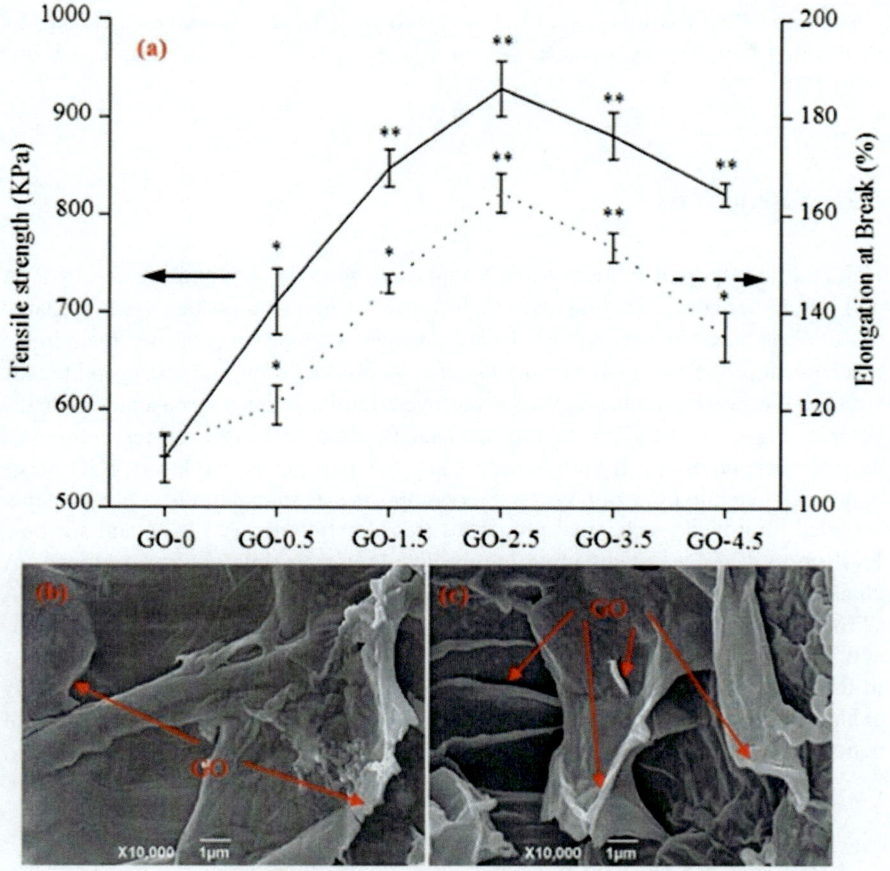

Fig. 17 **a** Tensile strength (solid line, left axis) and elongation at break (dotted line, right axis) of GO-0, GO-0.5, GO-1.5, GO-2.5, GO-3.5 and GO-4.5, **b**, **c** the tensile fracture surface of (**b**) GO-2.5 and **c** GO-4.5. *$P < 0.05$ and **$P < 0.01$ (compared with GO-0), n ¼ 6

6 Various Other Applications

Graphene reinforced PVA nanocomposites have many other applications. These applications are discussed in the following section.

6.1 Optical Films

Morimune et al. explained simple casting method for GO/PVA composites synthesis with improved transparency (490%) of nanocomposite films. Thus, marking these composites to be applied as optical films such as polarizer. Additionally, for many nanocomposites, transparency decreases with increasing GO content. Hence, by controlling GO content, transparency to opacity is controllable in nanocomposite films [50].

6.2 Safety Glass

Hajian et al. reported fabrication of Poly (vinyl butyral) via condensation of PVA and butyraldehyde. PVB/graphene (0.1–0.6 wt%) nanocomposites were prepared via solution blending process. Analysis confirmed a remarkable enhancement in the mechanical properties such as modulus of samples containing graphene and tensile strength. These nanocomposites have improved rigidity and increased adhesion to the surface of glass and can be used to form safety glass utilizing in stores, museums, automobiles, commercial building etc. [51]. Compton et al. explained rGO paper and paper-like thin films of PVA/Gr nanocomposites as shown in Fig. 18. rGO paper revealed improved mechanical properties (tensile strength: 125 MPa and stiffness level up to 40 GPa,) maintaining highly ductile and flexible nature. Adding PVA chains in host matrix improved hydrogen-bonding. Grafting a small amount of PVA (22 wt%) into the graphene oxide paper structure serving as a softening agent interacting strongly with rigid graphene oxide (rGO) nano sheets. Addition of PVA chains in the inter sheet space of rGO paper, reduced the role of water and dramatically enhanced the rigidity of composite papers. This resulted in the enhanced rigidity of nanocomposites as per observed in many natural materials, such as nacre, natural

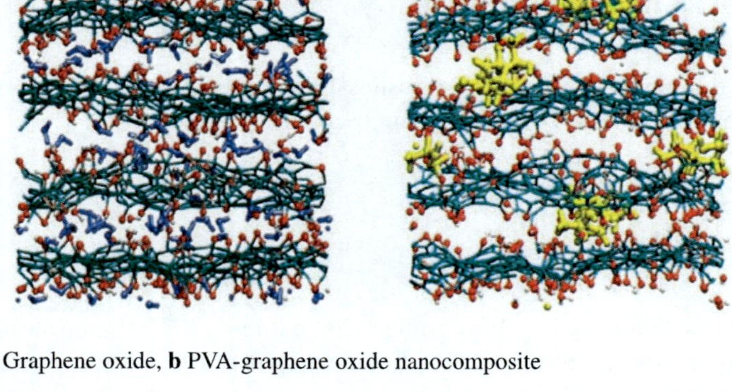

Fig. 18 **a** Graphene oxide, **b** PVA-graphene oxide nanocomposite

bio composites (e.g. collagen and spider silk), having flexible protein (lustrin and chitin), also molecular water and inorganic material (aragonite platelets) [52].

6.3 Filler Membranes

Dave et al. explained simple solution casting method for successful fabrication of rGO incorporated PVA composite membranes using DMSO as solvent. Analysis of PVA-GO composites compared to pristine (PVA) membrane, TGA results revealed enhanced thermal stability owing to the GO incorporation of PVA-GO composites. XRD results revealed an overall reduction in crystallinity of nanocomposites close to 42% in comparison to (PVA). At 2.5% acetic acid concentration, maximum water flux of 0.22 m^3 (STP)/m^2 h was obtained. Increase in acetic acid content resulted in molar fluxes of acetic acid and water reduction in the feed mixture this increased the separation factor. At 20% acetic acid concentration, maximum separation factor of 62.2 was recorded in $(PVA\text{-}GO)_2$ composite [53]. Atomic force microscopy of these nanocomposites is shown in Fig. 19.

7 Conclusions

Graphene reinforced PVA nanocomposites are fascinating materials due to their unique properties. These properties include mechanical, electrical and biological properties. These properties of nanocomposites are dependent on the amount of filler used. Results showed that there is always an optimum amount of filler required to get high performance of these nanocomposites towards various applications. A lot of investigations have been done in order to develop various chemical routes for their synthesis and their applications. These applications include sensors, food packaging, biodegradability, antibacterial activity against various bacteria, energy storage devices and membranes. As per reported by hundreds of articles based upon the subject of Gr/PVA nanocomposites, it is concluded that graphene reinforced PVA nanocomposites exhibiting higher level of strength, stiffness, superior mechanical properties makes these composites compatible contender for application in various fields of life. Gr/PVA nanocomposites with their exceptionally improved mechanical properties are being widely used in multiple areas of life, gaining worldwide attention. Although, exciting properties of these materials are already being explored but still there is a lot of room for further developments for the betterment of mankind.

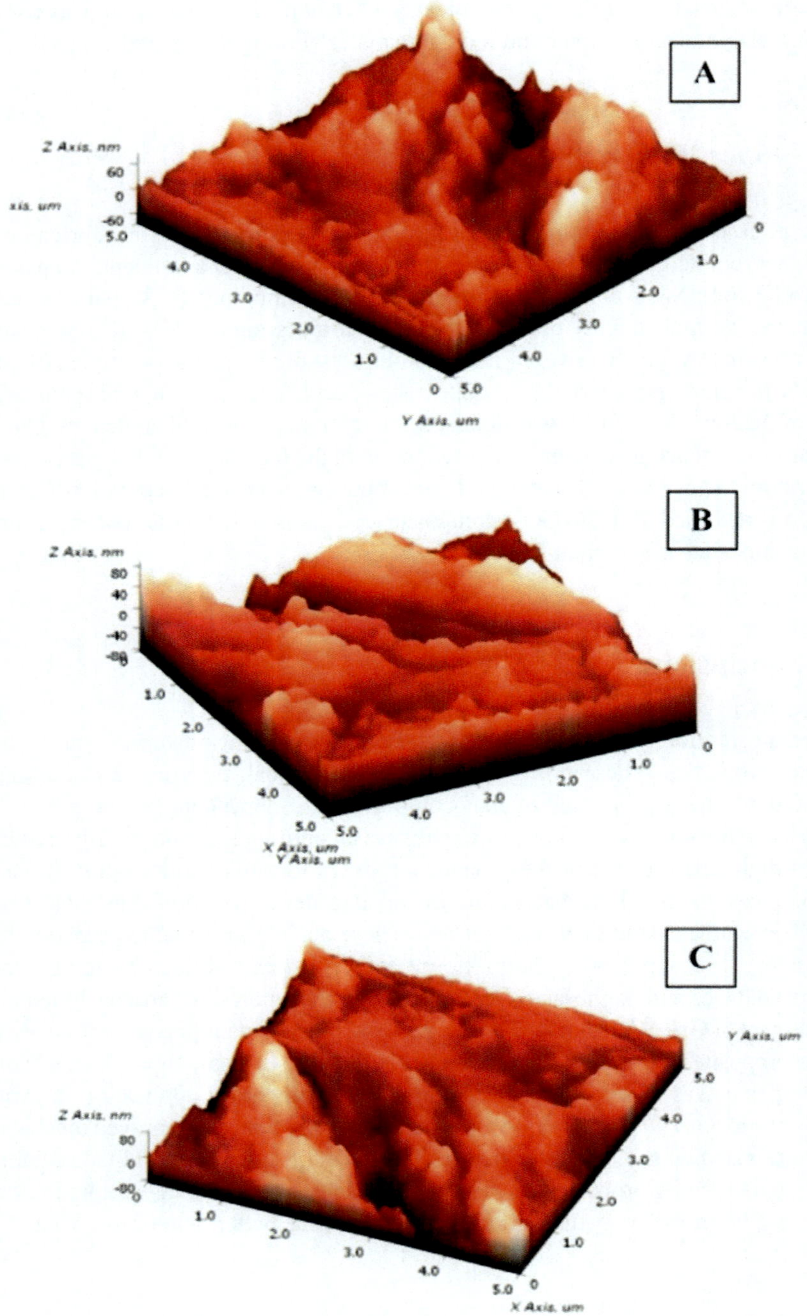

Fig. 19 Surface AFM images of the membranes used in the study. **a** $(PVA)_0$, **b** $(PVAGO)_1$, **c** (PVA-GO)

8 Future Prospects

Despite the fact that graphene reinforced PVA nanocomposites have shown fascinating properties but still there is a long way to overcome scientific and technological challenges. The major challenge is related to the control of their structures. If structures could be controlled then it is very helpful to understand underlying mechanisms. Another important aspect is the complete understanding of the interactions between graphene and PVA matrix. This understanding will help to prepare strength interface which is very critical in certain applications. One of the barriers understands of their long-term biocompatibility and potential toxicity as both of these factors are critical for their successful application in biomedical and pharmaceutical fields. It is proved experimentally that cytotoxicity is affected by the physicochemical properties of the nanocomposites. These physicochemical properties include their shape, size, distribution, purity, surface chemistry etc. Another barrier is related to their mechanical properties as these should be of high tensile strength and high elongation at break in order to withstand rupture. Deep understanding of their mechanical properties makes these nanocomposites more useful for packaging, sensors and energy storage applications.

Acknowledgements Hafeez Anwar acknowledges funding of Pakistan science foundation under Project No: PSF-NSF/ENG/P-UAF (05) and funding from higher education commission, Pakistan for national research program for university (NRPU), Project number 8545.

References

1. Adetayo A, Runsewe D (2019) Synthesis and fabrication of graphene and graphene oxide: a review. Open J Compos Mater 9:207–229. https://doi.org/10.4236/ojcm.2019.92012
2. Marcano DC, Kosynkin DV, Berlin JM, Sinitskii A, Sun Z, Slesarev A, Alemany LB, Lu W, Tour JM (2010) Improved synthesis of graphene oxide. ACS 4:4806–4814. https://doi.org/10.1021/nn1006368
3. Eletskii AV, Iskandarova IM, Knizhnik AA, Krasikov DN (2011) Graphene: fabrication methods and thermophysical properties. Phys Usp 54:227–258. https://doi.org/10.3367/ufne.0181.201103a.0233
4. Own SY, Me N (2011) Simple room-temperature preparation of high-yield large-area graphene oxide. Int J Nanomed 6:3443–3448. https://doi.org/10.2147/IJN:s26812
5. Yu H, Zha B, Chaoke B, Li R, Xing R (2016) High-efficient synthesis of graphene oxide based on improved hummers method. Nature 6:36143. https://doi.org/10.1038/srep36143
6. Kovtyukhova NI (1999) Layer-by-layer assembly of ultrathin composite films from micron-sized graphite oxide sheets and polycations. Chem Mater 11:771–778. https://doi.org/10.1021/cm981085u
7. Peng L, Xu Z, Liu Z, Wei Y, Sun H, Li Z, Zhao X, Gao C (2015) An iron-based green approach to 1-h production of single-layer graphene oxide. Nat Commun 6:1–9. https://doi.org/10.1038/ncomms6716
8. Chen J, Yao B, Li C, Shi G (2013) An improved Hummers method for eco-friendly synthesis of graphene oxide. Carbon N Y 64:225–229. https://doi.org/10.1016/j.carbon.2013.07.055

9. Hummers Jr WS, Offeman RE (1958) Preparation of graphitic oxide. J Am Chem Soc 80(6):1339–1339. https://doi.org/10.1021/ja01539a017
10. Staudenmaier L (1898) Method for the preparation of the graphite acid. Eur J Inorg Chem 31:1481–1487
11. Chen J, Li Y, Huang L, Li C, Shi G (2015) High-yield preparation of graphene oxide from small graphite flakes via an improved Hummers method with a simple purification process. Carbon N Y 81:826–834. https://doi.org/10.1016/j.carbon.2014.10.033
12. Su C, Xu Y, Zhang W, Zhao J, Tang X, Tsai C, Li L (2009) Electrical and spectroscopic characterizations of ultra-large reduced graphene oxide monolayers. 5674–5680. https://doi.org/10.1021/cm902182y
13. Rosillo-lopez M, Salzmann CG (2016) A simple and mild chemical oxidation route to high-purity nano-graphene oxide. Carbon N Y 106:56–63. https://doi.org/10.1016/j.carbon.2016.05.022
14. Brodie BC (1860) XXIII.—Researches on the atomic weight of graphite. Quart J Chem Soc London 12(1):261–268. https://doi.org/10.1098/rspl.1859.0007
15. Ranjan P, Agrawal S, Sinha A, Rao TR, Balakrishnan J, Thakur AD (2018) A low-cost non-explosive synthesis of graphene oxide for scalable applications. Sci Rep 8:1–13. https://doi.org/10.1038/s41598-018-30613-4
16. Shahriary L, Athawale A (2014) Graphene oxide synthesized by using modified Hummers approach. Int J Renew Energy Environ Eng 02:58–63
17. Shen J, Hu Y, Shi M, Lu X, Qin C, Li C, Ye M (2009) Fast and facile preparation of graphene oxide and reduced graphene oxide nanoplatelets. Chem Mater 21:3514–3520. https://doi.org/10.1021/cm901247t
18. Dong X, Huang W, Chen P (2011) In situ synthesis of reduced graphene oxide and gold nanocomposites for nanoelectronics and biosensing. Nanoscale Res Lett 6(1):1–6. https://doi.org/10.1007/s11671-010-9806-8
19. Huang K, Zhang J, Liu Y, Liu Y (2015) Science direct synthesis of reduced graphene oxide wrapped—copper sulfide hollow spheres as electrode material for supercapacitor. Int J Hydrogen Energy 1–10. https://doi.org/10.1016/j.ijhydene.2015.05.152
20. Haghighi B, Tabrizi MA (2013) Green-synthesis of reduced graphene oxide nanosheets using rose water and a survey on their characteristics and applications. RSC Adv 3(3):13365–13371. https://doi.org/10.1039/c3ra40856f
21. Shen J, Li T Long Y, Shi M, Li N, Ye M (2012) One-step solid state preparation of reduced graphene oxide. Carbon 50:2134–2140. https://doi.org/10.1016/j.carbon.2012.01.019
22. Habte AT, Ayele DW (2019) Synthesis and characterization of reduced graphene oxide (rGO) started from graphene oxide (GO) using the tour method with different parameters. Adv Mater Sci Eng 2019:9. https://doi.org/10.1155/2019/5058163
23. Kashyap S, Pratihar SK, Behera SK (2016) Strong and ductile graphene oxide reinforced PVA nanocomposites. J Alloys Compd 684:254–260. https://doi.org/10.1016/j.jallcom.2016.05.162
24. Wang J, Wang X, Xu C, Zhang M, Shang X (2011) Preparation of graphene/poly(vinyl alcohol) nanocomposites with enhanced mechanical properties and water resistance. Polym Int 60:816–822. https://doi.org/10.1002/pi.3025
25. Komarneni S (1992) Nanocornposites. 2:1219–1230
26. Yang Z, Peng H, Wang W, Liu T (2010) Crystallization behavior of poly(ε-caprolactone)/layered double hydroxide nanocomposites. J Appl Polym Sci 116:2658–2667. https://doi.org/10.1002/app
27. Sharma B, Shekhar S, Gautam S, Sarkar A, Jain P (2018) Nanomechanical analysis of chemically reduced graphene oxide reinforced poly (vinyl alcohol) nanocomposite thin films. Polym Test 70:458–466. https://doi.org/10.1016/j.polymertesting.2018.07.028
28. Yuan X (2011) Enhanced interfacial interaction for effective reinforcement of poly(vinyl alcohol) nanocomposites at low loading of graphene. Polym Bull 67:1785–1797. https://doi.org/10.1007/s00289-011-0506-z
29. Guo J, Ren L, Wang R, Zhang C, Yang Y, Liu T (2011) Water dispersible graphene noncovalently functionalized with tryptophan and its poly(vinyl alcohol) nanocomposite. Compos Part B Eng 42:2130–2135. https://doi.org/10.1016/j.compositesb.2011.05.008

30. Sellam C, Zhai Z, Zahabi H, Picot OT, Deng H, Fu Q, Bilotti E, Peijs T (2015) High mechanical reinforcing efficiency of layered poly(vinyl alcohol)–graphene oxide nanocomposites. Nanocomposites 1:89–95. https://doi.org/10.1179/2055033215Y.0000000001
31. Zhou T, Chen F, Tang C, Bai H, Zhang Q, Deng H, Fu Q (2011) The preparation of high performance and conductive poly (vinyl alcohol)/graphene nanocomposite via reducing graphite oxide with sodium hydrosulfite. Compos Sci Technol 71:1266–1270. https://doi.org/10.1016/j.compscitech.2011.04.016
32. Zhou TN, Qi XD, Fu Q (2013) The preparation of the poly(vinyl alcohol)/graphene nanocomposites with low percolation threshold and high electrical conductivity by using the large-area reduced graphene oxide sheets. Express Polym Lett 7:747–755. https://doi.org/10.3144/expresspolymlett.2013.72
33. Goumri M, Venturini JW, Bakour A, Khenfouch M, Baitoul M (2016) Tuning the luminescence and optical properties of graphene oxide and reduced graphene oxide functionnalized with PVA. Appl Phys a Mater Sci Process 122:1–8. https://doi.org/10.1007/s00339-016-9725-3
34. Muralidharan MN, Mathew S, Seema A, Radhakrishnan P, Kurian T (2016) Optical limiting properties of in situ reduced graphene oxide/polymer nanocomposites. Mater Chem Phys 171:367–373. https://doi.org/10.1016/j.matchemphys.2016.01.030
35. Thangamani JG, Deshmukh K, Sadasivuni KK (2017) White graphene reinforced polypyrrole and poly(vinyl alcohol) blend nanocomposites as chemoresistive sensors for room temperature detection of liquid petroleum gases. Microchim Acta 184(10):3977–3987. https://doi.org/10.1007/s00604-017-2402-1
36. Hwang S, Kang D, Ruoff RS, Shin HS. Polyvinyl alcohol reinforced and toughened with poly(dopamine)-treated graphene oxide, and its use for humidity sensing. 1–4
37. Wang Y, Shen C, Lou W, Shentu F (2016) Fiber optic humidity sensor based on the graphene oxide/PVA composite fi lm. Opt Commun 372:229–234. https://doi.org/10.1016/j.optcom.2016.04.030
38. Deshmukh K, Ahamed MB, Sadasivuni KK (2016) Graphene oxide reinforced polyvinyl alcohol/polyethylene glycol blend composites as high-performance dielectric material. J Polym Res. https://doi.org/10.1007/s10965-016-1056-8
39. Jose J, Al-harthi MA, Alma MA, Dakua JB, De SK (2015) Effect of graphene loading on thermomechanical properties of poly(vinyl alcohol)/starch blend. 41827:1–8 . https://doi.org/10.1002/app.41827
40. Chen J, Li Y, Zhang Y, Zhu Y (2015) Preparation and characterization of graphene oxide reinforced PVA film with boric acid as crosslinker. J Appl Polym Sci 132(22). https://doi.org/10.1002/app.42000
41. Yang X, Shang S, Li L (2011) Layer-structured poly (vinyl alcohol)/graphene oxide nanocomposites with improved thermal and mechanical properties. J Appl Polym Sci 120:1355–1360. https://doi.org/10.1002/app
42. He Y, Wang X, Wu D, Gong Q, Qiu H, Liu Y, Wu T, Ma J, Gao J (2013) Biodegradable amylose films reinforced by graphene oxide and polyvinyl alcohol. 142:1–11. https://doi.org/10.1016/j.matchemphys.2013.05.036
43. Cao Y, Wei W, Liu J, You Q, Liu F, Lan Q, Zhang C, Liu C, Zhao J (2015) The preparation of graphene reinforced poly(vinyl alcohol) antibacterial nanocomposite thin film. Int J Polym Sci 2015:7. https://doi.org/10.1155/2015/407043
44. Shang X (2011) Preparation of graphene/poly(vinyl alcohol) nanocomposites with enhanced mechanical properties and water resistance Jingchao Wang[a], Xianbao Wang[a, b,*], Chunhui Xu[a], Min Zhang[a]. 816–822. https://doi.org/10.1002/pi.3025
45. Liu L, Gao Y, Liu Q, Kuang J, Zhou D, Ju S, Han B (2013) High mechanical performance of layered graphene oxide/poly (vinyl alcohol) nanocomposite films. 2466–2472. https://doi.org/10.1002/smll.201300819
46. Jankovic A, Mitric M, Matic I, Juranic ZD, Rhee KY, Jin S, Ljiljana Z, Hui D (2016) The effect of graphene loading on mechanical, thermal and biological properties of poly(vinyl alcohol)/graphene nanocomposites. J Ind Eng Chem 34:250–257. https://doi.org/10.1016/j.jiec.2015.11.016

47. Hajeeassa KS, Hussein MA, Anwar Y, Tashkandi NY, Al-amshany ZM (2018) Nanocomposites containing polyvinyl alcohol and reinforced carbon-based nanofiller: a super effective biologically active material. 5:1–12. https://doi.org/10.1177/1849543518794818
48. Usman A, Hussain Z, Riaz A, Khan AN (2016) Enhanced mechanical, thermal and antimicrobial properties of poly(vinyl alcohol)/graphene oxide/starch/silver nanocomposites films. Carbohydr Polym 153:592–599. https://doi.org/10.1016/j.carbpol.2016.08.026
49. Shuai C, Feng P, Gao C, Shuai X (2015) RSC Advances nanocomposite sca ff olds for tissue engineering. RSC Adv 5:25416–25423. https://doi.org/10.1039/C4RA16702C
50. Morimune S, Nishino T, Goto T (2012) Poly (vinyl alcohol)/graphene oxide nanocomposites prepared by a simple eco-process. Polym J 44(10):1056–1063. https://doi.org/10.1038/pj.2012.58621
51. Hajian M, Reisi MR (2012) Preparation and characterization of polyvinylbutyral/graphene nanocomposite. J Polym Res 19(10):9966. https://doi.org/10.1007/s10965-012-9966-6623
52. Compton OC, Cranford SW, Putz KW, An Z, Brinson LC, Buehler MJ, Nguyen ST (2019) Tuning the mechanical properties of graphene oxide paper and its associated polymer nanocomposites by controlling cooperative intersheet hydrogen bonding. ACS nano 6(3):2008–2019. https://doi.org/10.6261021/nn202928w627
53. Dave HK, Nath K (2016) Journal of Water Process Engineering Graphene oxide incorporated novel polyvinyl alcohol composite membrane for pervaporative recovery of acetic acid from vinegar wastewater. J Water Process Eng 14:124–134. https://doi.org/10.1016/j.jwpe.2016.11.002

Graphene Grafted Chitosan Nanocomposites and Their Applications

Bhasha Sharma, Shashank Shekhar, Purnima Jain, Reetu Sharma, and K. K. D. Chauhan

Abstract Proliferation in the biopolymer nanocomposites domain is a manifestation of its fringe benefit, opportunity to generate reformulations of traditional nanofillers and polymer matrix. The diversification in conservative nanocomposites and environmental congenial is eminent attributes to the excessive contamination in biota. Chitosan, a naturally derived biopolymer obtained from the exoskeleton of crustaceans, fungi, animals, plants and vegetables contemplated as a promising biodegradable polysaccharide for a range of applications mainly in tissue engineering, electronics, biomedical devices, supercapacitors food packaging, bone substitutes, and water filtration. The limitation of chitosan is its low mechanical strength and extremely high hydrophilicity which could be ameliorated by the inclusion of graphene oxide (GO) nanoparticles which results in outstanding load transfer. Due to the outstanding mechanical and antibacterial properties of chitosan, it has found versatile applications in the industrial and biomedical fields. This robust indagation will cast light on fabrication, properties and various applications of graphene reinforced chitosan polymer nanocomposite.

Keywords Chitosan · Graphene oxide · Biomedical applications · Bioppolymers · Nanocomposites

1 Introduction

The innovation of high-performance bionanocomposites comprises depiction, fabrication, and manoeuvre of biopolymers with inclusion of nanomaterials. Biopolymer

---

B. Sharma (✉) · S. Shekhar · P. Jain · R. Sharma
Department of Chemistry, Netaji Subhas University of Technology, Delhi, India
e-mail: sharmabhasha@gmail.com

R. Sharma
e-mail: sharmabhasha@gmail.com

K. K. D. Chauhan
Bhashakaracharya College of Applied Science, University of Delhi, Delhi, India

B. Sharma and P. Jain (eds.), *Graphene Based Biopolymer Nanocomposites*, Composites Science and Technology,
https://doi.org/10.1007/978-981-15-9180-8_7

composites, class of new-fangled materials. Biopolymer nanocomposites have been used as an alternative over conventional polymers due to their environmental friendly, biocompatibility and non-toxic characteristics. Biopolymers have extremely high cost and subservient properties compared to other thermoplastic polymers which can be vanquished by embodiment of nanofiller. The inclusion of nanofiller like silicates, carbon nanotubes (CNTs), metal oxide nanoparticles and graphene based nanoplatelets can readily enhance the physical, thermal, mechanical and electrical properties of biopolymers. The subsequent advancement in the science and technology leads to demand of high strength and cost-efficient materials. The development of polymer has flourished in recent years due to its day to day utilization. The recent advances have shown that metal can be replaced by polymers. Polymers are very vital components composed of small molecules mainly monomers linked by repeating units. Synthetic and natural polymers have been employed in welfare and abetment of mankind. In addition, they are accountable for transportation, aerospace, clothing, biomedical devices, structural solicitation.

More than half of scientists all over the world are involved in the embellishment of polymers. GO readily changes the physical, chemical and thermal properties which eloquently affect the end product applications. Biopolymer nanocomposites with homogeneous distribution of nanoparticles are a formidable challenge for scientists attributes to the presence of strong cohesive forces and intermolecular interactions of nanoparticles in polymer matrix. GO evolved as an auspicious carbon based nanomaterial in which atoms are structured in a honeycomb lattice [1, 2]. GO, a single-layered, two dimensional, sp^2 hybridized building block of graphite has exceptional intriguing mechanical and electrical properties. GO derived nanomaterials incorporated into polymer matrix are formidable materials in biomedical devices, solar cells, packaging field, electronic applications, etc. [3, 4]. The exceptional properties such as high electrical conductivity, enhanced mechanical strength, improved thermal stability and highly chemical resistance have triggered quest among researchers. Its low cost and honeycomb structure make it extraordinary in contrary to CNTs. The accumulation of nanoparticles present within the matrix is due to the presence of *Van der Waal* forces which results in insolubility. Therefore, the formation of clusters ordinarily reduces the effectiveness of nanoparticles. In addition, high loading of nanoparticles and load transfer of GO can only is attained on uniform dispersion of nanoparticles into the polymer matrix.

2 Chitosan

Chitosan, another name of poly (*D*-glucosamine) is a natural carbohydrate polymer derived from the deacetylation of chitin is vibrant among the myriad of biopolymers. Chitosan, an efficient polysaccharide has attaining quest amidst research fraternity due to its non-toxic, biocompatibility and biodegradability properties. Chitosan has emerged out to be a potential candidate owing to its application in packaging, hydrogels, water etc. but the major drawback of chitosan is its lower molecular weight which obstruct its utilization [5, 6]. The concoction of reinforcing GO and chitosan produces high-performance biomaterials for utilization in industrial implications by remarkably enhancing characteristics for e.g. thermal, mechanical, electrical, physical, etc. To eradicate this issue, chitosan is grafted with GO nanoparticles. Inclusion of GO nanoparticles results in remarkable mechanical strength which can be attained at relatively lower loading of GO. Laterly, CNTS are contemplated to be an efficient candidate for polymers attributes to its exceptional properties and *1D* structure. Howbeit, the large scale production and higher cost of CNTs impede its industrial utilization. It is eminent to have strong interfacial adhesion between polymer and the nanofiller to achieve optimum properties. In chitosan, a crosslinking reaction takes place between epoxide linkage of GO and –OH, -COOH functional groups of chitosan. In addition, GO nanoparticles are negatively charged in aqueous solution which is a result of phenolic and carboxylic ionized groups. The dispersion in aqueous solution takes place as the result of electrostatic repulsion and hydrophilicity of GO nanoparticles. Being a hydrophilic biodegradable polymer, the interaction of GO and chitosan is favoured by acidic medium (polycationic material) involving –OH and $-NH_2$ functional groups [7].

2.1 Transformation of Chitin into Chitosan

Chitin, derivative of chitosan is inelastic, hard, nitrogeneous and white polysaccharide founded in outer skeleton of crabs, lobsters, shrimps etc. The waste produced from these natural polymers is a prime source of pollution on surface of coastal areas. Chitin is the plentiful acetylamino and natural polysaccharide estimated to be generated as much as cellulose annually [8]. Chitosan dervied from α $(1 \rightarrow 4)$-linked-2-amino-2-deoxy–D-glucopyranose, obtained from chitin through deacteylation. Figure 1 outlines the complete conversion of chitin into chitosan by involving organic acids followed by removal of proteins and further deacetylated in NaOH (40%) for 1–3 h at 120 °C. They are insoluble in water and extremely hydrophobic.

The degree of deacetylation of chitosan elucidates the number of amino groups present with the chains which is evaluated as the ratio of *D*-glucosamine units to the sum of *N*-acetyl- *D*-glucosamine and *D*-glucosamine. 60% of *D*-glucosamine residues should be present in deacetylated chitin to be considered as chitosan. The deacetylation method of chitin is conducted through chemical process under enzymatic hydrolysis or various alkaline conditions. Oligomers of chitosan can be utilized by the degradation of chitosan by employing specific enzymes and reagents like hydrogen peroxide. Henceforth, the present strategy comprises of analyzing the evolving properties and composition of products rather than directing predetermined characteristics by regulating the production process.

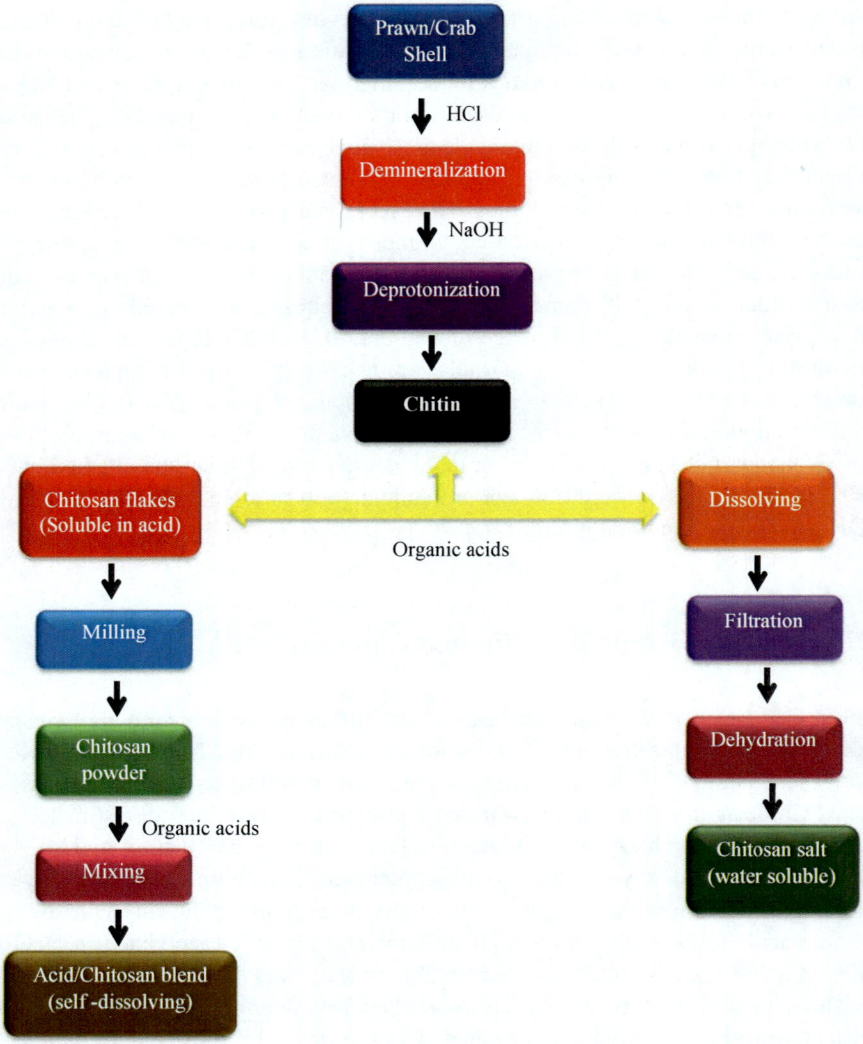

Fig. 1 Transformation of chitin into chitosan

2.2 *Graphene and Its Derivatives*

Graphene, an intriguing component with prodigious potential for many enormous applications. Graphene, constitutes a monolayer of sp^2 hybridized carbon atoms arranged in two dimensional honeycomb lattice [9]. GO has been previously synthesized by employing Modified Hummer's method. The formation of chitosan nanocomposite from graphene and its derivatives is associated with epoxide linkage.

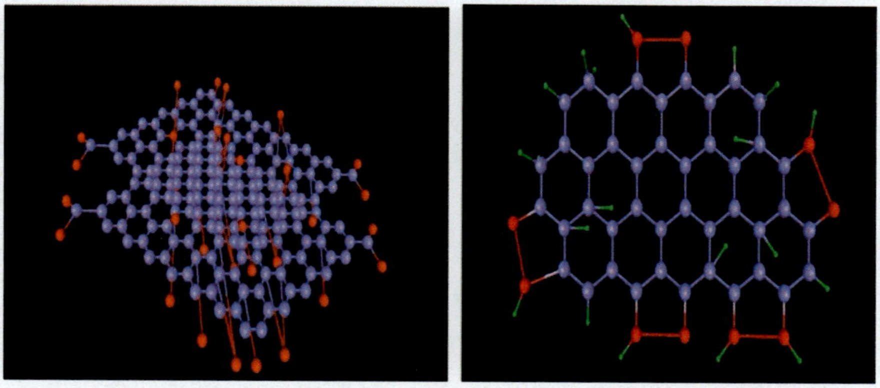

Fig. 2 3D Schematic illustration of GO and reduced GO

A small amount of loading of GO can enhance tremendous level of mechanical properties which can be attributed to the synergetic effect of strong hydrogen bonding interactions [10]. The hydroxylic & hydrophilic groups present in the chitosan matrix inhibit accumulation of GO nanosheets (Fig. 2).

3 Processing Method of Graphene Based Chitosan Nanocomposites

Solution Casting Approach- In this biom or granules have been dissolved in its respective solution i.e. acetic acid followed by chemical stirring method kept at certain temperature for several hours. In Chitosan, acetic acid is used as a processing solvent. Initially, nanoparticles are dispersed via ultrasonication process in the solvent and the resultant homogeneous solution was kept for magnetic stirring followed by addition of chitosan powder or flakes for 12 h [11, 12].

Electrospinning Technique—It is the most versatile technique currently employed for the production of nanofibers. It utilizes high voltage to produce electrically charged jet of polymeric melt or solution which facilitates the formation of fibers. The fibers produced by this technique usually have high specific surface area and high porosity which can be efficiently used in biomedical applications (Fig. 3).

4 Advantages of Incorporation of Graphene in Chitosan

The integration of GO into the chitosan not only upgrade its properties but also economically viable. The main key factors in production of polymer composites are degree of crystallinity, *Tg*, crystallization temperature, curing kinetics, use of

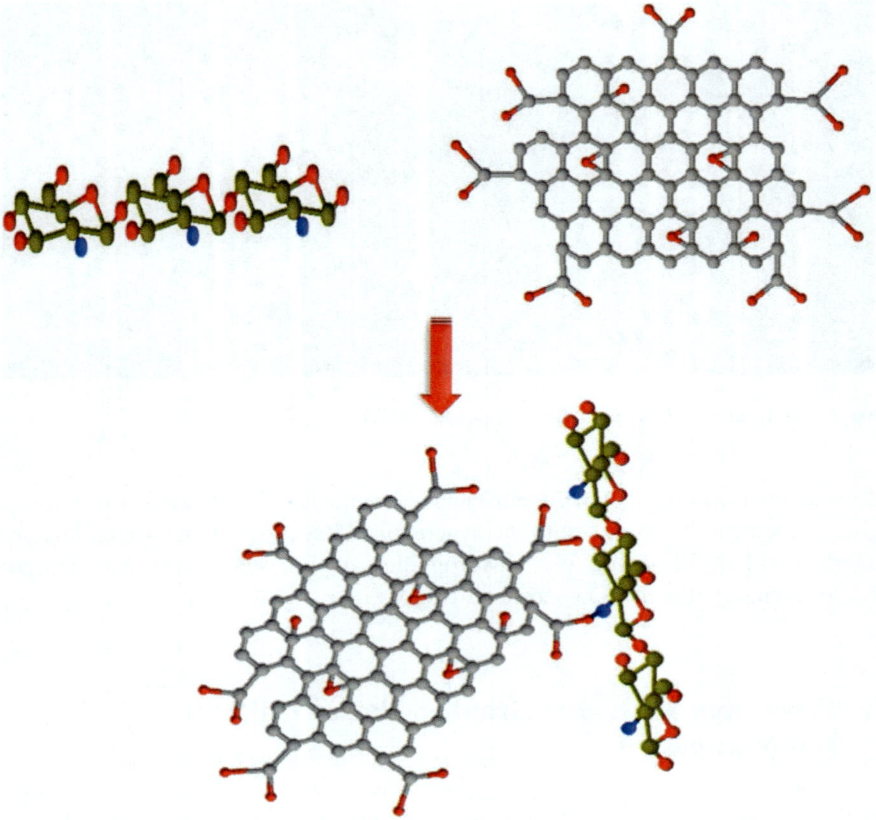

Fig. 3 Mechanism interaction of chitosan with graphene oxide

solvent etc. The reinforcement enhances the physical and mechanical properties of the matrix. Chitosan nanocomposites have outstanding properties compared to virgin polymers which depend on particle size, shape and surface area of reinforcing nanofiller. Following are the characteristics of inclusion of nanoparticle into polymer composites:

- Chitosan based nanocomposites own stronger and stiffer discontinuous fiber than continuous phase.
- GO reinforcements impart rigidity and strength into the polymer matrix.
- Composite materials are anisotropic in nature.
- They are resistant to corrosion and heat compared to other materials.
- They are light weight and have high resilience.
- GO reinforced high performance composites have lengths greater than cross sectional diameter.
- GO reinforcements transfer efficient load to the matrix, thereby making it mechanically strong.

5 Versatile Applications of Graphene Reinforced Chitosan

Due to the remarkable properties of chitosan, it can be extensively used in several biomaterials for instance in damaged tissues and organs, hydrogels, wound dressing, proliferation and cell attachment [13]. Chitosan based hydrogels can be implemented by lending it with other non-ionic polymers that are water soluble for e.g. poly (vinyl alcohol). In addition, thermosensitive hydrogels can be fabricated by using polyol salts [14]. As mentioned in Table 7.1, Chitosan can be utilized as artificial skin for individuals who go through extensive losses of skin which shows structural characteristics considered for generating substratum for replacement of skin. Chitosan can also be employed as an effective agent in coagulation of suspended solids for applications in range of food processing waste [15]. GO reinforced in chitosan can be a potential candidate for orthopaedic applications like bone tissue regeneration, cartilage regeneration, bone cements etc. attributes to the low cytotoxicity [16]. GO contains intrinsic antibacterial properties against gram positive and gram negative bacteria which makes it a crucial element in medication of resistant bacterial infections [17]. Jing and co-workers prepared a self-healing, self-adhesive hydrogel made of GO grafted chitosan implemented for applications in electroactive tissue engineering [18]. The multifunctional moieties attached to chitosan enable the nanomaterials to get easily attached. In addition, intramolecular and intermolecular hydrogen bonding molecules in chitosan stabilizes the packing structure which increases melting point [19–21]. Hieu et al. fabricated an aerogel used GO as adsorbent for removal of dye from water [22, 23]. Divband et al. examined chitosan based nanohybrid for antimicrobial applications result in enhanced crystallinity and low toxicity [23]. Tovar and co-workers designed GO grafted chitosan and poly (vinyl alcohol) antimicrobial biofilm for biomedical applications [24].

6 Limitations of Utilizing GO in Chitosan

- It has low porosity. Being a natural polymer, chitosan may be adulterated by inorganic and organic impurities.
- Chitosan displays broad polydispersity index that is why it is obligatory to evaluate viscosity rather than directing molecular mass.
- The disadvantage of solution casting is the brittle nature of nanocomposite films as scaffolds based on pristine chitosan are very brittle which can limit its utilization. To address the obstacle, nanofillers have been introduced to improve physicochemical properties. Hence, other nanocomposite preparation methods like *In-situ* polymerization process can be opted.
- The major disadvantage is that chitosan is only soluble in an acidic medium which makes investigations difficult to examine.
- To achieve mechanical strength and for applications in water purification, electrospun graphene based polymer nanocomposites can be produced.

Table 7.1 Elucidates diverse area of applications of derivatives of graphene reinforced chitosan reported by various researchers

Material	Properties	Applications	References
GO grafted chitosan	Resultant nanohybrid scaffold showed potential to regenerate bone at a rapid rate	Bone regeneration and controlled drug delivery	[25]
Morphed graphene reinforced chitosan	Improvement in impact resistance and high elasticity with 25% of hardness	Structural application such as packaging	[26]
Magnetic chitosan-GO	Removal of methyl orange and decrement in adsorption capacity with increasing pH values	Antibacterial and dye removal applications	[27]
Chitosan/reduced GO/Curcumin	Increased entrapment efficiency and enhanced cytotoxicity	Targeted delivery of anti-cancerous agents	[28]
Chitosan/GO aerogel	Increased adsorption and regeneration capabilities	Treatment of azo-dyed polluted water	[29]
GO and amine modified GO with chitosan	Improved mechanical properties and biomineralization ability	Promising tissue engineering applications as scaffold	[30]
GO nanosheets dispersed in chitosan	Enhanced printing accuracy and dielectric properties	Scaffolds for potential biomedical and 3D printing applications	[31]
Chitosan grafted GO aerogel	Increased adsorption capacity which reduced anthropogenic CO_2 gas	Adsorption of carbon dioxide gas at low cost	[32]
Chitosan grafted GO	104.16 mg/g of adsorption capacity was achieved at 2.0 pH	Adsorption of Chromium (VI) on nanocomposite	[33]
GO nanoparticles reinforced Chitosan	Increased proliferation with augmentation and enhanced physical, mechanical properties	Cartilage tissue engineering	[34]
CS/GO aerogels	Improvement of dye removal ability of nanocomposite	Potential adsorbents for water purification applications	[35]
Hydrothermally reduced GO dispersed in Chitosan	Enhanced electrical and mechanical properties followed by improved antioxidant property	Promising application in food and biological field	[36]
Blending of Chitosan with GO	Superior efficiency and reliability of electrochemical response towards glucose	Enzymatic Glucose Sensing	[37]

(continued)

Table 7.1 (continued)

Material	Properties	Applications	References
GO grafted chitosan	Enhancement in density of apatite like layer and degradation time	As bone substitute	[38]
Chitosan reinforced GO	Increased antibacterial properties, porosity and degradation rate	Bone cement in orthopaedic applications	[39]
GO grafted CS	Enhanced potential for detection of glucose level, increase in proliferation of cell health	Flexible and ultrathin sensor for diabetic wound monitoring	[40]
Magnetic GO functionalized chitosan	Increased efficiency of nanoadsorbent	Alkaloids separation in hotspot	[41]
GO/CS combined with cyclodextrin nano-sponge	Act as phase transfer agents and enhance	Catalyst recyclability	[42]
Chitosan coated with reduced GO	Excellent thermal insulation, 86% of evaporation efficiency and outstanding mechanical flexibility	Conversion of solar energy efficiency	[43]
GO/Chitosan/PVA nanofiber	Enhanced antibacterial activity and outstanding cytocompatibility	Applications in wound dressing	[44]
GO/Chitosan/Zinc oxide	Enhanced photodegradation activity and removal of dye under UV light irradiation	Photodegradation of acid orange 7	[45]
Chitosan/GO/Cellulose nanowhiskers	Displayed pH dependent exposure of drug in solution and improved flexibility was observed	Hydrogel applications to pH, temperature and external stimuli	[46]
Chitosan/GO	Low cost remote control device under sunlight and infrared light illumination	Lights stimuli actuator used for photoresponsive hydrogels induced water gradients	[47]
Magnetic GO derived chitosan	Fast kinetics, successful adsorbent recycling and high dye adsorption, 64% of efficiency of dye uptake was achieved	Dye removal	[48]
Chitosan/GO/PVA nanohybrid	Enhanced thermal stability and electrocatalytic activity	Efficient energy storage devices	[49]

- One of the major concerns is the reduction of graphene oxide to regain its electrical and mechanical properties; the usage of the reducing agent is the greatest challenge. Hence, an eco-friendly and non-toxic approach needs to be contemplated.
- Surprisingly, the aqueous solubility of chitosan depends on the pH which permits process ability under mild conditions which employs its applications in cosmetics also [50, 51].

7 Conclusion and Perspectives

The ascend in demand for polymer nanocomposites poses vital possibility due to advancement in recent creations in a sustainable manner attributes to its exceptional thermal, electrical and mechanical properties. In recent years, due to consumption of petroleum based products exploiting environment led to the utilization of biodegradable polymers in various applications. The substitution of conventional polymers with biodegradable polymers via biomimetic approach not only renders combative advantage but also proffer magnificent properties which have an innocuous effect. The voluminous utilization of traditional plastics in versatile commercial applications manifests exorbitant atrocities to the biota and sources of fossil fuels. GO grafted chitosan can be efficiently fabricated which improves thermal, mechanical properties assisted by the crosslinking reaction between hydroxyl and amine moieties exist within the nanoparticles and polymer matrix respectively. Further, thermosensitive application may be employed in future applications. Chitosan reinforced GO should be utilized in supercapacitors applications to impede the issue of e-waste generated in biota. Chitosan based GO polymer nanocomposite is an evolving technology amidst scientists attributes to the use of petroleum resources which obstructs the environmental policy.

Acknowledgements Authors would like to pay gratitude to Department of Chemisitry, Netaji Subhas University of Technology, Delhi, India.

References

1. Sharma B, Malik P, Jain P (2018) Biopolymer reinforced nanocomposites: a comprehensive review. Mater Today Commun 16:353–363
2. Sharma B, Malik P, Jain P (2019) To study the effect of processing conditions on structural and mechanical characterization of graphite and graphene oxide-reinforced PVA nanocomposite. Polym Bull 76(8):3841–3855
3. Sharma B, Shekhar S, Gautam S, Jain P (2018) Dynamic shear rheology behavior and long term stability kinetics of reduced graphene oxide filled poly (vinyl alcohol) biofilm. Polym Testing 69:583–592
4. Sharma B, Shekhar S, Malik P, Jain P (2018) Study of mechanism involved in synthesis of graphene oxide and reduced graphene oxide from graphene nanoplatelets. Mater Res Express 5(6):065012

5. Zuo PP, Feng HF, Xu ZZ, Zhang LF, Zhang YL, Xia W, Zhang WQ (2013) Fabrication of biocompatible and mechanically reinforced graphene oxide-chitosan nanocomposite films. Chem Cent J 7(1):39
6. Shao L, Chang X, Zhang Y, Huang Y, Yao Y, Guo Z (2013) Graphene oxide cross-linked chitosan nanocomposite membrane. Appl Surf Sci 280:989–992
7. Zhang H, Yan T, Xu S, Feng S, Huang D, Fujita M, Gao XD (2017) Graphene oxide-chitosan nanocomposites for intracellular delivery of immunostimulatory CpG oligodeoxynucleotides. Mater Sci Eng C 73:144–151
8. Kurita K, Yoshida A, Koyama Y (1988) Studies on chitin. 13. New polysaccharide/polypeptide hybrid materials based on chitin and poly (. gamma.-methyl L-glutamate). Macromolecules 21(6):1579–1583
9. Sharma B, Shekhar S, Gautam S, Sarkar A, Jain P (2018) Nanomechanical analysis of chemically reduced graphene oxide reinforced poly (vinyl alcohol) nanocomposite thin films. Polym Testing 70:458–466
10. Sharma B, Gautam S, Shekhar S, Sharma R, Rajawat DS, Jain P (2019) Facile synthesis of poly (vinyl alcohol) nanocomposite & its potential application to enhance electrochemical performance. Polym Testing 74:119–126
11. Jain P, Shekhar S, Chaudhary V (2019) Synergistic Reinforcement of Graphene Oxide-polyvinyl Alcohol Bionanocomposite to enhance mechanical performance. Res Rev J Phys 8(3):20–26
12. Bhasha SS, Jain P Thin films prepared using Sol Gel dip coating method. Metal Chalcogenide Nanostructures: Characteristics Synthesis 86
13. Jagur-Grodzinski J (2002) Electronically conductive polymers. Polym Adv Technol 13(9):615–625
14. Ruel-Gariepy E, Chenite A, Chaput C, Guirguis S, Leroux JC (2000) Characterization of thermosensitive chitosan gels for the sustained delivery of drugs. Int J Pharm 203(1–2):89–98
15. Croisier F, Jérôme C (2013) Chitosan-based biomaterials for tissue engineering. Eur Polymer J 49(4):780–792
16. Paz E, Forriol F, Del Real JC, Dunne N (2017) Graphene oxide versus graphene for optimisation of PMMA bone cement for orthopaedic applications. Mater Sci Eng C 77:1003–1011
17. Palmieri V, Papi M, Conti C, Ciasca G, Maulucci G, De Spirito M (2016) The future development of bacteria fighting medical devices: the role of graphene oxide. Expert Rev Med Devices 13(11):1013–1019
18. Jing X, Mi HY, Napiwocki BN, Peng XF, Turng LS (2017) Mussel-inspired electroactive chitosan/graphene oxide composite hydrogel with rapid self-healing and recovery behavior for tissue engineering. Carbon 125:557–570
19. Guo B, Sun Y, Finne-Wistrand A, Mustafa K, Albertsson AC (2012) Electroactive porous tubular scaffolds with degradability and non-cytotoxicity for neural tissue regeneration. Acta Biomater 8(1):144–153
20. Gupta P, Sharan S, Roy P, Lahiri D (2015) Aligned carbon nanotube reinforced polymeric scaffolds with electrical cues for neural tissue regeneration. Carbon 95:715–724
21. Shin SR, Jung SM, Zalabany M, Kim K, Zorlutuna P, Kim SB, Nikkah M, Kabhiry M, Azize M, Kong J, Wan KT, Palacios T, Dokmeci MR, Bae H, Tang X, Khademhossaini A (2013) Carbon-nanotube-embedded hydrogel sheets for engineering cardiac constructs and bioactuators. ACS Nano 7(3):2369–2380
22. Tu TH, Cam PTN, Phong MT, Nam HM, Hieu NH (2019) Synthesis and application of graphene oxide aerogel as an adsorbent for removal of dyes from water. Mater Lett 238:134–137
23. Khatamiana M, Divband B, Daryana M (2016) Preparation, characterization and antimicrobial property of ag+-nano Chitosan/ZSM-5: novel Hybrid Biocomposites. Nanomedicine J 3(4):268–279
24. Ruiz S, Tamayo JA, Delgado Ospina J, Navia Porras DP, Valencia Zapata ME, Hernandez M, Valencia CH, Zuluaga F, Grande Tovar CD (2019) Antimicrobial films based on nanocomposites of chitosan/poly (vinyl alcohol)/graphene oxide for biomedical applications. Biomolecules 9(3):109

25. Mahanta AK, Patel DK, Maiti P (2019) Nanohybrid Scaffold of Chitosan and Functionalized Graphene Oxide for Controlled Drug Delivery and Bone Regeneration. ACS Biomater Sci Eng 5(10):5139–5149
26. Velazquez-Meraz O, Ledezma-Sillas JE, Carreño-Gallardo C, Yang W, Chaudhari NM, Calderon HA, Rusakova I, Hernandez-Robles FC, Herrera-Ramirez JM (2019) Improvement of physical and mechanical properties on bio-polymer matrix composites using morphed graphene. Compos Sci Technol 184:107836
27. Jiang Y, Gong JL, Zeng GM, Ou XM, Chang YN, Deng CH, Zhang J, Liu HY, Huang SY (2016) Magnetic chitosan–graphene oxide composite for anti-microbial and dye removal applications. Int J Biol Macromol 82:702–710
28. Dhanavel S, Revathy TA, Sivaranjani T, Sivakumar K, Palani P, Narayanan V, Stephen A (2020) 5-Fluorouracil and curcumin co-encapsulated chitosan/reduced graphene oxide nanocomposites against human colon cancer cell lines. Polym Bull 77(1):213–233
29. Lai KC, Hiew BYZ, Lee LY, Gan S, Thangalazhy-Gopakumar S, Chiu WS, Khiew PS (2019) Ice-templated graphene oxide/chitosan aerogel as an effective adsorbent for sequestration of metanil yellow dye. Biores Technol 274:134–144
30. Olad A, Hagh HBK (2019) Graphene oxide and amin-modified graphene oxide incorporated chitosan-gelatin scaffolds as promising materials for tissue engineering. Compos B Eng 162:692–702
31. Ahmed J, Mulla M, Maniruzzaman M (2019) Rheological and dielectric behavior of 3D-Printable Chitosan/Graphene Oxide Hydrogels. ACS Biomater Sci Eng
32. Hsan N, Dutta PK, Kumar S, Bera R, Das N (2019) Chitosan grafted graphene oxide aerogel: synthesis, characterization and carbon dioxide capture study. Int J Biol Macromol 125:300–306
33. Samuel MS, Bhattacharya J, Raj S, Santhanam N, Singh H, Singh NP (2019) Efficient removal of Chromium (VI) from aqueous solution using chitosan grafted graphene oxide (CS-GO) nanocomposite. Int J Biol Macromol 121:285–292
34. Shamekhi MA, Mirzadeh H, Mahdavi H, Rabiee A, Mohebbi-Kalhori D, Eslaminejad MB (2019) Graphene oxide containing chitosan scaffolds for cartilage tissue engineering. Int J Biol Macromol 127:396–405
35. de Luna MS, Ascione C, Santillo C, Verdolotti L, Lavorgna M, Buonocore GG, Castaldo R, Filippone G, Xia J, Ambrosio L (2019) Optimization of dye adsorption capacity and mechanical strength of chitosan aerogels through crosslinking strategy and graphene oxide addition. Carbohyd Polym 211:195–203
36. Barra A, Ferreira NM, Martins MA, Lazar O, Pantazi A, Jderu AA, Neumayer SM, Rodriguez BJ, Enachesceu M, Ferraira P, Nunes C (2019) Eco-friendly preparation of electrically conductive chitosan-reduced graphene oxide flexible bionanocomposites for food packaging and biological applications. Compos Sci Technol 173:53–60
37. Poletti F, Favaretto L, Kovtun A, Treossi E, Corticelli F, Gazzano M, Palermo V, Zanardi C, Melucci M (2020) Electrochemical sensing of glucose by chitosan modified graphene oxide. J Phys: Mater 3(1):014011
38. Tavakoli M, Karbasi S, Soleymani Eil Bakhtiari S (2020) Evaluation of physical, mechanical, and biodegradation of chitosan/graphene oxide composite as bone substitutes. Polymer-Plastics Technol Mater 59(4):430–440
39. Valencia Zapata ME, Hernandez M, Herminsul J, Grande Tovar CD, Valencia Llano CH, Diaz Escobar JA, Vasquez-Laza B, Roman JS, Rojo L (2019) Novel bioactive and antibacterial acrylic bone cement nanocomposites modified with graphene oxide and chitosan. Int J Mol Sci 20(12):2938
40. Kafi MA, Paul A, Vilouras A, Hosseini ES, Dahiya RS (2019) Chitosan-Graphene Oxide based ultra-thin and flexible sensor for diabetic wound monitoring. IEEE Sens J
41. Tang T, Cao S, Xi C, Li X, Zhang L, Wang G, Chen Z (2020) Chitosan functionalized magnetic graphene oxide nanocomposite for the sensitive and effective determination of alkaloids in hotpot. Int J Biol Macromol
42. Sadjadi S, Heravi MM, Raja M (2019) Composite of ionic liquid decorated cyclodextrin nanosponge, graphene oxide and chitosan: a novel catalyst support. Int J Biol Macromol 122:228–237

43. Wang F, Wei D, Li Y, Chen T, Mu P, Sun H, Zhu Z, Liang W, Li A (2019) Chitosan/reduced graphene oxide-modified spacer fabric as a salt-resistant solar absorber for efficient solar steam generation. J Mater Chem A 7(31):18311–18317
44. Yang S, Zhang X, Zhang D (2019) Electrospun Chitosan/Poly (Vinyl Alcohol)/Graphene Oxide Nanofibrous Membrane with Ciprofloxacin Antibiotic Drug for Potential Wound Dressing Application. Int J Mol Sci 20(18):4395
45. Sheshmani S, Nejabat Ghamsari H (2019) Photodegradation of acid orange 7 from aqueous solution under visible light irradiation using nanosized ZnO/chitosan/graphene oxide composite. Int J Environ Anal Chem 1–10
46. Omidi S, Pirhayati M, Kakanejadifard A (2020) Co-delivery of doxorubicin and curcumin by a pH-sensitive, injectable, and in situ hydrogel composed of chitosan, graphene, and cellulose nanowhisker. Carbohyd Polym 231:115745
47. Duan J, Liu F, Kong Y, Hao M, He J, Wang J, Wang S, Liu J, Sang Y (2020) Homogeneous Chitosan/Graphene Oxide nanocomposite hydrogel-based actuator driven by efficient photothermally induced water gradients. ACS Appl Nano Mater
48. de Figueiredo Neves T, Dalarme NB, da Silva PMM, Landers R, Picone CSF, Prediger P (2020) Novel magnetic chitosan/quaternary ammonium salt graphene oxide composite applied to dye removal. J Environ Chem Eng 103820
49. Shekhar S, Anjana S, Bhasha S, Purnima J (2019) Electrochemical evaluation of functionalized graphene oxide filled PVA-chitosan biohybrid for supercapacitor applications. J Appl Polymer Sci 48610
50. Rinaudo M (2006) Chitin and chitosan: properties and applications. Prog Polym Sci 31(7):603–632
51. Shekhar S, Sharma R, Sharma S, Sharma B, Sarkar A, Jain P (2020) An exploration of electrocatalytic analysis and antibacterial efficacy of electrically conductive poly (D-Glucosamine)/graphene oxide bionanohybrid. Carbohydrate Polym 116242

Graphene Oxide—Plant Gum Nanocomposites for Sustainable Applications

Vinod V. T. Padil and Miroslav Černík

Abstract Nanotechnology is increasingly being promoted to adopt "greener" technology that will improve the environmental performance of existing industries, reduce consumption of resources and energy, and enable the environmentally benign economic expansion. A novel approach in the field of sustainable/green nanotechnology is to develop advanced bio-composite materials with desirable properties using renewable and benign materials as exemplified herein by plant gum-based resources. The present chapter focuses on the fabrication and characterization of bioplastic fibers/films based on plant gums or other natural polymers via electrospinning or solution casting methods. Additionally, the integration of graphene oxide as a potential additive to blend in the plant gum based films/fibers to improve their physicochemical, tensile, thermal and antibacterial properties are investigated. Further, how the green technologies such as plasma, and γ-ray irradiation treatments on plant gum fibers/films have affected the enhancement of their structural, mechanical properties and biodegradability are also epitomized with suitable examples. The bioplastic fibers/films produced from bio-based materials played a noteworthy role in numerous imperative sectors, such as environmental bioremediation, medical, food carrying bags and packaging are appraised. This chapter would be of interest to the broad readership of material scientists, chemists, engineers and researchers associated with the sustainable developments and foresee the progress of economic and ecological products based on plant gums.

Keywords Plant gums · Graphene oxide · Electrospinning · γ-ray irradiation · environmental bioremediation

---

V. V. T. Padil (✉) · M. Černík
Department of Nanomaterials in Natural Sciences, Institute for Nanomaterials, Advanced Technology and Innovation, Technical University of Liberec, Studentská, 1402/2, Liberec 1 461 17, Czech Republic
e-mail: vinod.padil@tul.cz

B. Sharma and P. Jain (eds.), *Graphene Based Biopolymer Nanocomposites*, Composites Science and Technology,
https://doi.org/10.1007/978-981-15-9180-8_8

1 Introduction

Bio-based products such as bioplastic fibers and films are getting prominent rooms for the development of greener and sustainable progression [1–6]. Improvement of functionalities such as thermal, mechanical, barrier and other physicochemical properties are the major goal to fulfil the requirement of bio-based biodegradable packing in comparison to the existing system in this field [7–12]. Currently using plastic derived materials (for e.g., LDPE (low-density polyethylene), HDPE (high-density polyethylene), PET (polyethylene terephthalate), PP (polypropylene), PBS (poly butylene succinate), PVC (poly vinyl chloride) and PPC polypropylene carbonate) etc. generates potential health hazard for environment via the production green gas effect and at the same time exhausting natural fossil resources for their production [13–15]. Bioplastics are categorized into various divisions such as non-biodegradable, biodegradable and biodegradable-petroleum and natural polymer based. Among them, bio-based biodegradable types (PLA (poly lactide), PHA (polyhydroxyalkanoates), PGA (poly glycolic acid), PBS (poly butylene succinate), TPS thermoplastic starch (TPS) and biodegradable petroleum based (PBAT (poly (butylene adipate-co-terephthalate), and PCL (poly ε-caprolactone) etc. are expensive, challenging process chemistry and would not to be able to fulfil the high demand for their applications in food, environmental and other industrial applications [9]. There are many natural biopolymers (silk, cellulose, chitosan, and starch), and agricultural vegetable wastes have been used for the production of bioplastic films for food packaging applications [15].

Developments of novel technology for the production of bioplastics from renewable materials are the key point for the ecological and sustainable community. The predicted total plastic production will be approximately 400 million tons in 2020 and the contribution of bio-plastic production may range closely 12 million tons. Bioplastics by biodegradable polymers/composites are believed to be key alternatives to synthetic plastics in the near future. Biodegradable polymer fibers, plasticizers and organic/inorganic fillers and processed by various methods and can be suitable to placate too many products for specific use. The biodegradable biopolymer nanocomposites are to create fibers, self-standing films for food packaging, environmental applications, medically related products (bandages) and antibacterial-based sanitary materials. Plant gums are bio-based biodegradable natural polysaccharides and their uses in many versatile fields such as food additives, pharmaceutical constituents, biomedical (tissue engineering, drug carrier, and biosensor) and biosorbent for environmental pollutants etc. have been broadly studied [16]. The non-toxicity, biodegradability, availability, superior physico-chemical performances form plant gums meet bio-based constituents for food wrapping, environmental and energy harvesting zones. However, applicability of them is restricted by their low strength, hydrophilicity, easy brittle, reduced mechanical strength, and amorphous nature, especially in contact to petroleum based plastics. Upgrading in the above properties is desired to make plant gums more suitable for packaging, environmental and energy harvesting applications. In this context, an environmental friendly reinforcing agent such as microporous carbon [17], graphene oxide (GO) [18], metal/metal oxide nanoparticles [19], clay [20] and other ingredients have been usefully demonstrated an effective filler to fabricate bioplastic films/fibers with natural gums with high performance and functionalities.

Electrospinning is one of the simplest, most economical and flexible choices (which is also amenable to being scaled up) applicable to polymers that need to be spun according to micro or nanoscale architecture [21, 22]. Polysaccharides display variable intrinsic characteristics such as molecular weight distribution, degree of deacetylation, purity, charged functional groups and crystallinity, all of which depend on their source of extraction [22]. There are many unique factors to electrospinning systems and their corresponding process parameters. The most significant polymer properties (viz. molecular weight distribution, glass-transition temperature and solubility) and solution characteristics (viz. viscosity, viscoelastic behavior, concentration, surface tension and electrical conductivity)—together with other inherent attributes (viz. substrate behavior, solution feed rate, field strength, geometry of electrode, vapor pressure of the solvent and relative humidity)—all serve to control and stabilize the electrospinning process [21, 23]. The development of 'green electrospinning'technique requires non-toxic solvents, environmental friendly polymers and materials [24]. Plant based materials and proteins are suitable choice to develop 'green' electrospun fibers utilizing water or other non-toxic solvents for electrospinning. Recently, plant gum based polymers have been demonstrated for making electrospun fibers and their applications in removal of environmental contaminants, catalysis, antibacterial and packing [25–29].

In this chapter, we focus on the fabrication of self-standing films/fibers based on plant gums, specially emphasis on gum arabic (GA), gum karaya (GK) and kondgogu gum (KG), with graphene oxide (GO) as a reinforcing filler (Fig. 1). Furthermore, other natural or synthetic biodegradable polymer incorporated with GO also demonstrates and compare their potential applications in environmental, biomedical, and food packaging arenas. Various affluence of the resulted fibers/films before and after interaction with GO are interpreted with their morphological, interfacial interaction, thermal, mechanical, hydrophobicity, and barrier (oxygen and water vapor) properties.

2 Plant Gums

Gums are hydrophilic polymers in nature and extracted from various sources such as botanical, animal, bacteria, trees, algae etc. Gum arabic (GA), gum karaya (GK), gum tragacanth (GT), kondagogu gum (KG) and gum ghatti (GG) are the major commercial gums extracted from plants. The structural sugars molecules of the various types of plant gums described in this chapter are presented in Fig. 2. Gums and their chemically modified structures have been researched for past 25 years including the extraction procedures from various sources, molecular fractionalization, compositional and structural elucidation, and their potential applications in food, environmental, energy

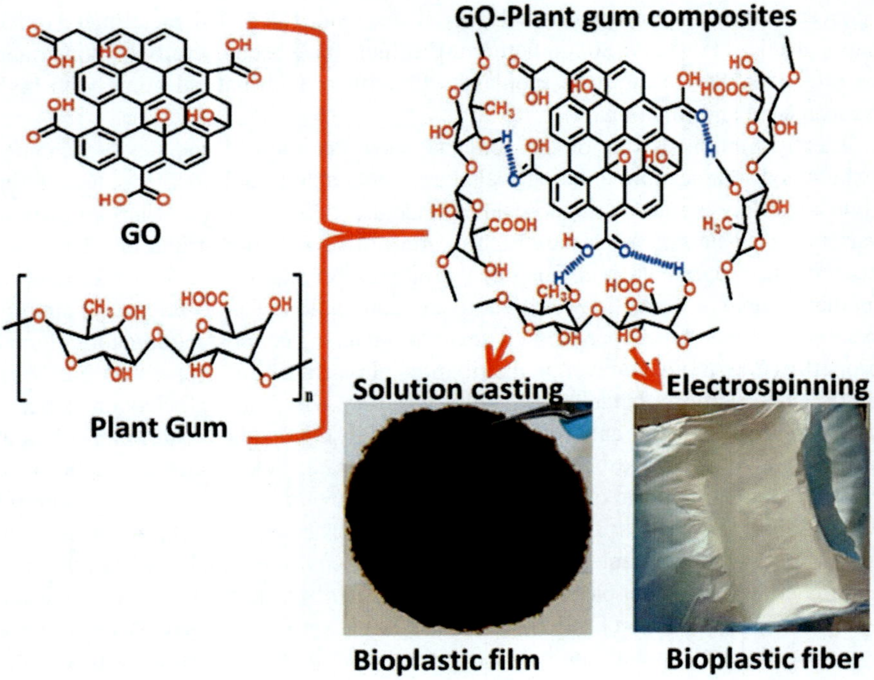

Fig. 1 Schematic representation of interaction between GO with plant gums and their electrospun fibers and self-standing film [18]. Reproduced with permission from Ref.

and biomedical aspects [16, 30–33]. Plant gums are complex polysaccharides with the advantages of harbouring several different types of bulky organic functional groups and having high molecular weights. They also have significant amounts of uronic acids and neutral sugars, their major functional groups being –OH, $-OCH_3$, –CO, and –COOH. The zeta potentials of these gums indicate that they contain negatively charged groups [16].

3 Electrospinning of Plant Gums

Electrospinning have been known for more than ninety years, since this technique was popularised after the discovery of electron microscope, and remarkable research has been carried out by many scientists and demonstrated that numerous organic/inorganic materials, polymers, composite materials including nanoparticles, drug molecules, proteins could be electrospun into fibers [34–37]. The development of electrospun fibres and membranes based on natural renewable materials for energy, environmental, and biomedical applications are still underway. Fabrication of electrospun fibres and membranes via "Green" electrospinning technology depend on the

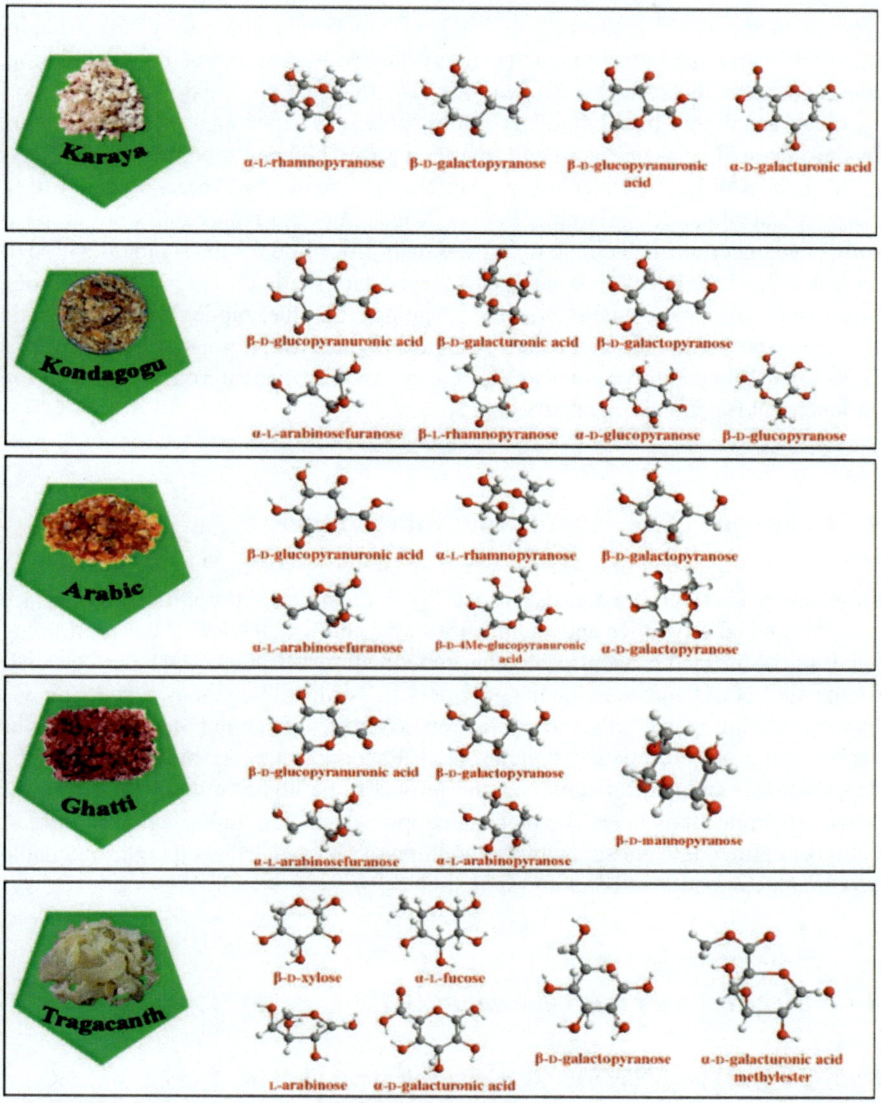

Fig. 2 Various commercial plant gums with their sugar structural moieties. Reproduced with the permission from [16]

development of non-toxic, inexpensive and environmentally friendly polymers and solvents (such as water or ionic solvents). Electrospun natural biodegradable polymers and their composites have shown countless flourishing in many fields such as the advance of water and air filtration membranes; environmental catalytic nanofibers; medical biotechnology (tissue/nerve repairing and engineering, drug encapsulation and distribution, and biosensors) and textiles [38–44]. Electrospun nanofibers derived from plant gums are a pioneering assembly of commercial and ecologically friendly materials with impending uses in the energy harvesting, environmental remediation, biosensor, catalytic and medical meadows [16, 45–47].

Electrospun bio-based biodegradable polymers such as chitin, chitosan, hyaluronic acid, and guar gum and their composite blended mixtures with copolymers such as poly vinyl alcohol or poly ethylene oxide have been prepared [48]. Lately, plant gums such as Arabic, Karaya, Kondagogu and Tragacanth were successfully have undertaken electrospinning and their procedure for the standardization of large scale production of their nanofibers were established [27, 49, 50]. Furthermore, electrospun composite nanofibers of plant gums with other biodegradable synthetic polymers (poly vinyl alcohol or poly ethylene oxide) using other additives such as DDSA, and other additives such as Ag, Au, or nano zero-valent iron NPs (to develop antibacterial functional membranes) [25].

4 Graphene Oxide-Plant Gum Fibers/films

Graphene oxide (GO) is a non-hazardous, bio-compatible, hydrophilic, great surface area, biological properties and chemically stable and which could act as reinforcing agent [51]. The rich oxygen containing functionalities (R–OH, –OH and –COOH) of the GO could interacted with polymers to form homogenous dispersion via hydrogen bonding, electrostatic interaction, assembling, van der waals interactions and bringing or coordination bonding [52]. Recently, gum Arabic (GA) could be dispersed GO and carbon nanotube and proved to be highly effective medium for potential applications [49]. The mechanical properties of cellulose and starch/PLA composite film could enhanced by the addition of GO and achieved high stretchable power and elongation at break [53, 54].

4.1 Electrospun Fibers Based on GO

Many polymers including natural or synthetic types were incorporated into GO for producing electrospun nanofibers. The major advantages of adding GO into electrospinning solution of polymers are to enhance the mechanical, electrical, thermal, morphological and porosity, controlling fiber diameters etc. of the electrospun fibers. The electrospun fibers of GO with natural/synthetic polymers and their potential applications are described below. GA, GO and PVA (poly vinyl alcohol) blend were prepared and demonstrated to electrospun into fibers via 'green' electrospinning [27]. The blended electrospinning solution of GA, GO and PVA and their produced electrospun fibers are presented in Fig. 3.

The electrospun fibers of GA/PVA (poly vinyl alcohol) are generated by electrospinning of blend mixtures of both PVA to GA in various ratios ranging from 50/50 to 100/0. Further, the best proportion of the blended mixture of PVA/GA was found

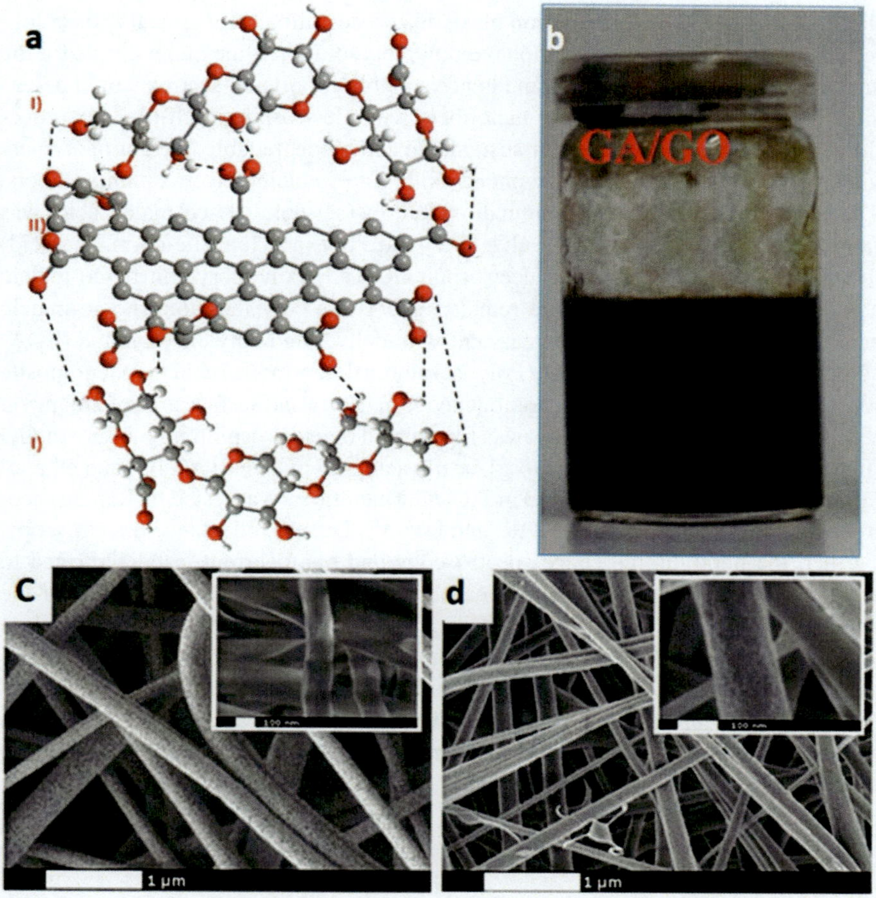

Fig. 3 The scheme depicting **a** the blended mixture of (I) GA, and (II) GO; **b** shows the electrospinning blend solution of GA/GO; and electrospun fibers of **c** GA/PVA; and **d** GO (05 wt.%)/GA/PVA visualized by scanning electron microcopy; Inset picture of **c**, **d** shows enlarged SEM images of electrospun fibers. Reprinted with permission from Ref. [27]

to be 70/30 to 90/10 provided uniform size and bead-less nanofibers. Various process parameters of the electrospinning such as conductivity, surface tension and viscosity of the blended mixture of PVA and GA were recorded [55]. Addition of GO into GA/PVA blend resulted into wrinkled, rough surface, and significantly smaller and thinner sized nanofibers in comparison to bare GA/PVA fibers without GO, as visualized by SEM image (Fig. 3d). The GO could certainly confirm in the electrospun nanofibers using Raman spectroscopy and additionally the amount of its content (wt. %) through TGA analysis.

The major advantage using GA for the preparation of GA/GO nanofibers are the eco-friendly use of tree polymer (eg; GA), electrospinning solvent (water) and the dispersibility of GO in GA solution make this process more economical and ecological. The polydispersity of the graphene/polymers solution is the major hurdle for the smooth electrospinning process and hence suitable copolymer section would necessitate the production of uniform nanofibers with electrospinning [56]. This procedure could be employed for other sustainable and biodegradable tree gum polymers via incorporating graphene or graphene oxide for producing electrospinning fibers. The nanocomposite of GO/Titanium dioxide/Poly (acrylonitrile-co-maleic acid) was fabricated by electrospinning of GO, TiO_2 and Poly (acrylonitrile-co-maleic acid) in DMF solvent [57]. From the electrospun GO/TiO_2/Poly (acrylonitrile-co-maleic acid) fibers were reduced to get reduced GO/TiO_2/ Poly (acrylonitrile-co-maleic acid) nanocomposite fibers by treatment with dehydrated ethylene diamine (EDA) solution. The fibers of RGO/TiO_2/ Poly (acrylonitrile-co-maleic acid) nanocomposite were 300–600 nm in diameters, homogeneous in nature and surface area of 28.4m^2/g, respectively. TiO_2/GO nanofibers was formulated by electrospinning mixture of GO (0–5.0wt.%), tetra butyl titanate (4 g), acetic acid (4.0 mL), PVP (1.4 g) and ethanol (13.0 mL) [58]. The amount of GO in TiO_2/GO nanofibers was tested by Raman spectroscopy, XPS analysis, XRD, FTIR, and UV–Vis Diffuse reflectance Spectroscopy. Further, the photoluminescence spectroscopy and photocurrent tests confirmed to examine the effectiveness of charge carrier trapping and transfer behaviour of the photo-induced electrons and holes. The degradation of organic dye (methylene blue) and 4-chlorophenol (an organic toxic compound used for preparation of organic dyes and drugs), under visible light irradiation using photocatalytic activity of the TiO_2/GO composite nanofibers were demonstrated. Polyelectrolyte electrospun nanofibers based on sulfonated PVA, GO and sulfosuccinic acid (cross-linking agent) were contrived by electrospinning [59]. The produced nanofibers were characterized by FE-SEM, FTIR, TGA, DSC, electrical conductivity and stability studies.

5 Surface Treatments to Improve the Functionality of Plant Gum Bioplastic Fibers

5.1 Plasma Treatment

Various types of plasma treatment techniques such as etching and sputtering, layer deposition, grafting, polymerization, and spray drying are used for the surface modification of nanofibers or metal plate or other polymer substances [66]. Plasma treatment enhances the surface wettability (hydrophobic or hydrophilic) of the materials without altering their bulk properties [66]. For improving the surface functionality, chemical modification and surface roughness, different types of gases such as oxygen, argon, helium, neon, nitrogen, sulphur hexafluoride and methane, etc. can be used as plasma source [67]. Plasma technology is an environmental friendly method which generates different gases to produce grafting of functionality such –OH, –CO, –COO and surface alternations (roughness, deposition, curing and polymerization) thereby modify the water contact angle, surface area and surface texture of the materials [68–73].

5.2 Plasma Treated Plant Gum Fibers

Electrospun fibers of arabic, kondagogu and karaya gums were treated with methane plasma and their physicochemical, thermal, mechanical properties were demonstrated [74, 75]. Generally, the water contact angle, specific surface area, surface roughness and water stability of the prepared GK, GK and KG are found to be enhanced after methane plasm treatment [76, 77]. The morphology of the plasma treated and non-treated fibres are presented in Fig. 4. The description of the plasma treatment conditions and instrument details are presented in our publications [27, 73]. Electrospun GA bioplastic fibers prepared were undergone methane plasma treatment and found that the enhancement of most of the parameters such as fibre diameter (250 ± 80 nm), BET surface area ($10.6 \pm 0.6\ m^2g^{-1}$), water contact angle (130°), and mechanical properties (tensile strength = 52.3 ± 9.5 mpa; elongation at break = 20.1 ± 1.8%) and thermal stability compared with untreated fibres [7]. Further, higher surface roughness and surface area of the methane plasm treated GA fibres showed potential antibacterial efficiency towards pathogenic bacterium

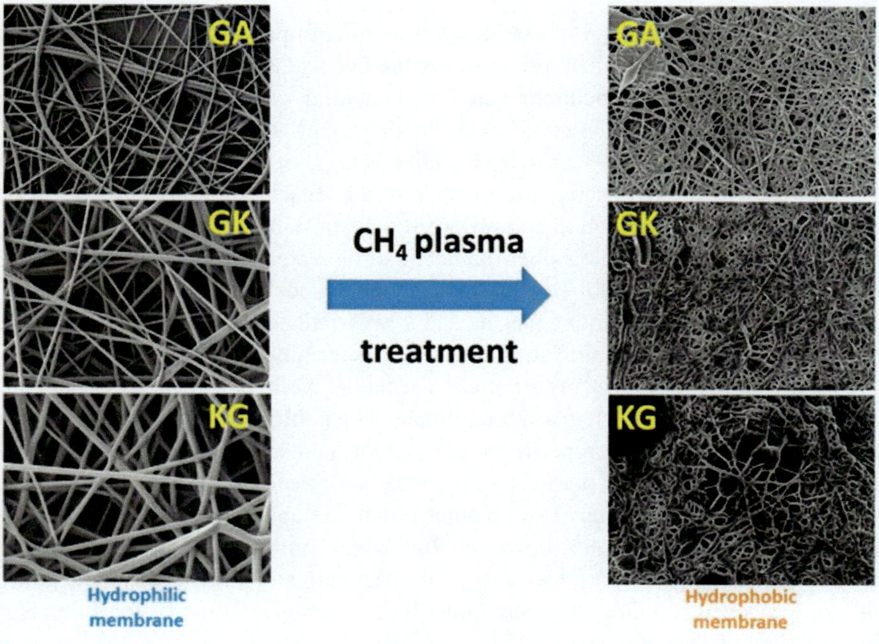

Fig. 4 Methane plasma treated gum fibers of GA, GK and KG, representing the fibres before and after plasma treatment and their morphological changes. Reprinted with permission from Ref. [55]

(*S. aureus and E. coli and P. auregenosa*). The plasma modified plant gum based fibres can be developed as potential antibacterial and food wrapping material.

Plasma modified electrospun GK fibers showed higher BET surface area (9.5 ± 0.6 m^2/g), porosity (79–82) and water contact angle (110.6 ± 0.8°) compared to unmodified fibers [19, 55]. Furthermore, this plasm treated fibers were usefully used for the potential application for the bioremediation of nanoparticles (Au, Ag, Pt, CuO and Fe_3O_4) sequestration from aqueous environment [19]. The major mechanisms of adsorption of various nanoparticles onto plasma-modified nanofibers were found to be physical interaction, functional group interface, high surface energy and surface area of the plasma fibers. The physicochemical properties of plasma treated gum kondagogu fibers showed exceptionally higher water contact area (112.6 ± 0.6°), porosity (82–84), BET surface area (9.8 ± 0.8 m^2/g) and greater water stability compared to plasma amended Arabic and Karaya fibers [55]. These fibers were efficiently applied for the waste-water purification, which are contaminated with nanoparticles [28]. With the recent research outputs, the plasma modification of "green" polymeric electrospun membranes based on plant gums has a pronounced prospective for many important areas such as water purification, food and medical strategies.

5.3 γ-ray Irradiation Treatment

Like plasma technology, γ-ray irradiation is also an important 'green' technology to modify the physicochemical properties of the fibers and membranes. The surface texture of the materials, including roughness, etching, crosslinking, functionality, drafting, and polymerization could easily modify using γ-ray irradiation [78]. The main radioactive source is ^{60}Co (a radioactive isotope of cobalt with a half-life of 5.2714 years with a radioactive dose of 5.9 kGy/h. Electrospun fibers of GA was modified with γ-ray irradiation and observed that substantial alternations has been observed with respect to their fiber diameter, surface area, stability, contact angle, surface roughness, hydrophilicity, thermal and mechanical properties [7]. There are considerable differences among plasma and γ-ray irradiated GA fibres, among their physicochemical and structural attributes noted, especially the hydrophilicity of GA fibers has improved after γ-ray irradiation whereas the hydrophobicity behaviour was enhanced after methane plasma treatment. The antibacterial potential, antioxidant and biodegradability properties of the γ-ray irradiated GA fibers were found to be greater in comparison to plasma modified fibers. Both plasma and γ-ray irradiated techniques are complementary to each other and the suitability of any method among them is appropriate to natural fibers or films which fully depends on the specific application. Additionally, food wrapping, antibacterial, antioxidant, and biodegradable of the developed GA bioplastic fibers would be suitable choice for a greener and sustainable development (Fig. 5).

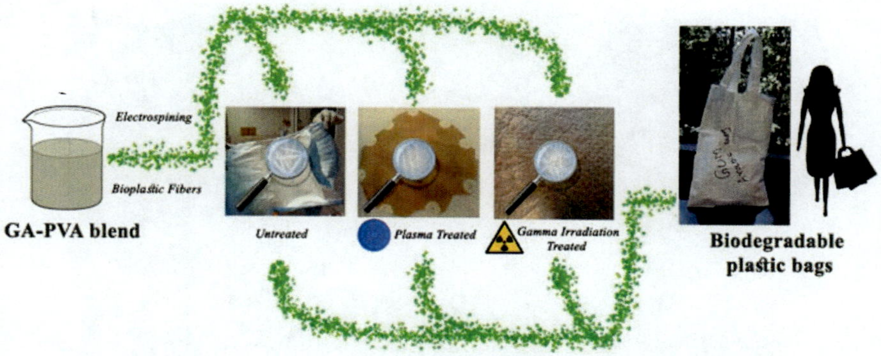

Fig. 5 Electropsun bioplastic fibers of GA with various treatments and their application to develop biodegradable shopping bags. Reproduced with permission from Ref. [7]

6 Bioplastic Films

Biopolymer based films derived from polysaccharides, proteins or lipids are getting more prominent role in food packaging applications due to their excellent possessions such as biodegradability, controlled gas and moisture barrier assets, economical, and antibacterial effect. There are in numerous natural resources available for the fabrication of bioplastic films such as starch, cellulose, chitosan/chitin and proteins like casein and gluten, wood, sugar, wheat, corn, rice and potatoes; vegetable waste materials, polylactate and natural oils and polyhydroxyalkanoates [79]. To replace or eliminate petroleum-based polymers, the natural based biodegradable materials have been established in a challenge to meet the mounting demand for food and pharmaceutical applications [80]. Even though, there are many advantage such as biodegradability, biocompatibility, non-toxic, and non-toxic waste generation with natural biopolymers, the higher cost and complicate extraction procedures are becoming hurdle for their application in food packaging and pharmaceuticals industries [81]. In this context, more economical alternative bio-based materials such as plant gums, vegetable wastes and silk sericin (waste from silk industry) etc. has been developed for food packaging, energy harvesting and other biomedical applications [17, 68].

6.1 Plant Gums-GO Bioplastic Films

The bioplastic films of major plant exudate gums such as Arabic, Karaya and Kondagogu with various amounts of GO (0.5% and 1.0%, respectively) were fabricated via solution casting method (Fig. 6). The prepared films showed exceptional thermal stability, mechanical attributes and oxygen and water vapor permeability to make them suitable for food packaging [18]. The addition of GO into plant gums (GA, GK and KG) improved their film forming along with physicochemical properties [18].

Fig. 6 Bioplastic film prepared from the plant gums/GO nanocomposites solution; **a** GA-GO; **b** GK-GO; and **c** KG-GO, respectively. Reprinted with permission from Ref. [18]

To further, establish films applicable to food packaging, the barrier properties such as oxygen and water vapour are key aspect designed for maintain the self-life of the products. Mostly, biopolymers are hydrophilic in nature and tend to swell when contact with water or in in elevation humidity conditions. However, upon bending with additives such as clay, GO and metal/organic or inorganic nanoparticles, the barrier properties are enhanced [69]. Intercalation of organic–inorganic nanomaterials are the pivotal to the advancement of biopolymer-based materials for food packaging with tailored barrier properties. Among various findings, nano silica based materials have been demonstred to be the ideal additives to intercalate the poymer matrices to develop hybrid organic–inorganic composites [70, 71]. The nanoscale plate morphology of layered silicates and other fillers often provides improved physical properties when incorporated in biopolymers, including enhanced mechanical, thermal stability, and barrier properties [71]. The oxygen transmission rate (OTR) studies of arabic, karaya and kondagou –GO nanocomposite films were carried out at a relative humidity (RH) of 65% (a standard method for food packaging—according to ISO 14,663–2) and a temperature of 23 °C, are presented in Fig. 7.

The oxygen permeability (OP) values were 1456, 1870 and 1487 $cm^3 \mu m\ m^{-2}\ day^{-1}\ atm^{-1}$ respectively for GA-GO, GK-GO, and KG-GO films, which are comparable to PET (polyethylene terephthalate) films and poly lactic acid (PLA) films [69,

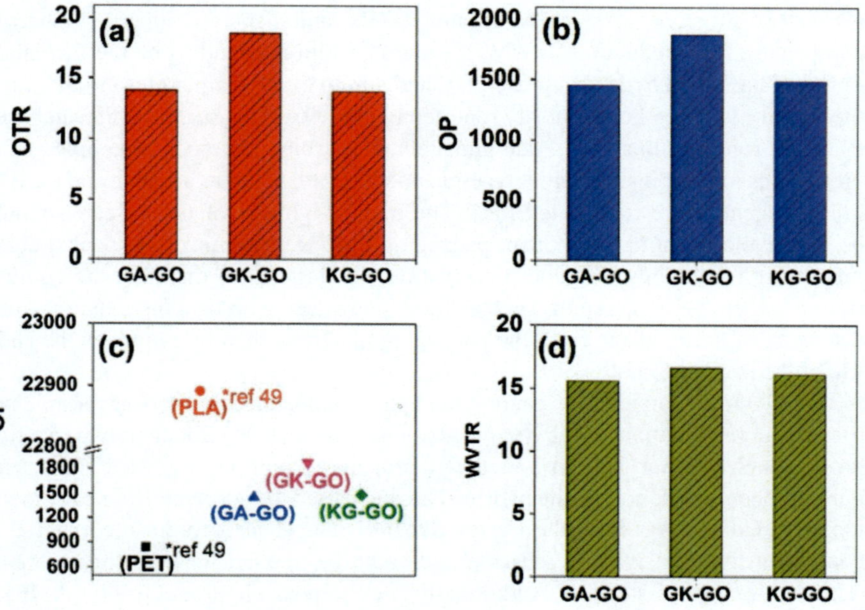

Fig. 7 The barrier properties of plant gums-GO composite films and their comparison with standard food packaging films based on PLA and PET available in the market **a** OTR (cm^3 m^{-2} day^{-1} atm^{-1}), **b** OP ($cm^3 \mu m$ m^{-2} day^{-1} atm^{-1}), **c** OP comparison with PLA and PET and **d** water vapor transmission rate (WVTR) (g m^{-2} day^{-1}). Reprinted with permission from Ref. [18]

72]. Many bio-based films based on pectin, starch, chitosan, xylan, and lignin interacted with GO or clay had higher barrier properties compared to their pristine samples which suggested that the importance of GO or clay with natural polymer and their composites would be a prosperous conjugate for formulating food packing materials [73]. An antibacterial film based on the combination of guar gum and chitosan was fabricated by solution casting method [74]. Polymer blending is one of the important benchmark to make suitable films with advance in functional properties and applications. In the case of wheat gluten proteins, paper fibers have been integrated to progress the mechanical properties of the resultant material composite [75]. For ensuring longer shelf life and preservation, the intrinsic antibacterial properties of guar gum-chitosan material as a packaging films has advantages to eliminate the risk of food-borne pathogens and thus enhance health related benefits due to cut down of adding other toxic preservatives in the food system. The fabricated guar gum-chitosan was shown lower OP, and higher mechanical properties for commercial use for food packaging.

The chitosan/PVA/ Ethyl vanillin blend films were prepared by solvent casting method and the effectiveness of ethyl vanillin for the improvement of this film with respect to their structural, tensile, optical, food compatibility and antibacterial efficacy were assessed [76]. The assimilation of Ethyl vanillin (EV) onto chitosan/PVA film further enhanced the barrier properties (both water vapour transmission rates

(WVTRs) and oxygen transmission rates (OTRs) and displayed superb UV barrier competence. The reduction in WVTR values of film is found to be the formation of Schiff base (aldehyde group from EV and amino group from chitosan) and intermolecular hydrogen bonds mostly reported in many systems such as chitosan based PVA and vanillin films [77]. Plant gums obtained from chia seeds and apple pectin blend films in combination with glycerol and glutaraldehyde (as additive and a cross-linking agent) were formulated [82]. The major highlight of this research would be the development of antioxidant packaging films. Such antioxidant packaging is innovating technology for food preservation which helps to maintain the equilibrium between the two systems such as food interaction with packing materials, and discharge of antioxidants from the packaging materials thereby improves the shelf life of the packed food [83].

The GO incorporated tragacanth gum also developed as bio-sorbent for an environmental remediation of heavy metals. For example, a radical polymerization product such as 2-Acrylamido-2-methyl-1-propanesulfonic acid)-grafted Tragacanth gum/graphene oxide composite hydrogel has been used the adsorption of heavy metal ions (Pb, Cd and Ag) from the wastewater [60]. The ensued composite material is competent, inexpensive and reusable and could be applied for other heavy metal removal from the wastewater. Additionally, the composite hydrogel [(poly (AMPS)-g-GT/GO)] with Ag nanoparticle could successfully apply as an antibacterial agent. Porous hydrogel beads based on GO and tragacanth gum (TG) by cross-linking with divalent ions (Ca^{2+} or Ba^{2+}) via ionic gelation technique was formulated for drug entrapment (Rivastigmine) and for controlled relesae study. The toxicity of the GO (0.5 wt % and concentration of GO-TG less than 125 μg/ml) by MTT assay has confirmed that the gum tagacanth–GO hydrogel is non-toxic. The development of tree gum based hydrogel have brilliant potential to be used in food packaging, anibacterial and biomendical (drug carriers and release study) fileds [84]. Recent studies on hydrogel based 'absorbed pads'have much importance for their ability to remove water thereby reducing the potential risk of microbial contamination while keeping the sensorial properties of the stored food [85]. In another work, GO interacted carboxymethyl cellulose–grafted poly(acrylic acid-co-acrylamide) hydrogel was synthesised and sucessfully applied for the adsorption of both cationic (methylene blue, neutral red and rhodamine B) and anionic (methyl orange, congo red and Indigo carmine) organic dyes [86].

7 Major Drawback to Develop Plant Gum-GO Based Bioplastics

Naturally occurring plant gums are known to be biodegradable, but with incorporation of GO, the biodegradation rate is comparatively slower than with plant gum based nanofibers/films. This can be attributed to several factors. GO in itself is non-degradable and hence does not contribute to the biodegradability of the films/fibers.

Further, the surface area of the fabricated material has direct correlation with the biodegradation rate. In addition, high interactions between the GO and the gums can also reduce the biodegradation rate. Incorporating GO into the gum matrix results in strong interactions between the gum matrix and the GO which results in pseudo crosslink effect significantly improving its mechanical stability, thus decelerating down the biodegradation rate. Additional investigations are required to establish the bio-degradation of GO-Gums films based on surface area accessibility for biodegradability. Further major challenges are focusing on large scale or industrial production of films/fibers and their cost effective for surface modification and GO or graphene or other additives integration into fibers to enhance the excellent properties, which are appropriate to various prospective applications in future.

8 Conclusions and Prospects

Many natural biomaterials and their electrospun fibers/films exhibited biocompatibility, non-toxicity, biodegradability and the availability of functional groups of medicinal value in their structures and have been extensively deployed in environmental, energy and medical applications. The important prospects for electrospun fibers/films based on tree gums relies on the commercialization of their products in the arena of food packaging/carrier bags, tissue engineering, wound dressing and disposable sanitary products. In this context, both academic and commercial organisations should be networked to increase interaction between them, in order to promote the nanofiber/films based products. Furthermore, the 'green' electrospinning process harnessing plant gums for the creation of nanofibers and composite films with potential additives such as GO, clay or nanoparticles are currently at the progressive stage and ample new exploratory investigation in this meadow is projected in the coming years. The impending uses of plant gums and their composite nanofibers/films developed as antibacterial agents, food wrapping and disposable ecological products (sanitary pads, carrier bags, towels and cups etc.) are likely to be exceedingly promising.

Besides, indications are that natural, biomaterial-based electrospun fibers have a favourable future as replacements of petroleum-based plastics which have become the scourge of the environment. The expansion of bioplastics from biodegradable, bio-based electrospun fibers/films would be challenging and researches relating to their application in many important areas are imperative if the viability and value of bioplastics is to be firmly established. Plant gums can thus serve as a renewable and exceedingly cost-effective bio-resource for a greener prospect. Regarding the chemical modification of biomaterials and their electrospinning parameters, the important aspects to be considered are the incorporation of additives or bioactive compounds into the nanofibers and the development of greener solvents for the electrospinning process, especially in applications of nanofibers for the biomedical and food technology sectors, in view of safety concerns. Additionally, the improving of physicochemical, structural and morphological by surface treatments through the adoption of several environmental friendly technologies such as atmospheric plasma, γ-ray irradiation and laser treatment with other procedures involving chemical grafting, heat treatment and functionalization will help generate inexpensive, viable and biogenic products with enhanced physicochemical and mechanical properties and processability for the future. In conclusion, plant gums and their composite fibers/film materials are making huge impact on the progression of greener technologies and products. These include water and air purification (progressive filters for the removal of contaminants from both polluted air and wastewater); renewable energy (harvesting, transformation and storage); electrodes for rechargeable batteries; catalysis; biomedical research [regeneration of various types of tissues (e.g. nerve, skin, heart, blood vessel]; the food industry (packaging, carrier bags, disposable food wraps) and medical biotechnology (antibacterial filtration membranes) Table 1.

Table 1 GO based electropsun composite fibers and their applications for the remediation of environmental contaminants

Electrospun composite fibers	Environmental pollutant	Mechanism of removal	% of removal efficiency/adsorption capacity	References
GA/PVA/GO	Methylene blue (MB)	chemisorption	90%; 5.8×10^{-5} mol/g	[27]
GT/GO	Pb^{2+}, Cd^{2+} and Ag^{+}	Adsorption into monolayer	142.5 pb(II); 112.50 Cd(II) and 132.12 Ag(I) mg g^{-1}	[60]
RGO/TiO_2/ Poly (acrylonitrile-co-maleic acid)	Malachite green (MG) and Leucomalachite green (LMG)	Adsorption and photocatalytic degradation	90.6% (MG); and 93.7% (LMG)	[57]
TiO_2/GO composite nanofibers	Methylene Blue (MB) and 4-chlorophenol (4-CP)	Photocatalytic degradation via photogenerated electron–hole pairs	~100% for 50 min (MB) and 80% (4-CP)	[58]
GO/BiOCl/ PAN nanofibers	Rhodamine B (RhB)	Photocatalytic degradation	% of removal of RhB (38.2%) in 1 h 30 min	[61]
PVA (poly(vinyl alcohol)/poly(acryli acid)/carboxylate graphene oxide nanosheet@polydopamine	Congo red (CR), Rhodamine B (RhB), and methylene blue (MB)	Strong π-π stacking, electrostatic interactions and abundant amino and hydroxyl groups interaction with dye molecule	Adsorption capacity of 31.3 mg/g for methylene blue; 8.4 mg/g for RbB and 12.9 mg/g for CR, respectively	[62]
maghemite (γ-Fe_2O_3) and GO/polyacrylonitrile (PAN)	As(V)	Electrostatic attraction and surface complexation	36.1 mg/g	[53]
Poly(lactic acid) (PLA)/Zeolitic imidazole framework/graphene oxide hybrid	MB	Adsorption and photocatalytic degradation	90%	[54]
$Cu_{2-x}Se$/rGO/PVP nanofibers	4-nitrophenol (4-NP)	Synergetic effect	~ 98.5%	[63]
polyvinylidene fluoride-graphene oxide (PVdF-GO) nanofibrous membranes	Methylene blue (MB), Neutral red (NR), Basic fuchsin (BF), Rhodamine B (RB) and Methyl orange (MO)	Electrostatic attraction–repulsion	99.2% (MB/MO); 98.4% (NR/MO); 98.1% (NF/MO), and 97.6% (RB/MO)	[64]

(continued)

Table 1 (continued)

Electrospun composite fibers	Environmental pollutant	Mechanism of removal	% of removal efficiency/adsorption capacity	References
Reduced graphene oxide (rGO) / Titanium dioxide (TiO_2) composite nanofibers	Methyl orange (MO)	Photocatalytic degradation	90%	[65]

Acknowledgements The authors acknowledge the assistance provided by the Research Infrastructure NanoEnviCz, under Project No. LM2018124, supported by the Ministry of Education, Youth and Sports of the Czech Republic and the European Union—European Structural and Investment Funds in the frames of Operational Program Research, Development and Education—Project Hybrid Materials for Hierarchical Structures (HyHi, Reg. No. CZ.02.1.01/0.0/0.0/16_019/0000843). This work also supported under the Inter Excellence—Action programme within the framework of Czech-Bavarian project "Tree Gum Polymers and their Modified Bioplastics for Food Packaging Application" (registration number LTAB19007), and additional project "Bio-based Porous 2D Membranes and 3D Sponges Based on Functionalized Tree Gum Polysaccharides and their Environmental Application" (registration number LTAUSA19091—TUL internal No. 18309/136) by the Ministry of Education, Youth and Sports in the Czech Republic under the "Inter Excellence—Action programme.

References

1. Lambert S, Wagner M (2017) Environmental performance of bio-based and biodegradable plastics: the road ahead. Chem Soc Rev 46:6855–6871. https://doi.org/10.1039/C7CS00149E
2. O'Dea RM, Willie JA, Epps TH (2020) 100th Anniversary of macromolecular science viewpoint: polymers from lignocellulosic biomass. current challenges and future opportunities. ACS Macro Lett 476–493. https://doi.org/10.1021/acsmacrolett.0c00024
3. Ibn Yaich A, Edlund U, Albertsson A-C (2017) Transfer of biomatrix/wood cell interactions to hemicellulose-based materials to control water interaction. Chem Rev 117:8177–8207. https://doi.org/10.1021/acs.chemrev.6b00841
4. Chen M-J, Li R-M, Zhang X-Q et al (2017) Homogeneous transesterification of sugar cane bagasse toward sustainable plastics. ACS Sustain Chem Eng 5:360–366. https://doi.org/10.1021/acssuschemeng.6b01735
5. Siracusa V, Rocculi P, Romani S, Rosa MD (2008) Biodegradable polymers for food packaging: a review. Trends Food Sci Technol 19:634–643. https://doi.org/10.1016/j.tifs.2008.07.003
6. Bayer IS, Guzman-Puyol S, Heredia-Guerrero JA et al (2014) Direct transformation of edible vegetable waste into bioplastics. Macromolecules 47:5135–5143. https://doi.org/10.1021/ma5008557
7. Padil VVT, Senan C, Wacławek S et al (2019) Bioplastic fibers from gum arabic for greener food wrapping applications. ACS Sustain Chem Eng 7:5900–5911. https://doi.org/10.1021/acssuschemeng.8b05896
8. Tedeschi G, Guzman-Puyol S, Ceseracciu L et al (2020) Multifunctional bioplastics inspired by wood composition: effect of hydrolyzed lignin addition to Xylan-Cellulose matrices. Biomacromol 21:910–920. https://doi.org/10.1021/acs.biomac.9b01569
9. Zhao X, Cornish K, Vodovotz Y (2020) Narrowing the gap for bioplastic use in food packaging-an update. Environ Sci Technol acs.est.9b03755. https://doi.org/10.1021/acs.est.9b03755
10. Turan D, Gunes G, Kilic A (2018) Perspectives of bio-nanocomposites for food packaging applications. In: Jawaid M, Swain SK (eds) Springer International Publishing, Cham, pp 1–32
11. Bouwmeester H, Hollman PCH, Peters RJB (2015) Potential health impact of environmentally released micro- and nanoplastics in the human food production chain: experiences from nanotoxicology. Environ Sci Technol 49:8932–8947. https://doi.org/10.1021/acs.est.5b01090
12. Papadopoulou EL, Paul UC, Tran TN et al (2019) Sustainable active food packaging from poly(lactic acid) and cocoa bean shells. ACS Appl Mater Interf 11:31317–31327. https://doi.org/10.1021/acsami.9b09755
13. Jawaid M, Swain SK (2018) Bionanocomposites for Packaging Applications. Springer International Publishing, Switzerland AG

14. Bhardwaj U, Dhar P, Kumar A, Katiyar V (2014) Polyhydroxyalkanoates (PHA)-Cellulose based nanobiocomposites for food packaging applications. ACS Symp Ser 1162:275–314. https://doi.org/10.1021/bk-2014-1162.ch019
15. Wang Q, Cai J, Zhang L et al (2013) A bioplastic with high strength constructed from a cellulose hydrogel by changing the aggregated structure. J Mater Chem A 1:6678. https://doi.org/10.1039/c3ta11130j
16. Padil VVT, Wacławek S, Černík M, Varma RS (2018) Tree gum-based renewable materials: Sustainable applications in nanotechnology, biomedical and environmental fields. Biotechnol Adv 36:1984–2016
17. Venkateshaiah A, Cheong JY, Shin S-H et al (2020) Recycling non-food-grade tree gum wastes into nanoporous carbon for sustainable energy harvesting. Green Chem 22:1198–1208. https://doi.org/10.1039/C9GC04310A
18. Venkateshaiah A, Cheong JY, Habel C et al (2020) Tree gum-graphene oxide nanocomposite films as gas barriers. ACS Appl Nano Mater 3:633–640. https://doi.org/10.1021/acsanm.9b02166
19. Padil VVT, Černík M (2015) Poly (vinyl alcohol)/gum karaya electrospun plasma treated membrane for the removal of nanoparticles (Au, Ag, Pt, CuO and Fe3O4) from aqueous solutions. J Hazard Mater 287:102–110. https://doi.org/10.1016/j.jhazmat.2014.12.042
20. Jamróz E, Kulawik P, Kopel P (2019) The effect of nanofillers on the functional properties of biopolymer-based films: a review. Polymers (Basel) 11:675. https://doi.org/10.3390/polym11040675
21. Ewaldz E, Brettmann B (2019) Molecular interactions in electrospinning: from polymer mixtures to supramolecular assemblies. ACS Appl Polym Mater 1:298–308. https://doi.org/10.1021/acsapm.8b00073
22. Schiffman JD, Schauer CL (2008) A review: electrospinning of biopolymer nanofibers and their applications. Polym Rev 48:317–352. https://doi.org/10.1080/15583720802022182
23. Hernández-Vargas J, González-Campos JB, Lara-Romero J, Ponce-Ortega JM (2014) A mathematical programming approach for the optimal synthesis of nanofibers through an electrospinning process. ACS Sustain Chem Eng 2:454–464. https://doi.org/10.1021/sc400360k
24. Jiang S, Hou H, Agarwal S, Greiner A (2016) Polyimide nanofibers by "green" electrospinning via aqueous solution for filtration applications. ACS Sustain Chem Eng 4:4797–4804. https://doi.org/10.1021/acssuschemeng.6b01031
25. Padil Vinod VT, Wacławek S, Senan C et al (2017) Gum karaya (Sterculia urens) stabilized zero-valent iron nanoparticles: characterization and applications for the removal of chromium and volatile organic pollutants from water. RSC Adv 7:13997–14009. https://doi.org/10.1039/C7RA00464H
26. Padil VVT, Senan C, Černík M (2015) Dodecenylsuccinic anhydride derivatives of gum Karaya (Sterculia urens): preparation, characterization, and their antibacterial properties. J Agric Food Chem 63. https://doi.org/10.1021/jf505783e
27. Silvestri D, Mikšíček J, Wacławek S et al (2019) Production of electrospun nanofibers based on graphene oxide/gum Arabic. Int J Biol Macromol 124:396–402. https://doi.org/10.1016/j.ijbiomac.2018.11.243
28. Padil VVT, Stuchlík M, Černík M (2015) Plasma modified nanofibres based on gum kondagogu and their use for collection of nanoparticulate silver, gold and platinum. Carbohydr Polym 121:468–476. https://doi.org/10.1016/j.carbpol.2014.11.074
29. Vinod VTP, Miroslav C (2014) Tree gum based electrospun nanofibre membranes: Process optimization, characterization and environmental application. In: NANOCON 2014—Conference Proceedings, 6th International Conference
30. Zhang S, Yu N, Zeng S et al (2018) An adaptive and stable bio-electrolyte for rechargeable Zn-ion batteries. J Mater Chem a 6:12237–12243. https://doi.org/10.1039/C8TA04298E
31. Chen J, Xie F, Li X, Chen L (2018) Ionic liquids for the preparation of biopolymer materials for drug/gene delivery: a review. Green Chem 20:4169–4200. https://doi.org/10.1039/C8GC01120F

32. Iravani S, Varma RS (2019) Plants and plant-based polymers as scaffolds for tissue engineering. Green Chem 21:4839–4867. https://doi.org/10.1039/C9GC02391G
33. Padil VVT, Senan C, Černík M (2019) "Green" polymeric electrospun fibers based on tree-gum hydrocolloids. In: Materials for biomedical engineering. Elsevier, pp 127–172
34. Luo CJ, Stoyanov SD, Stride E et al (2012) Electrospinning versus fibre production methods: from specifics to technological convergence. Chem Soc Rev 41:4708. https://doi.org/10.1039/c2cs35083a
35. Sun Y, Cheng S, Lu W et al (2019) Electrospun fibers and their application in drug controlled release, biological dressings, tissue repair, and enzyme immobilization. RSC Adv 9:25712–25729. https://doi.org/10.1039/C9RA05012D
36. Yan X, Yu M, Ramakrishna S et al (2019) Advances in portable electrospinning devices for in situ delivery of personalized wound care. Nanoscale 11:19166–19178. https://doi.org/10.1039/C9NR02802A
37. Zhang C-L, Yu S-H (2014) Nanoparticles meet electrospinning: recent advances and future prospects. Chem Soc Rev 43:4423. https://doi.org/10.1039/c3cs60426h
38. Xue J, Wu T, Dai Y, Xia Y (2019) Electrospinning and electrospun nanofibers: methods, materials, and applications. Chem Rev 119:5298–5415. https://doi.org/10.1021/acs.chemrev.8b00593
39. Wang J, Li Q, Liu D et al (2018) High temperature thermally conductive nanocomposite textile by "green" electrospinning. Nanoscale 10:16868–16872. https://doi.org/10.1039/C8NR05167D
40. Gao X, Gou J, Zhang L et al (2018) A silk fibroin based green nano-filter for air filtration. RSC Adv 8:8181–8189. https://doi.org/10.1039/C7RA12879G
41. Sridhar R, Lakshminarayanan R, Madhaiyan K et al (2015) Electrosprayed nanoparticles and electrospun nanofibers based on natural materials: applications in tissue regeneration, drug delivery and pharmaceuticals. Chem Soc Rev 44:790–814. https://doi.org/10.1039/C4CS00226A
42. Thavasi V, Singh G, Ramakrishna S (2008) Electrospun nanofibers in energy and environmental applications. Energy Environ Sci 1:205. https://doi.org/10.1039/b809074m
43. Yang X, Wang X, Yu F et al (2016) Hyaluronic acid/EDC/NHS-crosslinked green electrospun silk fibroin nanofibrous scaffolds for tissue engineering. RSC Adv 6:99720–99728. https://doi.org/10.1039/C6RA13713J
44. Kumar PS, Sundaramurthy J, Sundarrajan S et al (2014) Hierarchical electrospun nanofibers for energy harvesting, production and environmental remediation. Energy Environ Sci 7:3192–3222. https://doi.org/10.1039/C4EE00612G
45. Zare EN, Makvandi P, Borzacchiello A et al (2019) Antimicrobial gum bio-based nanocomposites and their industrial and biomedical applications. Chem Commun 55:14871–14885. https://doi.org/10.1039/C9CC08207G
46. Stevens LR, Gilmore KJ, Wallace GG, Het Panhuis M (2016) Tissue engineering with gellan gum. Biomater Sci 4:1276–1290
47. Pitkänen L, Heinonen M, Mikkonen KS (2018) Safety considerations of plant polysaccharides for food use: a case study on phenolic-rich softwood galactoglucomannan extract. Food Funct 9:1931–1943. https://doi.org/10.1039/C7FO01425B
48. Sun J, Perry SL, Schiffman JD (2019) Electrospinning nanofibers from chitosan/hyaluronic acid complex coacervates. Biomacromol 20:4191–4198. https://doi.org/10.1021/acs.biomac.9b01072
49. Cheng Q-Y, Zhou D, Gao Y et al (2012) Supramolecular self-assembly induced graphene oxide based hydrogels and organogels. Langmuir 28:3005–3010. https://doi.org/10.1021/la204558f
50. Debnath S, Maity A, Pillay K (2014) Magnetic chitosan–GO nanocomposite: Synthesis, characterization and batch adsorber design for Cr(VI) removal. J Environ Chem Eng 2:963–973. https://doi.org/10.1016/j.jece.2014.03.012
51. Wan C, Chen B (2012) Reinforcement and interphase of polymer/graphene oxide nanocomposites. J Mater Chem 22:3637. https://doi.org/10.1039/c2jm15062j

52. Goh K, Heising JK, Yuan Y et al (2016) Sandwich-architectured poly(lactic acid)–graphene composite food packaging films. ACS Appl Mater Interfaces 8:9994–10004. https://doi.org/10.1021/acsami.6b02498
53. Tripathy M, Hota G (2020) Maghemite and graphene oxide embedded polyacrylonitrile electrospun nanofiber matrix for remediation of arsenate ions. ACS Appl Polym Mater 2:604–617. https://doi.org/10.1021/acsapm.9b00982
54. Dai X, Li X, Zhang M et al (2018) Zeolitic imidazole framework/graphene oxide hybrid functionalized poly(lactic acid) electrospun membranes: a promising environmentally friendly water treatment material. ACS Omega 3:6860–6866. https://doi.org/10.1021/acsomega.8b00792
55. Padil VVT, Senan C, Wacsawek S, L·erník M (2016) Electrospun fibers based on Arabic, karaya and kondagogu gums. Int J Biol Macromol 91:299–309. https://doi.org/10.1016/j.ijbiomac.2016.05.064
56. Wacławek S, Padil VVT, Černík M (2018) Major advances and challenges in heterogeneous catalysis for environmental applications: a review. Ecol Chem Eng S 25:9–34. https://doi.org/10.1515/eces-2018-0001
57. Du F, Sun L, Huang Z et al (2020) Electrospun reduced graphene oxide/TiO_2/poly(acrylonitrile-co-maleic acid) composite nanofibers for efficient adsorption and photocatalytic removal of malachite green and leucomalachite green. Chemosphere 239:124764. https://doi.org/10.1016/j.chemosphere.2019.124764
58. Zhang L, Zhang Q, Xie H et al (2017) Electrospun titania nanofibers segregated by graphene oxide for improved visible light photocatalysis. Appl Catal B Environ 201:470–478. https://doi.org/10.1016/j.apcatb.2016.08.056
59. Gil-Castell O, Galindo-Alfaro D, Sánchez-Ballester S, et al (2019) Crosslinked Sulfonated Poly(vinyl alcohol)/Graphene Oxide Electrospun Nanofibers as Polyelectrolytes. Nanomaterials 9:397. https://doi.org/10.3390/nano9030397
60. Sahraei R, Ghaemy M (2017) Synthesis of modified gum tragacanth/graphene oxide composite hydrogel for heavy metal ions removal and preparation of silver nanocomposite for antibacterial activity. Carbohydr Polym 157:823–833. https://doi.org/10.1016/j.carbpol.2016.10.059
61. Zhang Y, Park M, Kim H-Y, Park S-J (2016) In-situ synthesis of graphene oxide/BiOCl heterostructured nanofibers for visible-light photocatalytic investigation. J Alloys Compd 686:106–114. https://doi.org/10.1016/j.jallcom.2016.06.004
62. Xing R, Wang W, Jiao T et al (2017) Bioinspired polydopamine sheathed nanofibers containing carboxylate graphene oxide nanosheet for high-efficient dyes scavenger. ACS Sustain Chem Eng 5:4948–4956. https://doi.org/10.1021/acssuschemeng.7b00343
63. Yang T, Zou HY, Huang CZ (2015) Synergetic Catalytic Effect of Cu 2–x Se nanoparticles and reduced graphene oxide coembedded in electrospun nanofibers for the reduction of a typical refractory organic compound. ACS Appl Mater Interfaces 7:15447–15457. https://doi.org/10.1021/acsami.5b03645
64. Ghaffar A, Zhang L, Zhu X, Chen B (2018) Porous PVdF/GO nanofibrous membranes for selective separation and recycling of charged organic dyes from water. Environ Sci Technol 52:4265–4274. https://doi.org/10.1021/acs.est.7b06081
65. Nasr M, Balme S, Eid C et al (2017) Enhanced visible-light photocatalytic performance of electrospun rGO/TiO_2 composite nanofibers. J Phys Chem C 121:261–269. https://doi.org/10.1021/acs.jpcc.6b08840
66. Chu P (2002) Plasma-surface modification of biomaterials. Mater Sci Eng R Rep 36:143–206. https://doi.org/10.1016/S0927-796X(02)00004-9
67. Daw R, Candan S, Beck AJ et al (1998) Plasma copolymer surfaces of acrylic acid/1,7 octadiene: surface characterisation and the attachment of ROS 17/2.8 osteoblast-like cells. Biomaterials 19:1717–1725. https://doi.org/10.1016/S0142-9612(98)00080-5
68. Perotto G, Ceseracciu L, Simonutti R et al (2018) Bioplastics from vegetable waste via an eco-friendly water-based process. Green Chem 20:894–902. https://doi.org/10.1039/C7GC03368K

69. Habel C, Schöttle M, Daab M, et al (2018) High-barrier, biodegradable food packaging. Macromol Mater Eng 303:1800333. https://doi.org/10.1002/mame.201800333
70. Walther A, Bjurhager I, Malho J-M et al (2010) Supramolecular control of stiffness and strength in lightweight high-performance nacre-mimetic paper with fire-shielding properties. Angew Chemie Int Ed 49:6448–6453. https://doi.org/10.1002/anie.201001577
71. Alexandre M, Dubois P (2000) Polymer-layered silicate nanocomposites: preparation, properties and uses of a new class of materials. Mater Sci Eng R Reports 28:1–63. https://doi.org/10.1016/S0927-796X(00)00012-7
72. Edlund U, Ryberg YZ, Albertsson A-C (2010) Barrier films from renewable forestry waste. Biomacromol 11:2532–2538. https://doi.org/10.1021/bm100767g
73. Vartiainen J, Vähä-Nissi M, Harlin A (2014) Biopolymer films and coatings in packaging applications—a review of recent developments. Mater Sci Appl 05:708–718. https://doi.org/10.4236/msa.2014.510072
74. Rao MS, Kanatt SR, Chawla SP, Sharma A (2010) Chitosan and guar gum composite films: preparation, physical, mechanical and antimicrobial properties. Carbohydr Polym 82:1243–1247. https://doi.org/10.1016/j.carbpol.2010.06.058
75. Rhim J-W, Lee J-H, Hong S-I (2006) Water resistance and mechanical properties of biopolymer (alginate and soy protein) coated paperboards. LWT Food Sci Technol 39:806–813. https://doi.org/10.1016/j.lwt.2005.05.008
76. Narasagoudr SS, Hegde VG, Vanjeri VN et al (2020) Ethyl vanillin incorporated chitosan/poly(vinyl alcohol) active films for food packaging applications. Carbohydr Polym 236:116049. https://doi.org/10.1016/j.carbpol.2020.116049
77. Tomadoni B, Ponce A, Pereda M, Ansorena MR (2019) Vanillin as a natural cross-linking agent in chitosan-based films: optimizing formulation by response surface methodology. Polym Test 78:105935. https://doi.org/10.1016/j.polymertesting.2019.105935
78. Svirachev DM (2005) Plasma treatment of polymer surfaces in different gases. Bulg J Phys 32:22–33
79. Scott G (2000) 'Green' polymers. Polym Degrad Stab 68:1–7. https://doi.org/10.1016/S0141-3910(99)00182-2
80. Jamshidian M, Tehrany EA, Imran M et al (2010) Poly-Lactic acid: production, applications, nanocomposites, and release studies. Compr Rev Food Sci Food Saf 9:552–571. https://doi.org/10.1111/j.1541-4337.2010.00126.x
81. Tokiwa Y, Calabia BP (2006) Biodegradability and biodegradation of poly(lactide). Appl Microbiol Biotechnol 72:244–251. https://doi.org/10.1007/s00253-006-0488-1
82. da Silva ISV, de Sousa RMF, de Oliveira A et al (2018) Polymeric blends of hydrocolloid from chia seeds/apple pectin with potential antioxidant for food packaging applications. Carbohydr Polym 202:203–210. https://doi.org/10.1016/j.carbpol.2018.08.061
83. López de Dicastillo C, Rodríguez F, Guarda A, Galotto MJ (2016) Antioxidant films based on cross-linked methyl cellulose and native Chilean berry for food packaging applications. Carbohydr Polym 136:1052–1060. https://doi.org/10.1016/j.carbpol.2015.10.013
84. Batista RA, Espitia PJP, Quintans J de SS, et al (2019) Hydrogel as an alternative structure for food packaging systems. Carbohydr Polym 205:106–116. https://doi.org/10.1016/j.carbpol.2018.10.006
85. Otoni CG, Espitia PJP, Avena-Bustillos RJ, McHugh TH (2016) Trends in antimicrobial food packaging systems: emitting sachets and absorbent pads. Food Res Int 83:60–73. https://doi.org/10.1016/j.foodres.2016.02.018
86. Dai H, Zhang Y, Ma L et al (2019) Synthesis and response of pineapple peelcarboxymethyl cellulose-g-poly (acrylic acid-co-acrylamide)/graphene oxide hydrogels. Carbohydr Polym 215:366–376. https://doi.org/10.1016/j.carbpol.2019.03.090

Graphene Functionalized Starch Biopolymer Nanocomposites: Fabrication, Characterization, and Applications

Ranjana Mishra and Ankit Manral

Abstract The discovery of Graphene as nanofiller has catalyzed a wide range of innovative areas of application. Many superior properties of graphene such as high surface area, lightweight, high mechanical strength, and high thermal conductivity as compared to other nanomaterials (layered silicates or carbon nanotubes) have garnered the attention for its application in the field of the composite. Bio-based polymers and their nanocomposites lead to very efficient interaction with the functional groups on graphene and GO due to the presence of hydroxyl, amino, or carboxyl groups in the biopolymers. Starch is the second abundant biopolymer after cellulose and has been of keen interest due to its excellent biodegradable property. Starch-biopolymers are widely used in several fields such as biomedical, pharmaceutical, food packaging, and agriculture. Incorporation of graphene into a starch biopolymer can broaden its application area by enhancing its properties. This chapter highlights the developed properties of graphene-starch nanocomposites and their fabrication techniques. It also discusses the characterization and application of graphene functionalized starch biopolymer nanocomposites (GFSBN).

Keywords Biopolymer · Graphene · Starch · Nanocomposites · Characterization · Application

1 Introduction

Biopolymer and their Nanocomposites have always been of keen interest for researchers as alternatives to synthetic polymers. Polymers derived from renewable resources such as cellulose, starch, and chitosan are called biopolymers. On the basis of methods of production biopolymers can be classified three main categories as shown in Fig. 1 [1]. Biopolymers possess many unique properties and some shortcoming simultaneously. Such defects can be replaced and reinforced with a wide range of fillers such as nanotubes, layered silicates in specific polymers, resulting in

R. Mishra (✉) · A. Manral
MPAE Division, Netaji Subhas University of Technology, Delhi, India
e-mail: mishra.ranjana2011@gmail.com

B. Sharma and P. Jain (eds.), *Graphene Based Biopolymer Nanocomposites*,
Composites Science and Technology,
https://doi.org/10.1007/978-981-15-9180-8_9

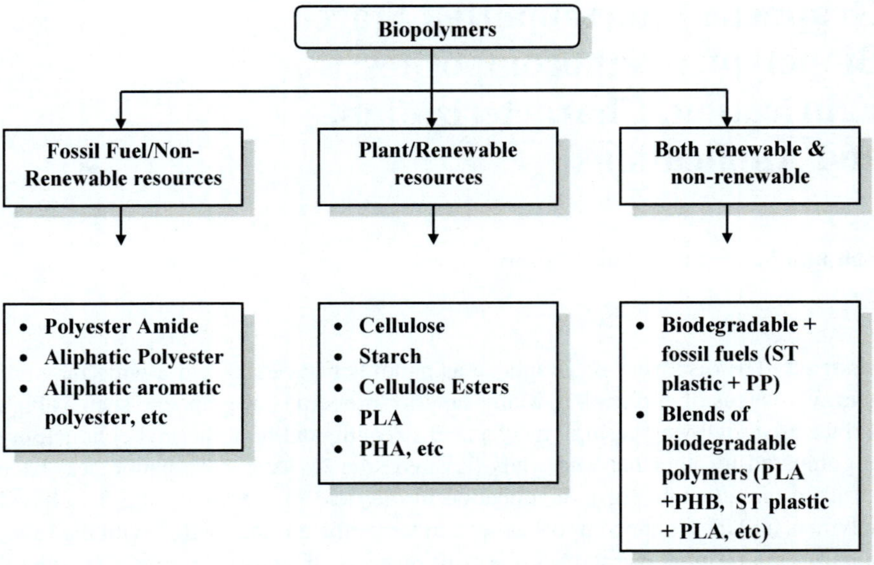

Fig. 1 Classification of bio-based polymers

a substantial improvement in their thermal, mechanical and other properties [2–4]. Various researchers who have worked in this area have reported that the addition of graphene as a polymer reinforcing agent can improve the overall properties and effectiveness of these composites. Graphene based polymers nanocomposites exhibit superior properties as compared to other nanomaterials such as layered silicates or carbon nanotubes [5–7].Graphene, the most basic form of carbon, is a 2D structure of sp^2 bonded carbon atoms arranged in hexagonal crystalline structure [8]. Graphene has a high surface area, due to which it boosts the interaction between the sheets and polymer material [9]. It also possesses high mechanical strength, electron mobility and thermal conductivity. On the other hand, graphene being the thinnest material in existence possesses quantum hall effect, high intrinsic power (130 GPa), high Young's modulus (1 TPa), high optical transmittance and high electronic transport as a single sheet [8, 10–13]. These promising properties mark its application in various fields of technologies such as electronics, biomedical, energy storage, etc.

Among all bio-based polymers, starch has garnered a lot of attention for the fabrication of environment-friendly biopolymers and nanocomposites. It has almost replaced synthetic plastic due to its potent properties such as biodegradability, abundance, renewability and low cost. Starch-based biopolymers are widely used in several fields such as biomedical, pharmaceutical, food packaging, and agriculture. But its application is affected by its barrier properties such as poor mechanical strength, hydrophilic property, and other limitations. Graphene and its derivatives like GO, RGO nanomaterials are incorporated to improve its barrier properties. This chapter contains a brief explanation of graphene functionalized starch biopolymer nanocomposites and their properties. It also highlights the various fabrication techniques of the development of these nanocomposites. To investigate the effect of graphene on starch biopolymer and to examine the morphology of the developed products, various characterization techniques have also been discussed in this chapter.

2 Starch

Starch is a natural polymer that is biodegradable, low cost and renewable due to which it is suitable for developing various sustainable material in different fields including the food, textile, adhesive, medical and packaging industries [14, 15]. Pure starch is a tasteless and odorless white powder commonly taken as a carbohydrate diet. The word "starch" means stiff or strong which comes from its German root. Starch is derived in granulated form from different plant tissues such as from wheat, potato, maize, rice, pea, amylomaizes, cassava, etc. Generally, Starch granules are between 1 and 100 μm and also available in variety of shapes [16]. This polysaccharide (starch) is composed of carbon, hydrogen and oxygen in ratio of 6:10:5 [$C_6H_{10}O_5$] [17]. Starch is the second most abundantly produced organic material after cellulose. The primary structure of starch polymer is much like cellulose; the difference lies in the secondary and tertiary structures which give different properties to these two materials [17].

2.1 Structure and Properties of Starch

Chemically, starch is a polymeric carbohydrate which contains two microstructures— linear microstructure amylose and branched microstructure amylopectin. Amylose is essentially a linear microstructure of α-1,4 linked D-glucose units [18, 19]. The length of this chain varies between 10^2 and 10^4 glucose units whereas amylopectin is a highly branched structure of short α-1,4 chains linked by α-1,6 bonds of which the length of this chain varies between 10^4 and 10^5 glucose units. The relative proportion of Amylose and Amylopectin in starch varies between 18 and 28% amylase and 72–82% amylopectin [18–20]. Table 1 summarizes the characteristics of different starches [21–29].

Starch composition is variable which solely depends upon their botanical origin. It usually exhibits some traces of protein, gluten, lipids or phosphate that can interfere with the properties of starch [16]. The chemical composition for different types of starch is summarized in Table 2.

Starch exhibits strong intermolecular linkages and hydrophilic properties through hydrogen bonding produced by hydroxyl groups [31]. Because of its hydrophilic characteristics, the starch's internal interaction and composition will be easily and effectively altered by water molecules, and hence its mechanical properties, glass transition temperature (T_g), and the dimension depend on the water content [31].

Table 1 Characteristic of different types of starch

Starch source	Amylose content (%)	Swelling power (g/g) (°C)	Solubility (%) (°C)	Light transmittance (%, at 650 nm)	References
Normal wheat	18–30	18.3–26.6 (100)	1.55 (100)	28	[21–23]
Normal potato	20.1–31	1159 (95)	82 (95)	96	[27–29]
Normal rice	5–28.4	23–30 (95)	11–18 (95)	24	[25, 26]
Normal corn	22.4–32.5	22 (95)	22 (95)	31	[21]
Waxy rice	0–2	45–50 (95)	2.3–3.2 (95)	–	[25, 26]
Waxy corn	1.4–2.7	–	–	46	[21]
High amylose corn	42.6–67.8	6.3 (95)	12.4 (95)		[21, 26, 29]

Table 2 Chemical composition of different types of starch [16]

Starch	Amylopectin content (%)	Amylose content (%)	Protein content (%)	Lipid content (%)	Phosphorus content (%)	Moisture content (%)
Waxy starch	99	<1	0.10	0.23	0.01	N. D
Potato	79–74	20–25	0.05	0.03	0.08	18–19
Wheat	72–73	26–27	0.30	0.63	0.06	13
maize	71–73	26–28	0.30	0.63	0.02	12–13
Amylomaize	20–50	50–80	0.50	1.11	0.03	N.D

Note N.D—Not Determined

For instance, if we reduce the moisture content (<20 wt%), the melting temperature tends to be quite close to the degradation temperature. Generally, glass transition temperature (T_g) of natural starch lies between 60 to 80 °C when the weight proportion of water content is in the range 0.12–0.14, which allows the successful injection moulding of starch to produce thermoplastic starch (TPS) polymers in the presence of water [31, 32].

2.2 *Modification of Starch*

Starch is poor in the dimensional stability and mechanical properties for its end products. That is why native starch cannot be used directly for many purposes. In order to mend the properties, starches are chemically and physically modified. Various modification techniques have been investigated by researchers such as cross-linking, esterification, pregelatinization and stabilization [34, 35]. Commonly used modification techniques and their effect on starch granules are illustrated in Fig. 2.

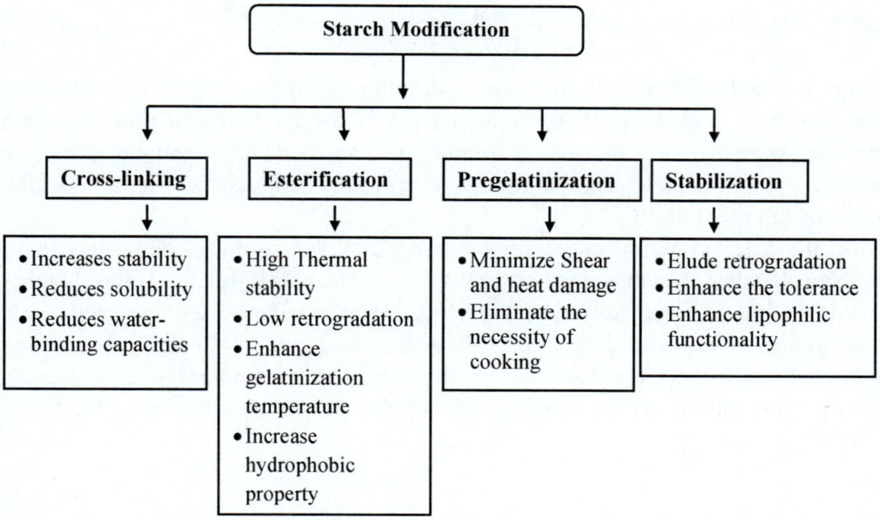

Fig. 2 Various modification techniques of starch and their effect

2.2.1 Cross-Linking

Cross-linking is one of the most commonly used chemical modifications that are applied to starch. In this process starch has been treated chemically to link the starch molecules together with cross bridges. In starch, most of the glucose units consist of one primary and two secondary hydroxyl groups, which easily react with many compounds such as ethylene compounds, organic chloro compounds, acid anhydrides, and epoxy [34].Increasing the number of cross-links makes a starch more heat resistant. The cross-linked starch becomes less soluble in water which results in higher stability. Due to these properties of cross-linked starch, we can use it effectively in various fields such as food packaging, adhesive, paper, and textile industry.

2.2.2 Esterification

Esterification is another effective method for altering and enhancing the structure of starch granules. In this process of starch modification, ester groups replace a few hydroxyl groups. There are many starch esters available, such as starch sulfates, starch alkenyl succinates, starch nitrates, starch phosphates and fatty acid starches. The purpose of this modification is to raise the hydrophobic property and enhance the water resistivity so that it can be used efficiently in the pharmaceutical and beverage industries. Esterification also enhances gelatinization temperature and thermal stability and decreases retrogradation.

2.2.3 Pregelatinization

Some researches show that in order to develop the appropriate properties many starches need cooking [34]. Pregelatinization is the modification process which removes the necessity of cooking. Mainly, two methods of Pregelatinization are widely used to achieve a large variety of cold thickening starch -spray cooking process and drum drying.

In the spray cooking process, the starch slurry gets atomized after entering a heating chamber. And then, the hot spray of water steam is injected into the chamber to cookout these atomized starches. This produces a standardized gelatinized starch with minimum heat and shear impact. In drum drying, the starch slurry is modified into cooked starch sheets. These starch sheets are then ground to the required size. This process minimizes shear and heat damage and eliminates the necessity of over cooking.

2.2.4 Stabilization

Native starch has some shortcomings such as poor thermal process-ability which can be improved by stabilization. Stabilization enhances starch tolerance to temperature changes and eludes retrogradation. Some bulky groups like octenylsuccinate are inserted onto the starch to avoid linear and scattered fragments to retrograde. The number of groups replaced determines the stabilization effect [34].

3 Starch-Based Biodegradable Polymers

Various polymers are used to prepare completely biodegradable starch-based composites. Polylactide (PLA) is the most promising biodegradable polyester having many useful properties. It possesses high strength and modulus, as well as good biocompatibility and processability [36, 37]. But due to weak interaction between hydrophobic PLA and hydrophilic starch granules this blend system is poorly compatible. To eliminate the incompatibility of this blend suitable compatibilizer should be added and the gelatinization method is applied which makes the biopolymer costly as PLA is considerably more costly than other industrial polymer [38–40]. Similarly, poly (ε—caprolactone) also called PCL is another synthetic biodegradable polymer derived from chemical polymerization that can be blended with starch to overcome properties of pure starch like high moisture sensitivity, low resilience, and high shrinkage [41–45]. On blending with PCL, dimensional stability and the impact resistance of starch is improved [46]. Poly vinyl alcohol (PVA) is also a biodegradable synthetic polymer that poses outstanding compatibility and mechanical properties with native starch [47]. The tensile strength of blend increases with increase in concentration of PVA [48]. PVA can also be used to increase the compatibility between starch and PLA which in result improves the mechanical property of starch/PLA blend [49, 50].

Starch can also be blended with biopolymers like chitosan and cellulose which are also biodegradable [51]. Blend exhibit good film forming property which is attributed to their inter and intra molecular hydrogen bonding that formed between hydroxyl groups and amino groups. The ratio of starch and chitosan affects the water barrier properties, mechanical properties, and miscibility of biodegradable blend [31, 52]. PLS and flax cellulose nanocrystals are cast together to obtain a starch-based nanocomposites film. The water resistance and mechanical properties are very much improved [53, 54].Although a variety of biopolymers are available for the development of composites, Graphene has excellent properties compared to others to meet the challenging requirement of advanced applications. In a neat biopolymer, only 1% of the polymer volume of graphene is required or sufficient for many electrical applications. The presence of hydroxyl, amino, or carboxyl groups in the biopolymers leads to very effective interaction with the Graphene and Graphene oxide functional groups [34].

3.1 Graphene Functionalized Starch Biopolymer Nanocomposites (GFSBN)

Mostly graphene is incorporated in nano form in the polymer, although, proper dispersion of graphene is needed for better properties of developed nanocomposites. Good distribution and adequate concentration of graphene in a biopolymer can also be a governing factor in influencing the properties of nanocomposites [34]. There are mainly three fabrication techniques to develop the graphene-based biopolymer nanocomposites: (1) In-situ intercalative polymerization (2) solution intercalation, and (3) melt intercalation. Besides these, there are other methods like drop-casting, wet-spinning, freeze-drying, and electro-spinning, etc. Figure 3 shows the schematic classification of fabrication techniques. In-situ polymerization method for production of nanocomposites generally involves the mixing of nanofillers in solution of monomers, accompanied by polymerization of monomers in the presence of dispersed filler. This process highly influenced the mechanical, electrical, and thermal conductive properties of developed nanocomposites. Solution Intercalation is a simple and most popular fabrication technique among all. It involves shear mixing or simple stirring of colloidal graphene heterogeneous mixture (suspensions) with a suitable and required biopolymer like starch, chitosan, PLA, cellulose [57]. Subsequently, evaporation of solvent causes the adsorbed polymer to assemble and form a sandwich between graphene and polymer which results in the fabrication of nanocomposites. This method is only applicable for thermoplastic biopolymer such are PLA, PHB etc.

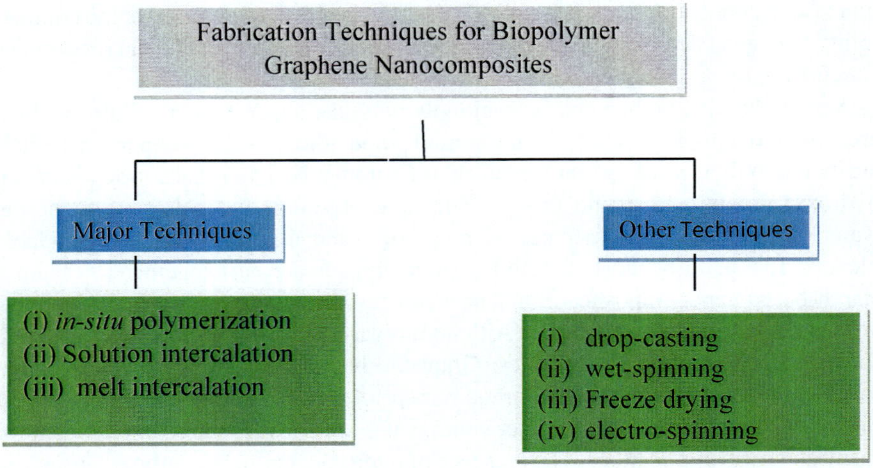

Fig. 3 Schematic classifications of fabrication techniques

In melt intercalation, a polymer melt and filler are mixed under high shear conditions. During this process, chemical functionalization with oxygen causes some matrix defects which lead to a restriction of electron flow and a reduction of electrical conductivity [34, 58].This process is introduced for a variety of biopolymers like Bio-polyamide (PA), cellulose, PLA, etc. for the development of graphene-based nanocomposites. The applications of nanocomposites fabricated by this method are used in electrical energy storage devices [58]. In summary, solution intercalation is a simple but highly effective fabrication technique of graphene–biopolymer nanocomposites manufacturing as it involves simple and efficient dispersion of graphite in polymer matrix. This method is widely used in the fabrication of graphene functionalized starch biopolymer nanocomposites (GFSBN).The study of properties and structural characterization of starch and chitosan-based graphene oxide (GO) nanocomposites has been conducted [59]. To prepare oxide Starch/Chitosan/Graphene Oxide biodegradable nanocomposites casting and solvent evaporation method was used by the researchers. It was concluded that oxide Starch/Chitosan/Graphene Oxide biodegradable nanocomposites films have higher tensile strength and lower elongation at break then the control CST/GO-n series [59].

Another research has been carried out on the fabrication of epoxy/starch-functionalized graphene oxide nanocomposites for coating applications [30]. For this study, graphene oxide (GO) nanoplatelets have been chemically functionalized with starch biopolymer to proceed with the opening of the epoxy ring. A high-performance epoxy nanocomposite was developed by this combination. The solution casting method was used in the fabrication of multi-functionalized graphene oxide/poly (lactide)/starch nanocomposites [33]. In this study Poly (lactide)-starch matrix was blended with multi-functionalized graphene oxide to develop the biopolymer nanocomposites. Various properties such as aging resistance, UV-shielding capacity, and surface hydrophobicity of these nanocomposites were enhanced by the inclusion of the f-GO (functionalized GO).

Starch-grafted graphene nanosheet and plasticized-starch composites were fabricated by a novel method. In this method Starch was grafted onto the graphene nanosheet (GN) surface by reduction of GO with hydrazine hydrate [55]. The composites were also produced using GN-starch in a plasticized-starch (PS) matrix as the filler. GN enhanced solubility and stability in water of GN-starch. A comparative study was conducted into the effect of GO and RGO (reduced graphene oxide) on the overall properties of starch [34]. For this study glycerol-plasticized starch (PS) nanocomposites were taken with GO and RGO fillers in presence of water as a solvent. Solution casting (SC) method is used to prepare the nanocomposites. Starch-GO nanocomposites exhibit better overall property then starch-RGO nanocomposites. Since GO has more polar groups than RGO which leads to better mechanical properties and better compatibility with PS than RGO. Besides that, GO has greater water dispersibility compared to RGO [60].Water vapour permeability (WVP) test show enhanced water vapour barrier properties for PS-GO nanocomposites than RGO. Several other fabrication methods are also used to develop these nanocomposites. According to a study, the Melt extrusion process was used to fabricate the glycerol-plasticized starch and graphene oxide hybrids [62]. Effect of Graphene oxide on glycerol-plasticized starch-GO hybrid was evaluated by using several characterization techniques such as X-ray diffraction (XRD), dynamic mechanical analysis (DMA), capillary rheometry, etc.

4 Analysis and Characterization Techniques

To analyze and understand the various facets of polymer nanocomposites morphology and properties, characterization is necessary. Characterization is generally done for the following purposes:

- To understand the effect of filler surface correction on composite properties and filler dispersion.
- To analyze the Quality of dispersion and orientation/alignment of fillers in the polymer matrix.
- To study the filler grafting interaction with the polymer chains and the chemical reactions between the two.
- In addition, to study the broad array of properties to ascertain the ability of nanocomposites for the application purpose.

There are generally three techniques of characterization of materials microscopic, macroscopic and spectroscopic.

Microscopic techniques of characterization examine the surface and sub-surface structure of a substance using electrons, photons, ions or physical cantilever tests to obtain data about the structure of any sample on a range of length scales.

Optical Microscope, Scanning Electron Microscope (SEM) and Transmission Electron Microscope (TEM) are the most commonly used instruments for the investigation of sample morphology. Macroscopic techniques are conducted to characterize various macroscopic properties of materials such as hardness, toughness, creep, fatigue, thermal, rheological and tribological etc. Trapezium-X Universal Testing Machine is mostly used to investigate the mechanical properties of composites. Mostly, Thermogravimetric analysis (TGA) and Differential scanning calorimetry (DSC) techniques are used for thermal characterization.

Chemical and functional properties such as photoelectric properties, chemical composition, composition variation, and crystal structure of composites are examined by using spectroscopic techniques. Fourier transform infrared spectroscopy (FTIR), X-ray diffraction (XRD), and X-ray photoelectron spectroscopy (XPS) are few commonly used techniques in functional and chemical characterization. Figure 4 shows the schematic diagram of characterization techniques. The effect of graphene on starch biopolymer has been evaluated in several studies which proved the use of graphene in the starch polymer is remarkable. Addition of graphene with starch can improve the overall property of starch for various applications. For instance, in a comparative study, it is concluded that the mechanical properties of starch-GO nanocomposites were increased 66% at 2 wt% GO loading whereas for starch RGO

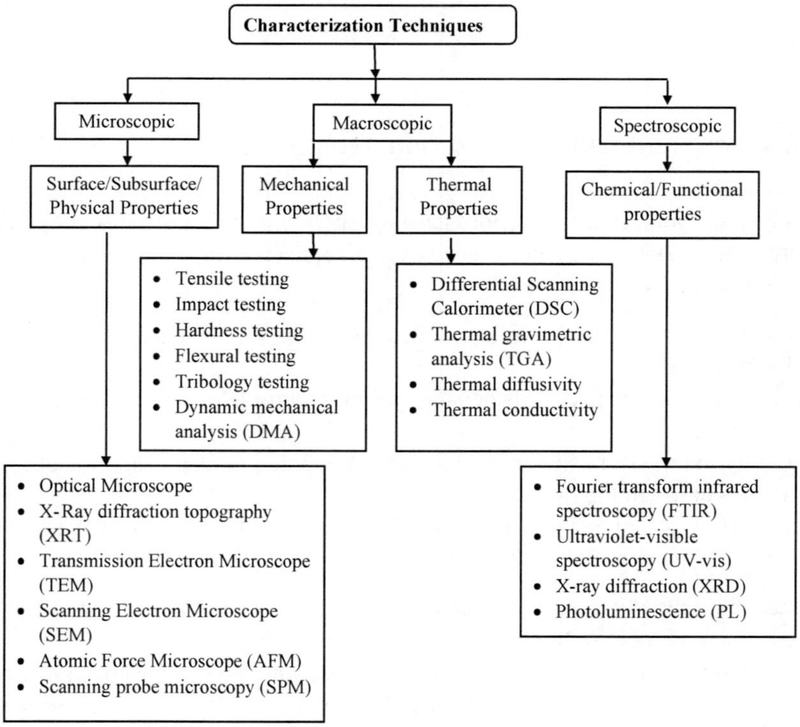

Fig. 4 Various techniques of characterization

nanocomposites tensile strength was increased by 67% for 6wt% RGO loading. This tensile test was carried out on the test metric AX M350-10 KN material testing machine at a crosshead speed of 50 mm/min. In case of thermal characterization, PS-GO nanocomposites degraded at 290 °C because of the about 27 wt% of oxygen functional group which weakens the interaction between PS matrix and GO filler. PS-RGO degraded at 314 °C having less percentage of oxygen containing groups (about 9 wt%). This shows that PS-RGO composites exhibit higher thermal stability than PS-GO nanocomposites. WVP (Water vapour permeability) tests were carried out by ASTM method to observe the effect of graphene on gas- and vapor-barrier properties of starch and WVP can be calculated as

$$WVP = \frac{\mathrm{WVTR}}{\mathrm{P(R1 - R2)}}.\mathrm{X}$$

where, P is the saturation vapour pressure in Pascal at the test temperature (25 °C). R1 is the relative humidity (RH) in the desiccator; R2, the RH in the permeation cell; and X is the thickness of the film (m).

Some of the tests show the improved water vapour properties for PS-4 wt% GO nanocomposites as compared to 8 wt% RGO loading PS-RGO nanocomposites. These microscopic, macroscopic and chemical/functional techniques clearly indicate the efficacy of graphene on the mechanical, thermal, chemical and other properties of different starch biopolymers nanocomposites which are shown in Tables 3, 4 and 5.

5 Applications of Graphene Functionalized Starch-Based Biopolymer Nanocomposites

Biopolymer nanocomposites are used in wide variety of fields such as in food industry they are used in majorly food packaging and edible films. Food packaging prevents food losses, keeps food fresh, provides food safety and maintains characteristics of food such as appearance, flavor, odor, etc. Due to increased antimicrobial potential and water resistance property of graphene functionalized starch biopolymer nanocomposites, it can be the most promising material for the food packaging industry. It can be used as a water filtration membrane and scaffold for cellular growth [56].These nanocomposites can be used as food serving items by converting them into trays or disposable dishes [34].GFSBN exhibits various enhanced properties, such as surface hydrophobicity, resistance to aging and UV shielding which marks it application in pharmaceutical industries and agriculture. In agriculture these biopolymer nanocomposites are used for mulch film, covering of greenhouse, and fertilizers-controlled release materials [63–65].

Starch-based biodegradable polymers are of great importance in medical field. They are commonly studied in the engineering of bone tissues. Addition of graphene

Table 3 Effect of graphene on the mechanical properties of different starch biopolymers nanocomposites

Matrix	Graphene	Preparation method	Tensile strength (MPa)	Young's modulus (GPa)	Other properties	References
OS/Chitosan/GO	0–2.0 wt%	SC	13.66–21.54	7.30	Elongation at Break—↓88.18–32.96%	[59]
PS-GO	2 wt% GO	SC	↑92	–		[61]
PS-RGO	6 wt% RGO	SC	↑54	–		[61]
PVA/GO/Ag/starch	0.5wt%GO	SC	↑26.81	↑145	Antimicrobial potential enhanced	[56]
Epoxy/starch-fGO	0.1 wt% GO	Solvent Exchange method	–	–	Fracture resistance enhanced	[30]
PLA-starch-f-GO	0.1–0.6 wt% f-GO	SC	–	–	– Increased the storage modulus – Increase in aging resistance properties	[33]
PS/GN-starch	1.774 wt% GN		25.4			[55]

Note PS—Plasticized starch, GO—Graphene Oxide, RGO—Reduced graphene oxide, SC—Solution casting, OS—Oxidized starch, fGO—functionalized graphene oxide

Table 4 Effect of graphene on the thermal properties of different starch biopolymers nanocomposites

Matrix–grapheme	Graphene	Preparation	Thermal property	References
OS/Chitosan/GO	0–2.0 wt% GO	SC	T_g 350 °C	[59]
PS-GO	27 wt% GO	SC	T_g 290 °C	[61]
PS-RGO	9 wt% RGO	SC	T_g 314 °C	[61]
PVA/GO/Ag/starch	0.5 wt% GO	SC	T_g 295 °C	[56]
PLA-starch-f-GO	0.1–0.6 wt% f-GO	SC	T_g 336.8–362.2 °C	[33]
PS/GN-starch	1.774 wt% GN	SC	K 9.7 × 10^{-4} S/cm	[55]

Where PS—Plasticized starch, GO—Graphene Oxide, RGO—Reduced graphene oxide, SC—Solution casting, T_g—Glass transition temperature, K—Thermal conductivity

Table 5 Effect of graphene on gas- and vapor-barrier properties/moisture uptake of different starch biopolymers nanocomposites

Matrix–graphene derivative	Graphene	Permeability/MU	References
PS–GO	4 wt% GO	WVP ↓ 43.7%	[61]
PS–RGO	8 wt% RGO	WVP ↓ 34.9%	[61]
OS/Chitosan/GO	0–2.0 wt % GO	MU ↓ 45.30%	[59]
PS/GN-starch	1.774 wt% GN	WVP ↓ from 1.5 × 10^{-10}–1.1 × 10^{-10} g/m s Pa	[55]

Where PS—plasticized starch, GO—Graphene Oxide, RGO—Reduced graphene oxide, WVP—Watervapour permeability, MU—Moisture uptake

enhances the mechanical strength of starch biopolymer, so it can increase its application in bone cement manufacturing. Bone cement is used in immediate structural support. They are also very useful in biomedical applications as they possess good biodegradability and biocompatibility. Products made by these nanocomposites are non-toxic and even possess better mechanical properties. These nanocomposites are of low-cost, light-weight, and high-reinforcement. Nanocomposites made from starch biopolymer–graphene derivatives are used as scaffolding materials, biosensor, fibers with wound healing capabilities, and for drug delivery [66, 67]. Graphene-starch epoxy nanocomposites possess admissible fracture resistance and curing potential that makes them appropriate for coating applications [30].

6 Conclusions

In this chapter, the processing of GFSBN, manufacturing techniques and its application in various fields like food industry, engineering, and medical components are discussed As the weight percentage of graphene in these nanocomposites is usually quite small (less than 5%), that emerge environment-friendly impact in several new applications. Limitations of starch biopolymer such as thermal resistance, oxygen/water vapor barriers, and low tensile strength are reduced to a considerable extent with the addition of graphene which accelerates the application of this biopolymer nanocomposite in all possible areas of application.

References

1. Babu RP, O'Connor K, Seeram R (2013) Current progress on bio-based polymers and their future trends. Prog. Biomater. 2(1):8
2. Lewitus D, McCarthy S, Ophir A, Kenig S (2006) The effect of nanoclays on the properties of PLLA-modified polymers part 1: mechanical and thermal properties. J Polym Environ 14(2):171–177
3. Luecha J, Sozer N, Kokini JL (2010) Synthesis and properties of corn zein/montmorillonite nanocomposite films. J Mater Sci 45(13):3529–3537
4. Mittal V (2008) Mechanical and gas permeation properties of compatibilized polypropylene–layered silicate nanocomposites. J Appl Polym Sci 107(2):1350–1361
5. Kuilla T, Bhadra S, Yao D, Kim NH, Bose S, Lee JH (2010) Recent advances in graphene based polymer composites. Prog Polym Sci 35(11):1350–1375
6. Soldano C, Mahmood A, Dujardin E (2010) Production, properties and potential of graphene. Carbon 48(8):2127–2150
7. Snook GA, Kao P, Best AS (2011) Conducting-polymer-based supercapacitor devices and electrodes. J Power Sources 196(1):1–12
8. Geim AK, Novoselov KS (2010) The rise of graphene. In: Nanoscience and technology: a collection of reviews from nature journals, pp 11–19
9. Rao CEE, Sood AE, Subrahmanyam KE, Govindaraj A (2009) Graphene: the new two-dimensional nanomaterial. Angewandte Chemie Int Edn 48(42):7752–7777
10. Dikin DA, Stankovich S, Zimney EJ, Piner RD, Dommett GH, Evmenenko G, Nguyen ST, Ruoff RS (2007) Preparation and characterization of graphene oxide paper. Nature 448(7152):457–460
11. Gadipelli S, Guo ZX (2015) Graphene-based materials: synthesis and gas sorption, storage and separation. Prog Mater Sci 69:1–60
12. Novoselov KS, Geim AK, Morozov S, Jiang D, Katsnelson, MI, Grigorieva I, Dubonos, SV, Firsov AA (2005) Two-dimensional gas of massless Dirac fermions in grapheme. Nature 438(7065):197–200
13. Stankovich S, Dikin DA, Dommett GH, Kohlhaas KM, Zimney EJ, Stach EA, Piner RD, Ngyuyen ST, Ruoff RS (2006) Graphene-based composite materials. Nature 442(7100):282–286
14. Araújo MA, Cunha AM, Mota M (2004) Enzymatic degradation of starch-based thermoplastic compounds used in protheses: identification of the degradation products in solution. Biomaterials 25(13):2687–2693
15. Teramoto N, Motoyama T, Yosomiya R, Shibata M (2003) Synthesis, thermal properties, and biodegradability of propyl-etherified starch. Eur Polymer J 39(2):255–261
16. Averous L (2004) Biodegradable multiphase systems based on plasticized starch: a review. Journal of Macromolecular Science Part C: Polymer Reviews 44(3):231–274
17. Ashori A (2014) Effects of graphene on the behavior of chitosan and starch nanocomposite films. Polym Eng Sci 54(10):2258–2263
18. Liu H, Xie F, Yu L, Chen L, Li L (2009) Thermal processing of starch-based polymers. Prog Polym Sci 34(12):1348–1368
19. Ramesh M, Mitchell JR, Jumel K, Harding SE (1999) Amylose content of rice starch. Starch-Stärke 51(8–9):311–313

20. Tomasik P, Schilling CH (2004) Chemical modification of starch. Adv Carbohydr Chem Biochem 59:175–403
21. Morrison WR, Milligan TP, Azudin MN (1984) A relationship between the amylose and lipid contents of starches from diploid cereals. J Cereal Sci 2(4):257–271
22. Whistler RL, BeMiller JN, Paschall EF (eds) (2012) Starch: chemistry and technology. Academic Press
23. Soulaka AB, Morrison WR (1985) The amylose and lipid contents, dimensions, and gelatinisation characteristics of some wheat starches and their A-and B-granule fractions. J Sci Food Agric 36(8):709–718
24. Lim ST, Kasemsuwan T, Jane JL (1994) Characterization of phosphorus in starch by 31P-nuclear magnetic resonance spectroscopy. Cereal Chem 71(5):488–493
25. Jane J, Kasemsuwan T, Chen JF, Juliano BO (1996) Phosphorus in rice and other starches. Cereal Foods World 41(11):827–832
26. Lii CY, Tsai ML, Tseng KH (1996) Effect of amylose content on the rheological property of rice starch. Cereal Chem 73(4):415–420
27. Kasemsuwan T, Jane JL (1996) Quantitative method for the survey of starch phosphate derivatives and starch phospholipids by 31P nuclear magnetic resonance spectroscopy. Cereal Chem 73(6):702–707
28. Kim YS, Wiesenborn DP, Orr PH, Grant LA (1995) Screening potato starch for novel properties using differential scanning calorimetry. J Food Sci 60(5):1060–1065
29. Leach HW, McCowen DL, Schoch TJ (1959) Swelling and solubility patterns of various starches and structure of starch granule. Cereal Chem 36:534–544
30. Yarahmadi E, Didehban K, Sari MG, Saeb MR, Shabanian M, Aryanasab F, Zarrinataj P, Reza Paran SM, Mozafari M, Rallini M, Puglia D (2018) Development and curing potential of epoxy/starch-functionalized graphene oxide nanocomposite coatings. Prog Org Coat 119:194–202
31. Lu DR, Xiao CM, Xu SJ (2009) Starch-based completely biodegradable polymer materials. Express Polym Lett 3(6):366–375
32. Stepto RFT (2003) The processing of starch as a thermoplastic. Macromol Symposia 201(1):203–212
33. Wang P, Tang S, Sheng F, Cai J, Fei P, Nawaz A, Walayat N, Javaid AB, Xiong H (2019) Crystallization, thermal stability, barrier property, and aging resistance application of multi-functionalized graphene oxide/poly (lactide)/starch nanocomposites. Int J Biol Macromol 132:1208–1220
34. Khan B, Niazi M, Samin G, Jahan Z (2017) Thermoplastic starch: A possible biodegradable food packaging material—A review. J Food Process Eng 40(3):e12447
35. Zhu F (2015) Composition, structure, physicochemical properties, and modifications of cassava starch. Carbohyd Polym 122:456–480
36. Jun CL (2000) Reactive blending of biodegradable polymers: PLA and starch. J Polym Environ 8(1):33–37
37. Mani R, Bhattacharya M (2001) Properties of injection moulded blends of starch and modified biodegradable polyesters. Eur Polymer J 37(3):515–526
38. Wang N, Yu J, Ma X (2008) Preparation and characterization of compatible thermoplastic dry starch/poly (lactic acid). Polym Compos 29(5):551–559
39. Park JW, Im SS, Kim SH, Kim YH (2000) Biodegradable polymer blends of poly (L-lactic acid) and gelatinized starch. Polym Eng Sci 40(12):2539–2550
40. Martin O, Avérous L (2001) Poly (lactic acid): plasticization and properties of biodegradable multiphase systems. Polymer 42(14):6209–6219
41. Li S, Vert M, Scott G, Gilead D (1995) Degradable polymers: principles and applications. Chapman & Hall, London, p 43
42. Vikman M, Hulleman SHD, Van D, Zee M, Myllärinen P, Feil H (1999) Morphology and enzymatic degradation of thermoplastic starch–polycaprolactone blends. J Appl Polym Sci 74(11):2594–2604

43. Dubois P, Krishnan M, Narayan R (1999) Aliphatic polyester-grafted starch-like polysaccharides by ring-opening polymerization. Polymer 40(11):3091–3100
44. Singh RP, Pandey JK, Rutot D, Degée P, Dubois P (2003) Biodegradation of poly (ε-caprolactone)/starch blends and composites in composting and culture environments: the effect of compatibilization on the inherent biodegradability of the host polymer. Carbohyd Res 338(17):1759–1769
45. Vertuccio L, Gorrasi G, Sorrentino A, Vittoria V (2009) Nano clay reinforced PCL/starch blends obtained by high energy ball milling. Carbohyd Polym 75(1):172–179
46. Averous L, Moro L, Dole P, Fringant C (2000) Properties of thermoplastic blends: starch–polycaprolactone. Polymer 41(11):4157–4167
47. Chiellini E, Corti A, D'Antone S, Solaro R (2003) Biodegradation of poly (vinyl alcohol) based materials. Prog Polym Sci 28(6):963–1014
48. Tudorachi N et al (2000) Testing of polyvinyl alcohol and starch mixtures as biodegradable polymeric materials. Polym Testing 19(7):785–799
49. Lawton JW (1996) Effect of starch type on the properties of starch containing films. Carbohyd Polym 29(3):203–208
50. Ke T, Sun XS (2003) Starch, poly (lactic acid), and poly (vinyl alcohol) blends. J Polym Environ 11(1):7–14
51. Zhai M, Zhao L, Yoshii F, Kume T (2004) Study on antibacterial starch/chitosan blend film formed under the action of irradiation. Carbohyd Polym 57(1):83–88
52. Bourtoom T, Chinnan MS (2008) Preparation and properties of rice starch–chitosan blend biodegradable film. LWT-Food Science and Technology 41(9):1633–1641
53. Cao X, Chen Y, Chang PR, Muir AD, Falk G (2008) Starch-based nanocomposites reinforced with flax cellulose nanocrystals. Express Polym Lett 2(7):502–510
54. Bao J, Xing J, Phillips DL, Corke H (2003) Physical properties of octenyl succinic anhydride modified rice, wheat, and potato starches. J Agric Food Chemi 51(8):2283–2287
55. Zheng P, Ma T, Ma X (2013) Fabrication and properties of starch-grafted graphene nanosheet/plasticized-starch composites. Ind Eng Chem Res 52(39):14201–14207
56. Usman A, Hussain Z, Riaz A, Khan AN (2016) Enhanced mechanical, thermal and antimicrobial properties of poly (vinyl alcohol)/graphene oxide/starch/silver nanocomposites films. Carbohyd Polym 153:592–599
57. Potts JR, Dreyer DR, Bielawski CW, Ruoff RS (2011) Graphene-based polymer nanocomposites. Polymer 52(1):5–25
58. Hu Y, Sun X (2013) Chemically functionalized graphene and their applications in electrochemical energy conversion and storage. Adv Graphene Sci 161–189
59. Ma J, Liu C, Li R, Wang J (2012) Properties and structural characterization of oxide starch/chitosan/graphene oxide biodegradable nanocomposites. J Appl Polym Sci 123(5):2933–2944
60. Li R, Liu C, Ma J (2011) Studies on the properties of graphene oxide-reinforced starch biocomposites. Carbohyd Polym 84(1):631–637
61. Ma T, Chang PR, Zheng P, Ma X (2013) The composites based on plasticized starch and graphene oxide/reduced graphene oxide. Carbohyd Polym 94(1):63–70
62. Ferreira WH, Andrade CT (2015) Characterization of glycerol-plasticized starch and graphene oxide extruded hybrids. Ind Crops Prod 77:684–690
63. Dilara PA, Briassoulis D (2000) Degradation and stabilization of low-density polyethylene films used as greenhouse covering materials. J Agric Eng Res 76(4):309–321
64. Dave AM, Mehta MH, Aminabhavi TM, Kulkarni AR, Soppimath KS (1999) A review on controlled release of nitrogen fertilizers through polymeric membrane devices. Poly-Plast Technol Eng 38(4):675–711
65. Guo M, Liu M, Zhan F, Wu L (2005) Preparation and properties of a slow-release membrane-encapsulated urea fertilizer with superabsorbent and moisture preservation. Ind Eng Chem Res 44(12):4206–4211
66. Qian C, Sun J, Yang J, Gao Y (2015) Flexible organic field-effect transistors on biodegradable cellulose paper with efficient reusable ion gel dielectrics. Rsc Adv 5(19):14567–14574

67. Li J, Zhao Y, Wang N, Ding Y, Guan L (2012) Enhanced performance of a MnO_2–graphene sheet cathode for lithium ion batteries using sodium alginate as a binder. J Mater Chem 22(26):13002–13004

Surface Functionalization of Graphene Based Polyhydroxyalkanoates Nanocomposites and Their Applications

Shreya Sharma, Shashank Shekhar, Anjana Sarkar, and Amit Kumar

Abstract Carbon nanotechnology coupled with polymer science has risen as an exceptionally encouraging device that could concurrently aggrandize all the biological and physical characteristics of polyhydroxyalkanoates. Nanoparticles with a grafted layer of polymers have become pervasive structures for implementation from the biomedical field to the production of sustainable energy. The design of carbon nanomaterials and surface modification delivers potentiality to invigorate logical endeavors in order to create a progressively ingenious community proficient of tackling the challenges. The inclusion of graphene in the matrix of polyhydroxyalkanoate manifests resistance against thermal degradation enhanced mechanical properties, lower gas permeability, and increment in electrical conductivity. Howbeit, several limitations restrict the competitiveness of graphene based polyhydroxyalkanoates with traditional synthetic nanocomposites or their implication as ideal biomaterials. To outmaneuver the drawbacks, modifications are a requisite to corroborate enhanced performance in distinct applications. Here, we detail the structure, function, review the functionalization routes, and discuss the current challenges looking forward toward future development directions.

Keywords Polyhydroxyalkanoates · Graphene · Surface functionalization · Nanocomposites

1 Introduction

One of the incredible accomplishments of science is the advancement of strategies which allow us to perceive the intended meaning of matter and alters its atomic structure. The alteration of several materials and their surfaces at the nanoscale has been

S. Sharma · S. Shekhar (✉) · A. Sarkar
Department of Chemistry, Netaji Subhas University of Technology, New Delhi, India
e-mail: shashanks634@gmail.com

A. Kumar
Department of Chemistry, University of Delhi, New Delhi, India

B. Sharma and P. Jain (eds.), *Graphene Based Biopolymer Nanocomposites*,
Composites Science and Technology,
https://doi.org/10.1007/978-981-15-9180-8_10

utilized in neoteric decades, particularly by nano scientists and technologists. Design of carbon nanostructures and surface modification delivers potentiality to invigorate logical endeavors in accordance to the creation of progressively ingenious community proficient for tackling the challenges. The implementation of carbonic nanostructures to several disciplines has been expedited by the functionalization of their surfaces and the significance of amended nano surfaces is very much contemplated in scientific work of the last decades. The superlative physiological chemistry of these functionalized nano structured materials is utilized for energy, cancer therapy, antimicrobial drug design and development, delivery in biological frameworks, biotechnological applications and aviation. The functionalization of nano structured materials with the essential moieties is reliant on the science of substructure utilized. In succession, the method of functionalization on nano surfaces essentially relies upon the essence of the issue and purposive utilization of the material. Lately, proficient strategies have been elaborated, the maximum of which are delegated either covalent or non-covalent functionalization [1].

Nanoparticles with a grafted layer of polymers have become pervasive structures for implementation from the biomedical field to the production of sustainable energy. The end-tethered polymer anchors are frequently utilized to alter the characteristics of nanoparticles, including, stability, dispersibility, compatibility towards solvent and assembly. Polymer chains fastened to latex nanoparticles speak to a specific challenge since both the canopy of end-fastened chains and the center of the nanoparticles are delicate and deformable. Consequently, determining the perfect polymer layer to improve the conduct of the nanoparticles for a particular application is conspicuous and standards are required to reasonably design such polymer canopy [2]. Graphene is a nanofiller with a sheet of a monolayer of carbon iotas implying sp^2 bonded atoms in a honeycomb crystal structure. The stiffest and most grounded studied material has Young's modulus and tensile strength of up to 1 TPa and 130 GPa, respectively. The measured electrical and thermal conductivities are approximately 7200 Sm^{-1} and 1800 Wm^{-1} K^{-1}, respectively [3, 4]. The oxidized form of graphene, graphene oxide comprises of carbonyl, epoxide, and hydroxyl groups moieties superficially that permit arrangement of hydrogen bonds. Reduced graphene oxide is synthesized synthetically or thermally, with a lower oxygen content comparative to GO [4]. Graphene and its subsidiary products have received consideration appreciably for their incredible physicochemical properties, including, high strength, ease of functionalization, elasticity, high surface area to volume ratio, and chemically non-reactive [5–7].

Polyhydroxyalkanoates (PHAs) constitute a class of biodegradable polyesters produced by a broad diversity of microbes for intracellular carbon and vitality stockpiling. PHAs are stored in the cytoplasm as granules ranging in size from 0.2 to 0.5 μm. PHAs are classified based on the number of carbon monomeric units into short-chain-length PHAs comprising of 3–5 carbon monomeric units, medium chain-length PHAs comprising of 6–14 carbon monomeric units, and long-chain-PHAs comprising of almost 14 carbon monomeric units in the 3-hydroxyalkanoate unit [8, 9]. Several undesirable characteristics limit the use of PHAs. For instance, poly (3-hydroxybutyrate) lacks sound mechanical properties due to the presence of large crystals. This renders poly (3-hydroxybutyrate) unacceptable for several uses, including biomedical and packaging applications. Additionally, the poor thermal stability of poly (3-hydroxybutyrate proffers incompatibility towards traditional thermal processing strategies and susceptible to thermal degradation. Lack of advanced functionality, high hydrophobicity, and significant expense has additionally confined further development of its usage. As to the biodegradability of PHAs, the profoundly hydrophobic nature of unmodified PHAs and slow rate of degradation are not preferred in several leading biomedical implications. Customization of PHAs to accomplish better minimal cost and better conduct including sound mechanical properties, crystallinity, amphiphilicity, degradation rate, and surface features is a requisite to satisfy explicit applications [9].

Carbon nanotechnology coupled with polymer science has risen as an exceptionally encouraging device that could concurrently aggrandize all the biological and physical characteristics of PHA. The inclusion of graphene manifests resistance against thermal degradation enhanced mechanical properties, lower gas permeability, and increment in electrical conductivity. Theoretical and experimental results have demonstrated graphene's remarkable prospect for supporting variety of functionalities in a class of polymers that is profoundly esteemed for its eco-accommodating and biocompatible characteristics [3, 10, 11]. The production of graphene derived poly (hydroxy alkanoates) production has been delineated in Fig. 1.

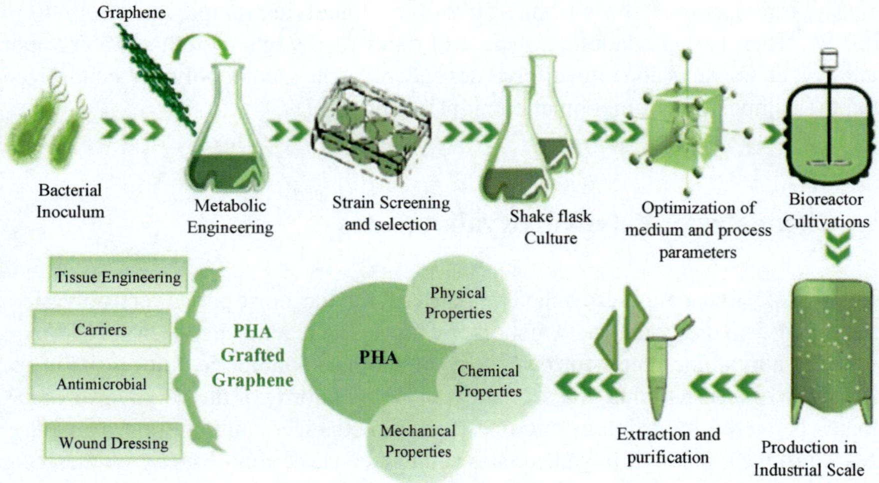

Fig. 1 Graphene grafted poly (hydroxyalkanoate) production

2 Polyhydroxyalkanoates

Polyhydroxyalkanoates are a broad class of linear thermoplastic polymers, an adherent of bio-based bioplastic category. Produced by different microorganisms as carbon and energy stocks, the monomers of varying length are produced. Structurally, the hydroxyacid monomers are found with R-pendant groups ranging from methyl to tridecyl. The most common way for production at industrial scale includes the utilization of pure cultures of natural or engineered microbial strains. Biosynthetic routes can be divided into two distinct steps including the generation of hydroxyacyl-CoA through inherent metabolism from structurally related carbon source or unrelated carbon source and polymerization of hydroxyacyl-CoA into polyhydroxyalkanoates by Polyhydroxyalkanoate synthase. However, the prerequisite of a sterile environment make the process expensive and complicated. Poly (hydroxybutyrate) is tantamount to polypropylene and demonstrates great protection from dampness and procure phenomenal hindrance properties to gases. In addition to their biocompatibility and biodegradability, polyhydroxyalkanoates are insoluble in water, have great protection from hydrolytic assault, impervious to ultraviolet rays, sink in water and, act as piezoelectric materials [12–14].

Biodegradation of such bio polyesters is determined by the kind and composition of the polymer, ecological constraints and the kind of microbes. Polyhydroxyalkanoates are dissolvable in chloroform and other chlorinated solvents. Their glass transition temperature shifts from −50 to 4 °C, liquefying temperature from 40 to 180 °C. Thermo degradation temperature, rigidity, Young's modulus, water vapor and oxygen transmission rate differ as dependent on the kind of polymer synthesized and the composition of monomeric unit [15, 16].

3 Chemistry of Polyhydroxyalkanoates

Polyhydroxyalkanoates researched to date are for the most part direct polyesters containing 3-hydroxy fatty acid monomeric units. The reaction between the carboxyl group of a monomer and hydroxyl group of another monomeric unit results in the formation of an ester bond. Subsequently, the vast majority of the polyhydroxyalkanoates delivered by different microbes are of rectus (R) configuration. Nevertheless, sinister (S) polyhydroxyalkanoates could likewise be combined by synthesizing copolymer with some stereo polymers. The hydroxyl subbed carbon particle is of the R configuration. However, exceptions may exist in a couple of situations where chirality is absent. There is incredible variety in the alkyl side chain, which might comprise of halogenated, aromatic, epoxidized or other expanded or other branched monomeric units. These varieties in the size and synthesis of side-chain substituents are liable for chemical alteration of polyhydroxyalkanoates and for their distinct applications [17, 18].

4 Graphene-Based Polyhydroxyalkanoates

Graphene, a 2D atomic sheet framed via carbon iotas orchestrated uniformly in hexagonal structures, with brilliant electrical and thermal conductivity has been investigated for the enhancement of characteristics exhibited by polyhydroxyalkanoates. Graphene is a truly steady compound and it has been demonstrated to be biocompatible, improving the cell adhesion of osteoblasts and mesenchymal cells. The wonder material has resistance multiple times better than that of steel and has extraordinary characteristics of versatility, flexibility, just as incredible mechanical properties. The graphene family comprises of materials emerging from chemical or physical modifications of graphene. For a variety of applications, it is vital to change over graphene into different mixes on the grounds that, for instance, unadulterated graphene has demonstrated a specific cytotoxicity, which could initiate cell apoptosis and a decreased cell adhesion. Graphene oxide has displayed very significant antibacterial activity against a wide range of bacteria. In this way, graphene oxide gives the chance of fortifying poly-(3-hydroxybutyrate-co-3-hydroxyvalerate (the most common member oh polyhydroxyalkanoate), going about as a nucleating operator for crystallization, which significantly enhances the mechanical characteristics of composite materials and expands the temperature of most extreme debasement. Moreover, GO doesn't meddle with the biodegradability of poly-(3-hydroxybutyrate-co-3-hydroxyvalerate), but could confine the versatility of polymer chains in the crystal growth process. Poly-(3-hydroxybutyrate-co-3-hydroxyvalerate) has likewise been strengthened with the incorporation of reduced graphene oxide. On account of its piezoelectricity and physiological characteristics, polyhydroxyalkanoate might be more qualified for biomedical use by consolidating it with graphene [19–21].

Besides, unsaturated polymeric chains of polyhydroxyalkanoates could be directly grafted to graphene (Fig. 1) with an active thiol by a click reaction between thiol and alkene bond, while graphene oxide could likewise be directly esterified with polyhydroxyalkanote comprising a hydroxyl gathering. On account of graphene's well circulation inside polymer lattice, this copolymer has better conductivity, which is of incredible centrality for the advancement of biodegradable conductive polyhydroxalkanoate-based materials [22]. Incorporation of thermally reduced graphene into medium chain length polyhydroxyalkanotes resulted into enhancement of melting temperature by 1–3 °C, modulus by 200−590%, and the electrical conductivity by >7 orders of magnitude. The results also inferred that solid interfacial interactions build the level of modulus improvement in these group of biopolyesters and the utilization of graphene oxide with its greater fraction of hydroxyl, carbonyl, and epoxide moieties might be an approach to additionally enhance interfacial fascination with medium chain length poly (hydroxyalkanotes) [10]. Incorporation of graphene nanoparticles in poly (3-hydroxybutyrate-co-4hydroxybutyrate) resulted in elevated storage modulus, non-isothermal melt crystallization peak temperature values and overall isothermal melt crystallization rates when compared to pristine polymer [23]. Upshift in glass transition rate, thermal stability, and modulus was reported on incorporation of graphene onto Poly-(R)-3 hydroxyoctanoate [5].

5 Surface Modification Techniques

5.1 Covalent Surface Functionalization

Significant approaches to alter the connection of bond are brought about by covalent functionalization. The change of hybridization from sp^2 to sp^3 distorts the translational evenness of graphene and the characteristics of nanofillers, as both the transport and electronic properties are affected. The covalent linkages are established between the functional units with the skeleton of the composite [6, 7].

The presence of appropriate functional groups is the absolute step towards developing polymers from the surface. Numerous polymerization techniques, for instance, reversible addition-fragmentation chain transfer polymerization (RAFT), atom transfer radical polymerization (ATRP), ring-opening polymerization (ROP), nitroxide mediated radical polymerization (NMRP), and anionic polymerization techniques have been utilized to functionalize graphene. In the course of recent years, a few examinations have been accounted for the utilization of hybrid polymers, permitting the accomplishment of alluring properties from the subsequent materials. Two fundamental methodologies, "grafting to" and "grafting from", have been accounted for the covalent grafting of polymers. The "grafting to" methodology depends on the binding of as-arranged or commercially accessible polymer atoms onto the graphene surface by an assortment of concoction responses, for instance, esterification, amidation, radical coupling, and so on. The polymer must have reasonable reactive functional gatherings for the design of nanocomposites. The "grafting to" methodology permits joining a polymer with the functional end/side gatherings (for instance, $-NH^2$, –OH, –COCl, and –COOH) (Fig. 2). However, easy to carry with both linear and dendritic polymers, the joining effectiveness is in every case low attributable to the steric deterrent of the pre-joined macromolecular chains. If occurrence of "grafting from" methodology arises, the polymer is bound to the graphene surface by in-situ polymerization of monomers in nearness of responsive graphene or graphene upheld initiators. The primary bit of leeway of this methodology is that the carbon nanomaterial based polymer nanocomposites can be set up with high joining thickness. This methodology has been utilized effectively to unite polymers through radical, cationic, anionic, and ring-opening polymerization. For down to earth applications, the functionalization of nanocomposites should be controlled. In this manner, there is a genuine requirement for unique concoction responses that permit the change of nanocomposites straightforwardly. Cycloaddition reactions have emerged as prominent routes [24–28].

Inferable from their magnetic, electrical, and, catalytic properties, metallo nanoparticles with characteristic shapes and sizes have acquired impressive consideration. Nanoparticles with a controlled shape or size are synthesized by diminishing metal forerunners within the sight of organic surface-topping operators. These surface-topping operators may frequently ruin further utilizations of the nanostructured materials, forbidding dispersion in several variety of solvent or on the other hand making surface functionalization progressively troublesome. Strategies that

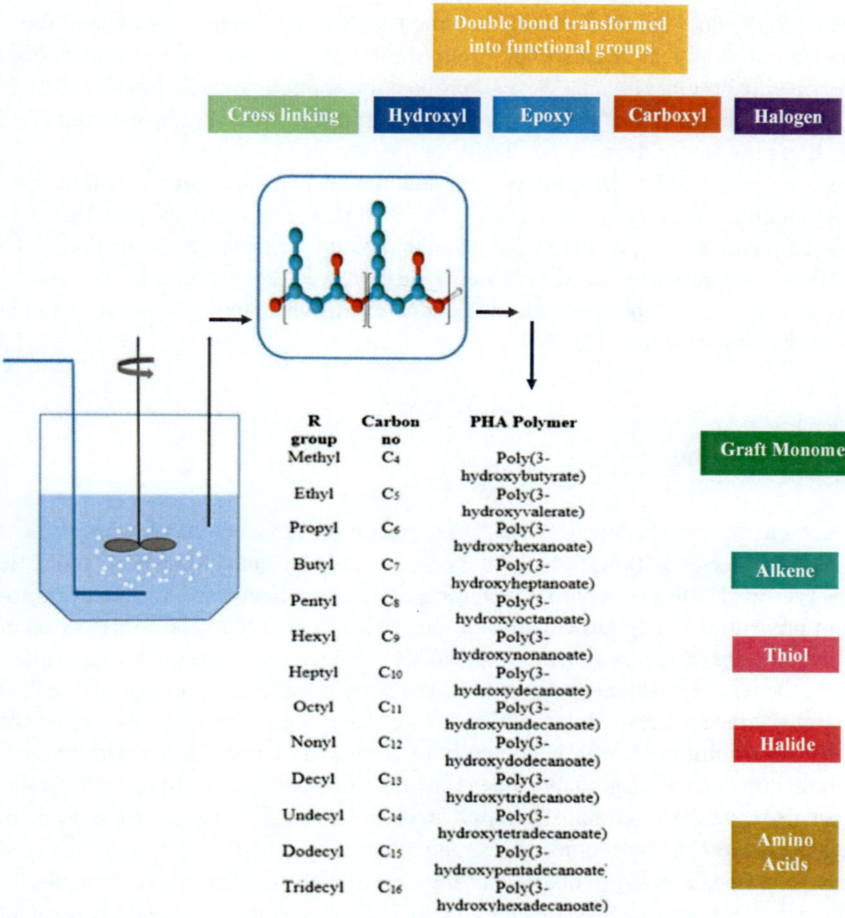

Fig. 2 Introduction of functional groups into variant structure of PHA

encourage basic and widespread surface modifications are critical to different implications. Specifically, the immobilization of shape-controlled nanostructured materials on aid can additionally expand the potentiality. Metal catalysts bolstered via carbon nanocomposites have astounding trademark properties, inferable from their high surface area, thermal stability, and permeable surface, which goes about as a platform to forestall the agglomeration of the immobilized metallic particles [29–31].

5.2 Non-covalent Surface Functionalization

The major drawback of covalent functionalization includes the permanent destruction of the structure that further influences huge changes in their physical properties.

Stacking and water-soluble bonding assume a significant job in non-covalent functionalization. Conjugation can occur through non-covalent associations by appending some specific optional ligands. Several propositions of non-covalent binding include electrostatic interaction, $\pi-\pi$ stacking and entrapping biomolecules in biocompatible films like phospholipids, polymer, DNA, and so on.

Surface coating likewise confers extra functionality to nanoparticle as it permits ligand/medication binding. As coating may disturb the magnetic/physical properties of nanocomposites, it ought to be taken in thought for imaging applications. The modification of the surface by deposition of the inorganic/organic layer is the epitaxial approach. Epitaxial approach gives a more extensive extent of multifunctional nanoprobes improvement [32–34].

6 Applications

Surface chemistry is critical to the biocompatibility and conduct of biomedical gadgets. Surface alteration strategies are consequently expected to stifle antagonistic impacts as well as trigger a particular biological reaction in vivo, for instance promotion or prevention of cell growth, prevention of blood clot, acceptance or rejection of explicit proteins, cell adhesion or satisfactorily arrangement of the growth of explicit cells in the tri-dimensional space. It is additionally beneficial to control the surface wettability, since modestly hydrophilic surfaces have been indicated ideal in view of its biocompatibility [35, 36]. Nilkamal et al. developed a non-toxic, antimicrobial, biocompatible and biodegradable magnetic Fe_3O_4/RGO-g-Poly(3-hydroxybutyrate-co-3-hydroxyvalerate) composite based porous 3D scaffold to design sensors in imaging area for tissue engineering applications. The functionality in the reduced graphene oxide inferred a marvelous for consolidation of magnetic particles to achieve implementations in biomedical field and gadgets, optics to electrochemical vitality capacity and conversion. The supermagnetic behavior of magnetic nanoparticles allows it to be utilized in MRI/multimodal imaging in biological implications [35].

Avossa et al. utilized porphyrin as a polymer doping agent substitute to functionalize the polymeric matrix of poly (hydroxybutarate). The remarkable sensing properties of porphyrin were exploited for the design and creation of a nanofibrous chemical sensor. The surface modification of poly (hydroxybutarate) grafted graphene with porphyrin produced chemical sensor with the ability to be applied for multifaceted environments [36]. Low strength, poor resistance against heat, vulnerability to microbial attack, etc. limit the utilization of polyhydroxyalkanoate polymer in packaging industries. However, high performance polymer nanocomposites with potent antibacterial activity have proven effective to overcome such problems. Graphene and its derivatives dispersed in a good extent can serve as an efficient filler to provide enhanced mechanical properties. Pengwu Xu et al. designed Go-grafted-poly (hydroalkanoate) nanocomposite coupled with surface alteration brought about by grafting long alkyl chain quaternary salts (LAQ). LAQ exhibits antibacterial as well as hydrophobic characteristics. Therefore, the multifunctional nanofiller in contemplation to the combined advantages provided resistance against heat, reduction of oxygen permeation by 86% and marvelous antibacterial and mechanical properties to be utilized in packaging applications [37–39].

7 Challenges and Limitations

The functionalization science of graphene grapfted polyhydroxyalkanoate has an incredible possibility and is gaining a significant energy. The functionalization by well-defined polymers with precise control over molecular weight, polydispersity has its ability to introduce different architectures onto the graphene surface [40–42]. Production of tailor made materials based on the detailed information of their complex structure property relationships for fulfilling the demands in optoelectronic, energy, and engineering materials can be achieved. Some of the major drawbacks are listed;

- The blending of polymers is complicated by interfacial interactions between the two polymers. Poor interfacial interactions often cause the deterioration of properties and subsequently poorer mechanical behavior.
- Substrate choice limits self-assembled monolayer choice which further limits chemical methodologies.
- Stability of chemically modified substrates. Reactivity, mechanisms, techniques and kinetics involved in the chemical modification would be useful for probing structure, function and stability relationship of the nanocomposite.
- Use of harsh chemicals to functionalize pose environmental concerns for generation of toxic waste when chemicals are applied on the industrial scale, risking cross-reactivity downstream.
- A major drawback concerns the control of both molecular weights and polydispersity of the immobilized chain. Controlled polymerization techniques and fabricating techniques are a requisite.
- Durability of the target surface to maintain a longer life opportunity is a major concern among the scientists working in the field.
- Covalent functionalization on graphene surface deteriorates the properties related to the transport of electrons or phonons due to the conversion of sp^2 carbon into sp^3 carbon.

8 Conclusion and Future Perspectives

The discovery of graphene as nanofiller has prompted the creation of minimal effort, light-weight, and high-support nanostructured composites, driving the path to a broad scope of inventive implications. The nearness of amino, carboxyl, or hydroxyl bunches in the biopolymers prompts a very productive interaction with the functional gatherings on graphene and graphene oxide. These materials have condition cordial effect in different novel applications that develop on the grounds that the measure of graphene in the film is generally exceptionally small 5%. Unlike synthetic polymers, the grafting procedures for consolidating graphene subsidiaries to biobased polymers have not been broadly researched. The territory additionally should be investigated as it might enhance the performance of nanocomposites contrasted with the solution blending and melt blending strategies.

Graphene grafted bacterial PHAs have an assortment of valuable physical properties making them potential materials for a variety of applications. Be that as it may, absence of functional moieties makes them badly assorted for utilization in several applications. Surface alteration strategies have been the subject of esteemed interest among scientists, with the point of lessening the undesirable properties. For the functionalization of graphene and GNs-based progressed nanoplatforms and biocomposites by means of surface functionalization, however exceptional advancement has been practiced, a considerable number of current methodologies still need to be improved. Covalent surface functionalization is preferred substantially. The soundness of non-covalently functionalized frameworks in physiological arrangement despite everything should be affirmed both in long haul and in vivo which may likewise move new ideas for the functionalization of graphene and its subordinates.

Acknowledgements Authors would like to acknowledge Netaji Subhas University of Technology, Delhi, India.

References

1. Punetha VD, Rana S, Yoo HJ, Chaurasia A, McLeskey JT Jr, Ramasamy MS, Sahoo NJ, Cho JW (2017) Functionalization of carbon nanomaterials for advanced polymer nanocomposites: a comparison study between CNT and graphene. Prog Polym Sci 67:1–47
2. Wichaita W, Kim YG, Tangboriboonrat P, Therien-Aubin H (2020) Polymer-functionalised polymer nanoparticles and their behaviour in suspensions. Polym Chem 11(12):2119–2128
3. Chee WK, Lim HN, Huang NM, Harrison I (2015) Nanocomposites of graphene/polymers: a review. RSC Adv 5(83):68014–68051
4. Shrivastava NK, Saidi MAA, Othman N, Zurina M, Hassan A (2019) Fillers and reinforcements for advanced nanocomposites. In: Bio-based polymers and nanocomposites, Springer, Cham, pp 29–48
5. Abdala A, Barrett J, Srienc F (2013) Synthesis of poly-(R)-3 hydroxyoctanoate (PHO) and its graphene nanocomposites. In: Green polymer chemistry: biocatalysis and materials II. American Chemical Society, pp 199–209
6. Usman A, Hussain Z, Riaz A, Khan AN (2016) Enhanced mechanical, thermal and antimicrobial properties of poly (vinyl alcohol)/graphene oxide/starch/silver nanocomposites films. Carbohyd Polym 153:592–599
7. Thiruppathi R, Mishra S, Ganapathy M, Padmanabhan P, Gulyás B (2017) Nanoparticle functionalization and its potentials for molecular imaging. Adv Sci 4(3):1600279
8. Raza ZA, Abid S, Banat IM (2018) Polyhydroxyalkanoates: characteristics, production, recent developments and applications. Int Biodeterior Biodegradation 126:45–56
9. Li Z, Yang J, Loh XJ (2016) Polyhydroxyalkanoates: opening doors for a sustainable future. NPG Asia Mater. 8(4):e265

10. Barrett JS, Abdala AA, Srienc F (2014) Poly (hydroxyalkanoate) elastomers and their graphene nanocomposites. Macromolecules 47(12):3926–3941
11. Rivera-Briso AL, Aachmann FL, Moreno-Manzano V, Serrano-Aroca Á (2020) Graphene oxide nanosheets versus carbon nanofibers: enhancement of physical and biological properties of poly (3-hydroxybutyrate-co-3-hydroxyvalerate) films for biomedical applications. Int J Biol Macromol 143:1000–1008
12. Gadgil BST, Killi N, Rathna GV (2017) Polyhydroxyalkanoates as biomaterials. MedChemComm 8(9):1774–1787
13. Mohapatra S, Maity S, Pati S, Dash A, Samantaray DP (2020) Polyhydroxyalkanoates: the green polymer. In: Frontiers in soil and environmental microbiology
14. Możejko-Ciesielska J, Kiewisz R (2016) Bacterial polyhydroxyalkanoates: still fabulous? Microbiol Res 192:271–282
15. Mathuriya AS, Yakhmi JV (2017) Polyhydroxyalkanoates: biodegradable plastics and their applications. In: Handbook of ecomaterials, pp 1–29
16. Kovalcik A, Obruca S, Fritz I, Marova I (2019) Polyhydroxyalkanoates: their importance and future. BioResources 14(2):2468–2471
17. Raza ZA, Riaz S, Banat IM (2018) Polyhydroxyalkanoates: properties and chemical modification approaches for their functionalization. Biotechnol Prog 34(1):29–41
18. Halevas EG, Pantazaki AA (2018) Polyhydroxyalkanoates: chemical structure. In: Biosynthesis, chemical structures and applications, p 133
19. Sun J, Shen J, Chen S, Cooper MA, Fu H, Wu D, Yang Z (2018) Nanofiller reinforced biodegradable PLA/PHA composites: current status and future trends. Polymers 10(5):505
20. Yu X, Cheng H, Zhang M, Zhao Y, Qu L, Shi G (2017) Graphene-based smart materials. Nat Rev Mater 2(9):1–13
21. Rivera-Briso AL, Serrano-Aroca Á (2018) Poly (3-hydroxybutyrate-co-3-hydroxyvalerate): enhancement strategies for advanced applications. Polymers 10(7):732
22. Luo Z, Wu YL, Li Z, Loh XJ (2019) Recent progress in polyhydroxyalkanoates-based copolymers for biomedical applications. Biotechnol J 14(12):1900283
23. Ning Z, Xin S, Wu X, Xu K, Han C, Dong L (2016) Crystallization behavior, mechanical properties, and enzymatic degradation of biosourced poly (3-hydroxybutyrate-co-4-hydroxybutyrate)/graphene nanocomposites. J Therm Anal Calorim 124(3):1705–1715
24. Hetemi D, Pinson J (2017) Surface functionalisation of polymers. Chem Soc Rev 46(19):5701–5713
25. Rodríguez-San-Miguel D, Zamora F (2019) Processing of covalent organic frameworks: an ingredient for a material to succeed. Chem Soc Rev 48(16):4375–4386
26. Shekhar S, Sharma R, Sharma S, Sharma B, Sarkar A, Jain P (2020) An exploration of electrocatalytic analysis and antibacterial efficacy of electrically conductive poly (D-glucosamine)/graphene oxide bionanohybrid. Carbohydr Polym 240:116242
27. Mosayebi M, Salehi Z, Doosthosseini H, Tishbi P, Kawase Y (2020) Amine, thiol, and octyl functionalization of GO-Fe_3O_4 nanocomposites to enhance immobilization of lipase for transesterification. Renew Energy 154:569–580
28. Shepherd C (2016) New routes for functionalised nanoparticles for polymer nanocomposites. Doctoral dissertation. University of Oxford
29. Selim MS, Shenashen MA, El-Safty SA, Higazy SA, Selim MM, Isago H, Elmarakbi A (2017) Recent progress in marine foul-release polymeric nanocomposite coatings. Prog Mater Sci 87:1–32
30. Miculescu M, Thakur VK, Miculescu F, Voicu SI (2016) Graphene-based polymer nanocomposite membranes: a review. Polym Adv Technol 27(7):844–859
31. Capek I (2019) Nanocomposite structures and dispersions, vol 23. Elsevier
32. Wang Y, Yang C, Mai YW, Zhang Y (2016) Effect of non-covalent functionalisation on thermal and mechanical properties of graphene-polymer nanocomposites. Carbon 102:311–318
33. Georgakilas V, Tiwari JN, Kemp KC, Perman JA, Bourlinos AB, Kim KS, Zboril R (2016) Noncovalent functionalization of graphene and graphene oxide for energy materials, biosensing, catalytic, and biomedical applications. Chem Rev 116(9):5464–5519

34. Yong-Gyun JUNG, Yoo HJ, Kim JS, Kim JS, Byeon CS (2020) U.S. patent no. 10,544,324. U.S. Patent and Trademark Office, Washington, DC
35. Pramanik N, De J, Basu RK, Rath T, Kundu PP (2016) Fabrication of magnetite nanoparticle doped reduced graphene oxide grafted polyhydroxyalkanoate nanocomposites for tissue engineering application. RSC Adv 6(52):46116–46133
36. Avossa J, Paolesse R, Di Natale C, Zampetti E, Bertoni G, De Cesare F, Scarascia-Mugnozza G, Macagnano A (2019) Electrospinning of polystyrene/polyhydroxybutyrate nanofibers doped with porphyrin and graphene for chemiresistor gas sensors. Nanomaterials 9(2):280
37. Xu P, Yang W, Niu D, Yu M, Du M, Dong W, Chen M, Lemstra PJ, Ma P (2020) Multifunctional and robust polyhydroxyalkanoate nanocomposites with superior gas barrier, heat resistant and inherent antibacterial performances. Chem Eng J 382:122864
38. Mahmood H, Pegoretti A, Brusa RS, Ceccato R, Penasa L, Tarter S, Checchetto R (2020) Molecular transport through 3-hydroxybutyrate co-3-hydroxyhexanoate biopolymer films with dispersed graphene oxide nanoparticles: gas barrier, structural and mechanical properties. Polym Testing 81:106181
39. Suvarnaphaet P, Sasivimolkul S, Sukkasem C, Pukesamsombut D, Tanadchangsaeng N, Boonyagul S, Pechprasarn S (2019) Biodegradable electrode patch made of graphene/PHA for ECG detecting applications. In: 2019 12th biomedical engineering international conference (BMEiCON). IEEE, Nov 2019, pp 1–5
40. Shekhar S, Sarkar A, Sharma B, Jain P (2020) Electrochemical evaluation of functionalized graphene oxide filled PVA-chitosan biohybrid for supercapacitor applications. J Appl Polym Sci 137(17):48610
41. Sharma B, Gautam S, Shekhar S, Sharma R, Rajawat DS, Jain P (2019) Facile synthesis of poly (vinyl alcohol) nanocomposite & its potential application to enhance electrochemical performance. Polym Testing 74:119–126
42. Sharma B, Malik P, Jain P (2019) To study the effect of processing conditions on structural and mechanical characterization of graphite and graphene oxide-reinforced PVA nanocomposite. Polym Bull 76(8):3841–3855

Natural Rubber/Graphene Nanocomposites and Their Applications

K. B. Bhavitha, Srinivasarao Yaragalla, C. H. China Satyanarayana, Nandakumar Kalarikkal, and Sabu Thomas

Abstract The current review focuses on giving a basic understanding of graphene structure, preparation methods, their role in fabricating and the mechanical performance of elastomer and biopolymer composites. Since the physical properties and the performance of graphene reinforced elastomer or biopolymer composites predominantly depend on the rate of dispersion of graphene in the matrix, the physical and chemical interaction of polymer chains with the graphene and the orientation of the graphene within the matrix, here, a thorough study of these topics is carried out. Moreover, the current challenges and future perspectives involving in graphene biopolymer/elastomer composites are also discussed.

Keywords Graphene · Natural rubber · Biopolymer composites

1 Introduction

Elastomers are one of the types of polymers that exhibit elasticity [1]. High elongation, visco elasticity and low modulus with weak intermolecular forces is common properties of most elastomers. Among other properties, elastomers can deform easily upon elongation and exhibit flexibility before mechanical failure. Due to the aforementioned characteristics, elastomers have been utilized for manifold applications

K. B. Bhavitha · N. Kalarikkal · S. Thomas (✉)
International and Inter-University Centre for Nanoscience and Nanotechnology, Mahatma Gandhi University, Kottayam, Kerala 686560, India
e-mail: sabuthomas@mgu.ac.in

S. Yaragalla (✉)
Smart Materials Group, Istituto Italiano Di Tecnologia, Via Morego 30, 16163 Genova, Italy
e-mail: yaragallasrinu@gmail.com

K. B. Bhavitha
Department of Physics, St Teresas's College, Ernakulam, Kerala 682011, India

C. H. China Satyanarayana
Department of Chemistry, S.K.B.R College, Amalapuram, India

B. Sharma and P. Jain (eds.), *Graphene Based Biopolymer Nanocomposites*,
Composites Science and Technology,
https://doi.org/10.1007/978-981-15-9180-8_11

such as industrial, health care and so on. Initially, people from America started using natural rubber a kind of biopolymer [2] but it is not become popularized until the research work on vulcanization in 1839 by Charles Goodyear and Thomas Hancock [3]. At present, polymer industries have been receiving huge industrial and academic interest due to the increase in demand of high quality and engineering materials, which cause a remarkable impact in global revenues, estimated to be 5600 crores Indian rupees by 2020 [4]. These materials were utilized for manufacturing and development of efficient polymeric materials, which have brought revolution in material science and engineering [5–11]. Biopolymer nanocomposites that can be originated from elastomers exhibit excellent improved properties. However, their synthetic methods are differing from the regular polymer nanocomposites; they should undergo crosslinking by means of vulcanization during preparation. Several research groups have produced efficient elastomeric nanocomposites where they introduced different fillers such as layered silicates, silica nanoparticles, carbon black, multi-walled carbon nanotubes [12–22]. These fillers significantly influence the final performance of the nanocomposite and each type of filler having its own impact on the end use applications as they differ from structural and geometrical properties. Importantly, a homogeneous distribution of the fillers inside elastomers is needed for enhanced physical properties such as mechanical, thermal and electrical. To attain this condition, chemical treatment of nanofillers is required to boost the interfacial interactions between filler and polymer as well as filler and filler [23].

Graphene is initially originated from graphite, is a single layer form of graphite and a two dimensional carbon allotrope. It contains hexagonal arrangements of carbon atoms in a honeycomb lattice. It became more popular during the twenty-first century and recognized by scientific and industrial groups throughout the world. Graphene can be utilized in various implications such as energy storage materials, and antistatic coatings and effective reinforcement in polymer materials due to its unprecedented mechanical and electrical properties. It can exhibit high modulus of elasticity, (1 TPa), high electron and thermal conductivity (up to 5000 W/mK) along with 98% optical transmittance [24–26]. Many reports had revealed that incorporation of small quantity of graphene into polymers significantly improve the physical performance of those materials [27–29]. Therefore, small amount of graphene is needed to explore physical performance of various polymers. Many research articles have been reported on graphene based elastomers, thermosets and thermoplastics polymer nanocomposites [30–36]. However, we attempted to review specifically advances regarding graphene based elastomer matrix nanocomposites. Special focus has been given here to target the techniques used in the preparation of graphene and its elastomer nanocomposites. Their mechanical properties with reference to the morphology and the applications of the final materials have been reviewed thoroughly.

2 About Graphene

Geim and Novoselov unveiled in their article [24] that graphene is the initial compound to generate other forms of carbon materials like 0D buckyballs, 1D nanotubes, and 3D graphite. It consists of hexagonal carbon atoms in which each carbon atom undergoes sp^2 hybridization, where the each carbon atoms contains one pure π electron, which is the responsible for electrical conductivity at room temperature. It is one layer structure of graphite; nevertheless, wrinkles can be generated during preparation process of graphene, these may affect the compatibility with polymers.

2.1 Preparation Methods of Graphene

Graphene has gained much interest for fabricating smart nanocomposites owing to its exceptional intriguing properties, which caused for the creating innovative methods for the production of low cost material in large quantities without compromising the quality. The structural features of the synthesized graphene are mainly dependent upon the synthetic method used. Most popular preparation methods for the production of graphene will be disclosed in the following section.

2.2 Mechanical Exfoliation from Graphite

Geim and coworkers employed this method for the first time to prepare monolayer graphene and it has gained much interest in the research community thereby throughout the world [37]. This method includes peeling of graphene layers from natural graphite using 'Scotch' tape. Using this method, the total number of graphene layers can be tuned and the size of the graphene layers in the order of microns and it depends on the grain size of the graphite. Major positive side of this method is that it can produce graphene with high quality properties. Nevertheless, this method cannot be used to produce graphene in large enough quantities for the manufacturing of nanocomposites [38, 39].

2.3 Solvent/Liquid-Phase Exfoliation

This method encompasses exfoliation of the graphite in presence of organic solvents such as dimethylsulfoxide [40], tetrahydrofuran [41], *N*-methyl-pyrrolidone [42], dimethyl-formamide [43], and perfluorinated aromatic solvent [44]) via sonication. This method is very beneficial since the solvent-graphene mixture could utilize the generated energy during the sonication process, which motivates the exfoliation of graphite. Therefore, selection of solvent with suitable surface tension, which is equal or higher than the graphene-graphene interaction, is very crucial to opt this method. Common organic solvents are more effective for producing graphene with this method, but Coleman et al. [45] revealed that aqueous-surfactant suspensions were also could be utilized successfully for exfoliating the graphite. A report revealed that the exfoliation of graphite can be done in the presence of aqueous media and inorganic salt mixtures such as $CuCl_2$ and NaCl [46]. The positive side of this method is that there are many organic solvents that can be utilized for the fabrication of quality graphene associated with commercial applications. Nevertheless, the problem associated with this method is the small yield efficiency. This method is suitable for preparing nanocomposites via solution blending, though usage of harmful solvents is not eco-friendly.

2.4 Thermal Exfoliation

This method involves thermal treatment of graphite to exfoliate graphene at higher temperatures. Rapid microwave heating of acid functionalized graphite results in graphite nanoplatelets with size of 10 nm thickness [47]. Trapped acid, epoxy, and hydroxyl groups among graphitic layers can be vaporized during microwave heating which results in expanded graphite. The main factor for the formation of graphene that needs to be consider for this method is that the competition associated with the faster degradation of the epoxy and hydroxyl groups of graphite oxide and diffusion rate of the evolving gases. The pressure generated during the evolution of gases can separate the graphene sheets efficiently by dominating the van der Waals forces present among the graphene layers [48]. To achieve complete exfoliation of graphitic layers, elevated temperatures (1050 °C), high pressure (700 $m^2 g^{-1}$) with high heating rates (2000 °C/min) are required [49, 50]. It is possible to produce graphene, which can exhibit some of the properties of single layer graphene. This method can be useful to produce graphene in a bulk quantity. This is why this method is used for the preparation of polymer nanocomposite materials for engineering applications.

2.5 Chemical Vapour Deposition (CVD) Process

First, this method started on Ni substrate through chemical vapour deposition for the manufacturing of graphene [51]. In this, gaseous hydrocarbon reactants are carefully targeted onto a substrate through the reaction reactor, where such reactants decompose and form the reactive carbon radicals on the metal substrate [52]. The reaction occurs when the gases encounter the substrate surface inside the reaction chamber, which leading to generate the material. The production quality of the graphene completely depends on the set temperature, gaseous byproducts during the course of the process. Various metal surfaces are being utilized for the graphene deposition [53, 54] and copper is one of the best substrates which are currently used. Several other substrates based on carbon are also been used [55]. There are several problems associated with this method though it can produce high-quality graphene. The main problem associated with this method is the extraction of the graphene, it is so complicated due to the structural damages occur which ultimately effect the final properties of the final product. Besides, this method is so expensive owing to the usage of metal-based materials and the consumption of more energy. This method is slightly changed by using roll to roll process to improve the graphene quantity, here the complete formation of graphene occurs on Cu rolls and it can be easily extracted with the help of a polymer film [56]. Graphene prepared by this method having high quality so that it is been used for many engineering applications as electronic vehicles, sensing materials and so on. The drawback of this method is that graphene cannot be produced in large quantity for manufacturing advanced polymeric materials. Some reports revealed the production of graphene in large quantity by CVD [57–59].

3 Characterization of Graphene

3.1 *Microscopy Techniques*

Transmission electron microscopy has one of the best techniques to understand the morphology; size and the number of layers of graphene (see Fig. 1). It estimates the size of the formed layers, crystalline nature and defects, which determine the characteristic properties (mechanical, thrermal, electronic) of the material [60–63]. Novoselov and his group reported the thickness of the graphene is 0.4 nm in one of their studies using atomic force microscopy (AFM) [37]. After this invention, AFM is widely used to estimate the thickness of graphene. Besides, scanning tunneling and electron microscopy techniques are also used to understand the surface morphology, and depth atomic structure of graphene [64–68].

3.2 *X-Ray Diffraction and Raman Spectroscopy Analysis*

X-ray diffraction (XRD) analysis is a smart technique that can be used to understand the crystalline nature and interlayer spacing of formed graphene during the preparation (see Fig. 2a). Graphite exhibits sharp characteristic peak related to hexagonal carbon skeleton at at $2\theta = 26°$ which broadens as number of layers decreases and completely disappears for single layer graphene [69]. XRD can give the information of graphene crystallinity and the exfoliation of graphene layers. Moreover, it can be utilized for tracing distribution of graphene inside polymer nanocomposites.

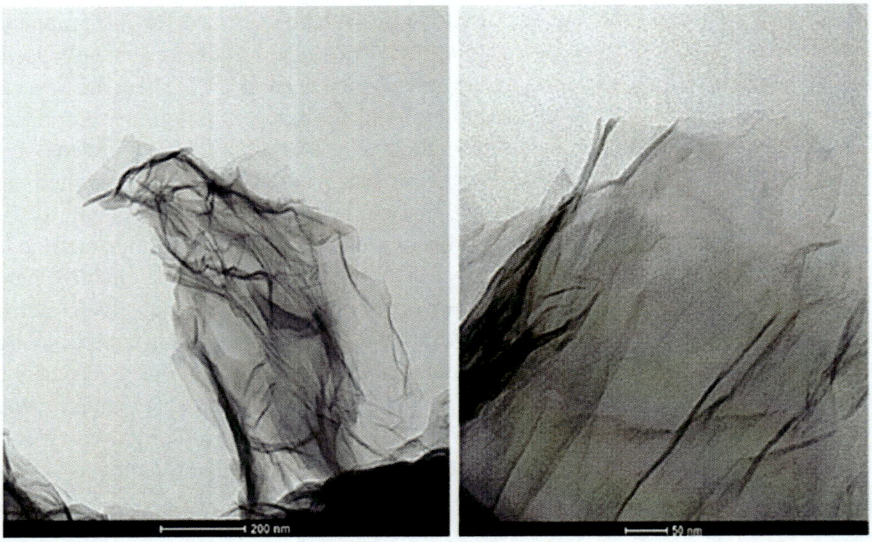

Fig. 1 TEM micrographs of graphene. Reference [27] Copyright 2020, reproduced from Elsevier

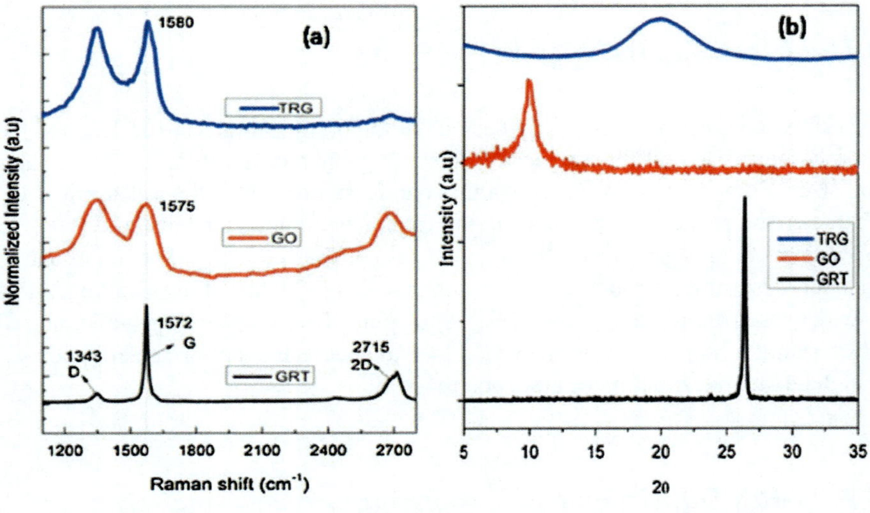

Fig. 2 Raman spectra and XRD of graphite (GRT) graphite oxide (GO) and Thermally exfoliated graphene (TRG). Reference [27] Copyright 2020, reproduced from Elsevier

Raman spectroscopy is one of the most powerful techniques to identify the electronic structure of graphene (see Fig. 2b) owing to the delocalization of free π electrons. Single layer graphene can produce a strong signal upon Raman scattering which is beneficial for the understanding the quality of the material [70, 71]. Graphite and graphene-associated materials can exhibit, the graphitic G band at ~1581 cm^{-1}, the deformation D band at ~1331 cm^{-1} and the G_0 (or 2D) band related to electronic structure at ~2651 cm^{-1}. This technique gives the clear picture related to the number of layers of graphene [72]. Dressel-haus et al. [73] have reported that the appearance of characteristic 2D band at different wavenumbers for graphene prepared by different methods discloses the numbers of graphene layers. The D band discloses the deformation and surface defects of the material [74]. The ratio between D and G band reveals the formation of graphene and functionalized graphene from the graphite. This ratio is crucial to estimate the quality of defects in the graphitic materials.

4 Graphene Reinforced Elastomer Nanocomposites

Many research articles disclosed related to the fabrication of graphene associated elastomeric materials and their polymer nanocomposites. Commonly, the orientation, dispersion of the filler inside the polymer can decide the physical performance and other characteristics of the resultant hybrid materials and nanocomposites. Several procedures including solution blending and melt mixing are being employed for the fabrication/preparation of such composites. However, each procedure has its own importance with reference to the distribution of the filler inside the polymer [75, 76]. Mixture of methods is needed to apply to prepare elastomeric-based nanocomposites, as it is not finished within a single step. A series of materials like crosslinking agents, catalysts and processing materials etc., are required in order to process elastomer mixture to provide a material with deserved properties. So that, processing of elastomer composites consists of various stages, in which, elastomers/polymers are mixed with fillers, crosslinking materials via two-roll mill [77–81]. Preparation techniques regarding to the elastomer/bio nanocomposites will be addressed in the coming sections.

4.1 Melt Mixing/Blending

This technique favored by industry, because it combines low cost and speed [82–84]. Initially, fillers are dispersed in the selected polymer matrix by applying a shear force during this process. The main usage of this method lies in the absence of utilization of various organic solvents. There are some problems associated with this technique despite the aforementioned advantages, because high temperature is required to disperse nanofillers inside elastomer, this can cause degradation of the material. Filler dispersion can be obstructed by the high amounts of filler and viscosity of the polymer and it requires high shear forces to control the viscosity, these can destroy the graphene layers. Poor dispersion of the filler inside polymer is noticed with this method and it was evidenced by the research through Kim et al. [75] where they investigated on thermally reduced graphene (TRG) filled thermoplastic

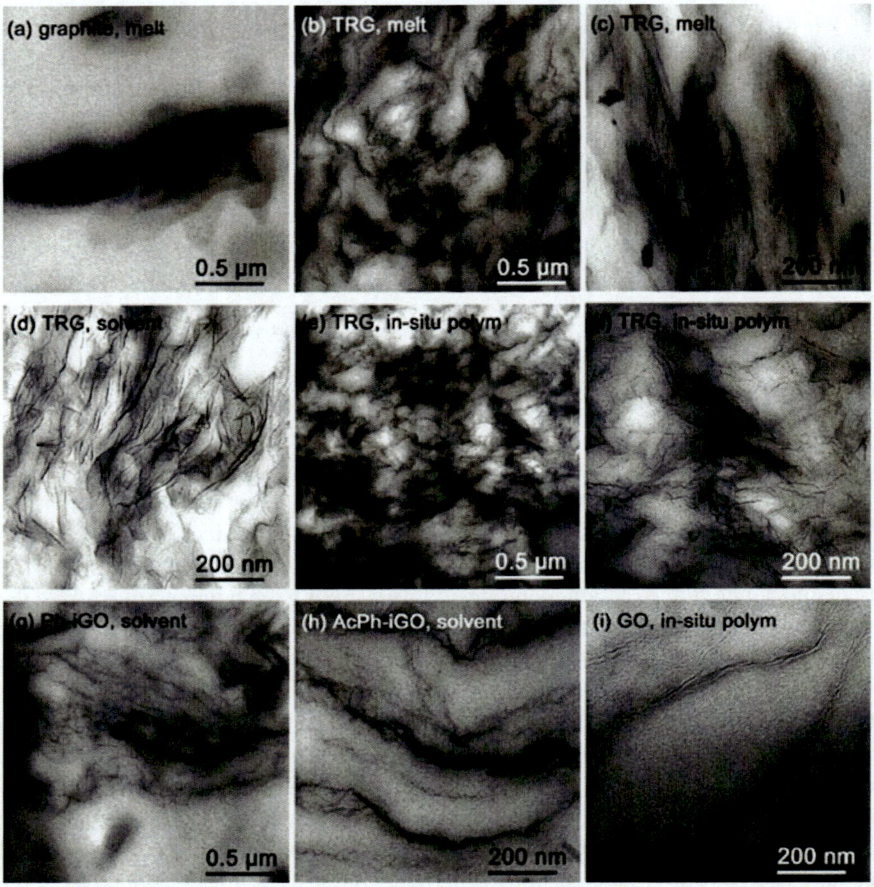

Fig. 3 Dispersion of graphite, graphene, GO, TRG inside polyurethane matrix with respect to comparison of the different processing methods. Reference [75] Copyright 2020, American Chemical Society

polyurethane (TPU) composites (see Fig. 3). It is very difficult to achieve highest homogeneous distribution of the filler inside polymer using this method.

4.2 Solution/Latex Blending

This method is generally reported as an effective fabrication technique due to the ease of processing graphite derivatives and graphene in various solvents and is generally used for the preparation of graphene based polymeric materials [75–81]. In this process, the mixing of graphene colloidal suspensions with the selected elastomer is done in a particular solvent by simple stirring or shear mixing. Sonication is performed in order to get better dispersion of graphitic fillers inside elastomer. However, agglomeration of graphene layers is possible even at lower loadings owing to the vander Walls forces. The restacking of the graphene platelets in solution mixing process could be prevented using organic surfactants, which have the great tendency to stabilize colloidal nanoparticle suspensions. Latex blending method is effective one to produce elastomer or biopolymer graphene based nanocomposites, the aggregation of the fillers inside biopolymer can be controlled kinetically during processing [79–81]. One of the major difficulties involving with this method is to eliminate or separate solvents used during preparation of biopolymer composites. Because of this issue, this method is limited only for certain applications, where homogeneous distribution of graphene filler could be achieved.

4.3 *In Situ Polymerization*

Graphene/biopolymer or elastomer composites have also been synthesized by in-situ polymerization methods [85]. In this technique, high level of dispersion of graphene could be achieved in comparison with solution mixing. Initially, the graphene/graphite or functionalized graphene's is added to the neat monomers. The interlayer distance of graphite greatly increases in this process. Kim et al. [85] have reported synthesis of TPU/graphene composites through in situ polymerization technique. This method is quite useful for the fabrication of graphite/graphene filled elastomer composites, but it has some limitations. It is not applicable if the polymerization takes place between the graphitic layers [86]. Moreover, it required many reagents for the polymerization purpose.

5 Mechanical Properties

Incorporation of carbon nanomaterials into elastomers results elastomer carbon nanocomposites, which exhibits excellent mechanical properties, compared to individual elastomers. Because of these fillers can transform their mechanical strength to the elastomer; consequently, high stress is required to break the resultant nanocomposite. The physical features of these nanomaterials are very important for tuning the physical performance of such nanocomposites. Mainly, homogeneous dispersion of the filler inside matrix could decide the final performance of the elastomer nanocomposites. In fact, agglomeration of nanomaterials or nanoreinforcement inside polymer matrix can suppress the mechanical strength of the resultant material. Natural rubber-graphene nanocomposites exhibited enhanced mechanical properties that mainly depend on the level of dispersion of graphene regardless of the preparation conditions [87]. Interfacial interactions between graphene and polymer could greatly influence the final properties of the composite. There have been many reports regarding the effect of nanofillers such as carbon nanotubes [16, 21], silica [8], clays [88, 89] on the physical performance of the polymer. Though these fillers do not show unique properties like graphene that is why graphene based composites received much attention among their counterparts.

5.1 Tensile Properties

Much research has disclosed regarding the physical performance of graphene reinforced elastomer composites [75–79]. Zhan and his co-workers had investigated the tensile properties of graphene reinforced NR composites and they observed significant improvement in modulus and tensile strength of neat NR [87]. Besides, decrease in elongation of neat NR was noticed with respect to graphene loading indicating that the graphene layers interacting with the NR chains via π-π interactions. High efficiency affordable composites with promising mechanical properties could be fabricated with graphene fillers [90]. Graphene reinforced elastomeric composites showed significant enhancement in modulus in comparison with other graphene/polymer composites, which validating the strong interfacial interactions anchoring among graphene and elastomer chains. In contrast to the significant increase in modulus, the tensile strength of elastomers is compromised by the addition of rigid fillers and the most of the cases elastomers- graphene composites often show a decline in tensile strength. Potts et al. [91, 92] have studied comprehensively the RGO reinforced natural rubber composites. They understood that dramatic improvement in tensile performance in nanocomposites prepared by latex blending method [91]. In another report [92], they explored the effect of thermally-exfoliated graphene oxide (TEGO) on the physical performance of natural rubber with the help of a two-roll mill. They found that TEGO has significantly improved the tensile properties of natural rubber as shown in Fig. 4. Transmission electron microscopy validates the homogeneous dispersion of TEGO inside elastomer, which causes for observing such an improvement in mechanical performance [92]. Tensile properties of the material are listed in Table 1. It can be understood that the increment in stiffness is less than the report of Potts et al. [91], but an improvement in extension/strain was

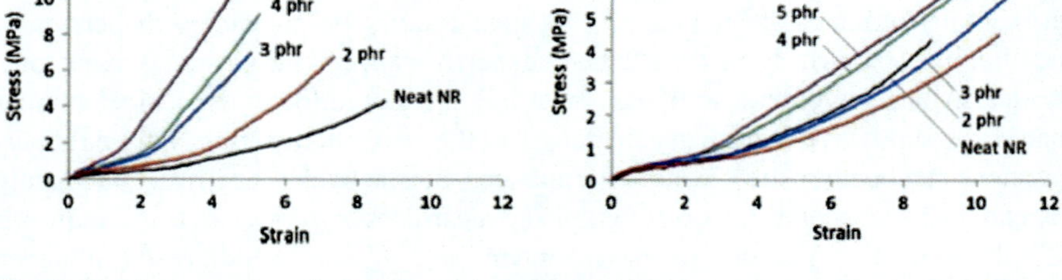

Fig. 4 Tensile properties of graphene and modified graphene reinforced NR composites **a** L-TEGO/NR **b** TEGO/NR. Reference [92] Copyright 2020, reproduced from Elsevier

Table 1 Mechanical properties of 5 phr of graphene, modified graphene, carbon black reinforced natural rubber composites

Filler	E_{100} (MPa)	E_{300} (MPa)	σ_f (MPa)	Strain at break
Neat NR	0.41	0.84	5.15	8.44
TEGO	0.43	1.28	6.05	8.96
L-TEGO	1.07	5.19	10.90	5.05
RGO	1.59	9.01	10.18	3.19
CB	0.51	1.19	6.28	8.34

Reference [92] Copyright 2020, reproduced from Elsevier

achieved. Overall, considerable improvement in tensile properties with graphene can be achieved using a two-roll mill method. It is also summarized in Table 1 that higher levels of strength and stiffness can be obtained with GO than with carbon black. Graphene oxide (1phr) reinforced natural rubber composite showed significant improvement in fatigue resistance under cyclic loading [93]. In fact, composites prepared by solution blending resulted the best mechanical performance. However, use of hazardous solvents making this method less applicable. Significant improvement in tensile properties of elastomer nanocomposites with graphene also has been noticed via ecofriendly two-roll mill method, which is industrially viable.

Mülhaupt and his group reported the modified graphene [94] and RGO [95] reinforced styrene-butadiene rubber (SBR) composites. Significant enhancement in tensile properties of SBR was noticed with the addition of both thermally-reduced GO (TRGO) and chemically- reduced GO (CRGO) to SBR (see Table 2). From the Table 2, one can notice that the improved mechanical properties with graphene-based fillers compared with other nanofillers. In another report, Zhang et al. [96] had noticed improvement in the tear strength (50%) and tensile strength (47%) for the natural rubber latex (NRL)/GE (Chemically Reduced Graphene, CRG-2 wt%) composites. In contrast, the low improvement in tensile properties of NR/MWCNT (multi-walled carbon nanotube) and NR/CB (carbon black) composites was noticed. The strong interfacial interactions among graphene and NR chains and great reinforcing action of graphene are responsible for their observed high mechanical properties [97]. Zhan

Table 2 Tensile properties of graphene, modified graphene, carbon black, CNT reinforced SBR composites

Type of filler	E_{300}	E_{50}	σ_f (MPa)	Maximum strain at break
Neat SBR	4.7	0.8	5.1	3.24
TRGO (25 phr)	–	10.9	17.5	0.88
CRGO (25 phr)	12.4	2.8	12.6	2.95
CB (25 phr)	6.7	1.2	8.2	3.65
CNT (25 phr)	–	2.1	9.2	2.60

Reference [94]

et al. [80] in another investigation revealed the improved tensile strength, modulus, in crosslinked graphene filled elastomer composites in comparison with uncrosslinked composites. In a conclusion, a well dispersed graphene and modified graphene elastomer system, way of preparation method, formation of interfaces between graphene and elastomer could greatly influence the mechanics of their composites.

5.2 Dynamic Mechanical Properties

Dynamic mechanical analysis can give the details about viscous and elastic nature of graphene reinforced elastomer composites [76–78, 81, 82]. In general, filler particles improve the storage modulus of a composite at a temperature above the glass transition temperature T_g in comparison with the neat elastomer [98]. Moreover, nanofillers can induce crystallinity of the whole composite, which further enhances the T_g of the composite owing to the suppressing of segmental mobility of elastomer chains around the graphitic filler. Araby et al. [76] had observed significant improvement in the elastic modulus at rubbery region of graphene incorporated styrene-butadiene rubber (SBR) (see Fig. 5). At another instance, huge increase in modulus was recorded by Potts et al. [91] with GO reinforced natural rubber composites owing to the increased interface region between the fillers and the matrix, which were boosted by the solution blending method. Nanocomposites prepared by Solution mixing method showed superior mechanical properties in comparison with melt mixing, two roll mill methods due to the weak interfacial interactions and agglomeration of nanomaterials. Araby et al. at another instance [76] had revealed the effect of graphene in enhancing the glass transition temperature of SBR composites owing to the fact that the elastomer chain mobility suppressed by the graphene. In contrast, a report revealed plasticizing nature of graphene where a decrease in glass transition can be observed in case of elastomer nanocomposites [78].

Fig. 5 Dynamic mechanical analysis of SBR composites with respect to graphene loading. Reference [76] Copyright 2020, reproduced from Elsevier

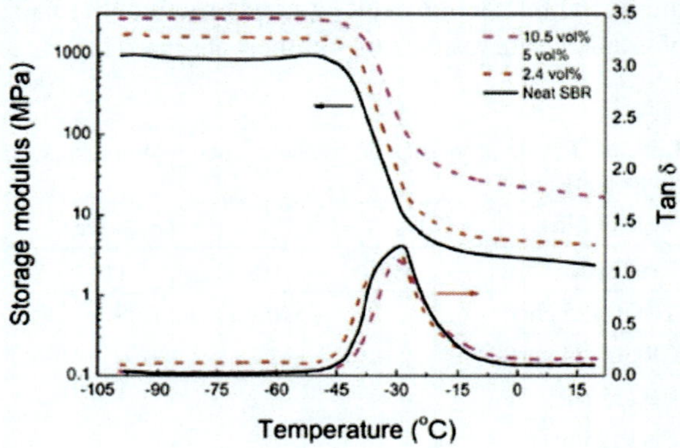

6 Conclusions and Perspectives

Structural features and synthesizing methods of graphene are summarized initial part of this review. The mechanical properties of the elastomer composites are highlighted with special focus on their physical/chemical interactions with respect to the graphene and modified graphene loadings. Graphene and its derivatives could greatly enhance the mechanical properties of elastomers even at extremely low loadings. This is because of huge surface area of graphene in comparison with other nanofillers. However, significant improvement in barrier, thermal, electrical properties already revealed for graphene-reinforced elastomer composites are not highlighted in this review. In addition, various preparation methods of graphene and elastomer nanocomposites have been addressed with highlighting advantages and disadvantages of such process to target the desirable applications of the material. Due to unique properties of graphene nanofillers, they show manifold advantages as compared to other nanofillers. In a conclusion, graphene can be utilized as successful reinforcement in elastomer nanocomposites in order to improve the mechanical properties such as tensile strength, elongation at break, modulus etc. as compared to carbon black and graphite filled composites. Raising the production of high-quality graphene is an important factor to be considered to develop not only elastomeric nanocomposites but also other polymer composites with enhanced physical properties. Processing techniques could greatly influence the dispersion graphene inside elastomer and thus ultimately affect the mechanical properties. The physical and chemical interactions also play a crucial role to decide the physical properties of the composites. These interactions could drastically lift the modulus and strength of elastomer composites. Mechanical and other physical properties of the elastomer composites can be directed by the way of preparation, alignment of the filler, and the homogeneous distribution.

References

1. Mc Naught AD, Wilkinson A (1997) Volume of activation, ΔV. In: IUPAC compendium of chemical terminology. IUPAC, Research Triagle Park, NC
2. Hosler D (1999) Prehistoric Polymers: rubber processing in ancient mesoamerica. Science (80)284:1988–1991. https://doi.org/10.1126/science.284.5422.1988
3. Heinrich G, Basak GC (2011) Advanced rubber composites. Springer, Berlin
4. Ceresana Research (2011) Protecting plastics and rubber: Ceresana analyzes the stabilizer market. Pigment Resin Technol 40:prt.2011.12940eaa.022. https://doi.org/10.1108/prt.2011.12940eaa.022
5. Sinha Ray S, Okamoto M (2003) Polymer/layered silicate nanocomposites: a review from preparation to processing. Prog Polym Sci 28:1539–1641. https://doi.org/10.1016/j.progpolymsci.2003.08.002
6. Paul DR, Robeson LM (2008) Polymer nanotechnology: nanocomposites. Polymer (Guildf)
7. Heinrich G, Klüppel M, Vilgis TA (2002) Reinforcement of elastomers. Curr Opin Solid State Mater Sci. https://doi.org/10.1016/S1359-0286(02)00030-X

8. Alexandre M, Dubois P (2000) Polymer-layered silicate nanocomposites: preparation, properties and uses of a new class of materials. Mater Sci Eng R Reports 28:1–63. https://doi.org/10.1016/S0927-796X(00)00012-7
9. Pavlidou S, Papaspyrides CD (2008) A review on polymer–layered silicate nanocomposites. Prog Polym Sci 33:1119–1198. https://doi.org/10.1016/j.progpolymsci.2008.07.008
10. Zou H, Wu S, Shen J (2008) Polymer/silica nanocomposites: preparation, characterization, properties, and applications. Chem Rev 108(9):3893–3957
11. Moniruzzaman M, Winey KI (2006) Polymer nanocomposites containing carbon nanotubes. Macromolecules 39:5194–5205. https://doi.org/10.1021/ma060733p
12. Koerner H, Price G, Pearce NA et al (2004) Remotely actuated polymer nanocomposites—stress-recovery of carbon-nanotube-filled thermoplastic elastomers. Nat Mater 3(2):115–120
13. Bokobza L (2007) Multiwall carbon nanotube elastomeric composites: a review. Polym (Guildf) 48:4907–4920. https://doi.org/10.1016/j.polymer.2007.06.046
14. Yaragalla S, Mishra RK, Thomas S, Kalarikkal N, Maria HJ (2018) Carbon-based nanofillers and their rubber nanocomposites 1st edition carbon nano-objects. Elsevier, p 402
15. Leblanc JL (2002) Rubber-filler interactions and rheological properties in filled compounds. Prog Polym Sci 27(4):627–687
16. Frogley MD, Ravich D, Wagner HD (2003) Mechanical properties of carbon nanoparticle-reinforced elastomers. Compos Sci Technol 63(11):1647–1654. https://doi.org/10.1016/S0266-3538(03)00066-6
17. Donnet JB (2003) Nano and microcomposites of polymers elastomers and their reinforcement. Compos Sci Technol 63(8):1085–1088. https://doi.org/10.1016/S0266-3538(03)00028-9
18. Boonstra BB (1979) Role of particulate fillers in elastomer reinforcement: a review. Polymer 20(6):691–704 (Guildf). https://doi.org/10.1016/0032-3861(79)90243-X
19. Edwards DC (1990) Polymer-filler interactions in rubber reinforcement. J Mater Sci 25:4175–4185. https://doi.org/10.1007/BF00581070
20. Joly S, Garnaud G, Ollitrault R et al (2002) Organically modified layered silicates as reinforcing fillers for natural rubber. Chem Mater 14(10):4202–4208. https://doi.org/10.1021/cm020093e
21. Ponnamma D, Sadasivuni KK, Grohens Y et al (2014) Carbon nanotube based elastomer composites—an approach towards multifunctional materials. J Mater Chem C 2:8446–8485. https://doi.org/10.1039/C4TC01037J
22. Araby S, Meng Q, Zhang L et al (2015) Elastomeric composites based on carbon nanomaterials. Nanotechnology 26:112001. https://doi.org/10.1088/0957-4484/26/11/112001
23. Basu D, Das A, Stöckelhuber KW et al (2014) Advances in layered double hydroxide (LDH)-based elastomer composites. Prog Polym Sci 39(3):594–626
24. Geim AK, Novoselov KS (2007) The rise of graphene. Nat Mater 6(3):183–191
25. Singh V, Joung D, Zhai L et al (2011) Graphene based materials: past, present and future. Prog Mater Sci 56(8):1178–1271
26. Novoselov KS, Geim AK, Morozov SV et al (2005) Two-dimensional gas of massless dirac fermions in graphene. Nature 438:197–200. https://doi.org/10.1038/nature04233
27. Yaragalla SAPM, Kalarikkal N, Thomas S (2015) Chemistry associated with natural rubber–graphene nanocomposites and its effect on physical and structural properties. Ind Crops Prod 74:792–802. https://doi.org/10.1016/j.indcrop.2015.05.079
28. Yaragalla S, Sarath Chandran C, Kalarikkal N et al (2015) Effect of reinforcement on the barrier and dielectric properties of epoxidized natural rubber-graphene nanocomposites. Polym Eng Sci 55. https://doi.org/10.1002/pen.24131
29. Yaragalla S, Sindam B, Abraham J et al (2015) Fabrication of graphite-graphene-ionic liquid modified carbon nanotubes filled natural rubber thin films for microwave and energy storage applications. J Polym Res 22:137. https://doi.org/10.1007/s10965-015-0776-5
30. Yaragalla S, Mishra RK, Thomas S, Kalarikkal N, Maria HJ (2019) Carbon-based nanofillers and their rubber nanocomposites, fundamental and applications, 1st edn. Elsevier, p 496
31. Yaragalla S, Hanum R (2013) Electrical properties of graphene filled natural rubber composites. Adv Mater 812:263–266

32. Sengupta R, Bhattacharya M, Bandyopadhyay S, Bhowmick AK (2011) A review on the mechanical and electrical properties of graphite and modified graphite reinforced polymer composites. Prog Polym Sci 36:638–670. https://doi.org/10.1016/j.progpolymsci.2010.11.003
33. Terrones M, Martín O, González M et al (2011) Interphases in graphene polymer-based nanocomposites: achievements and challenges. Adv Mater 23(44):5302–5310. https://doi.org/10.1002/adma.201102036
34. Cai D, Song M (2010) Recent advance in functionalized graphene/polymer nanocomposites. J Mater Chem 20:7906. https://doi.org/10.1039/c0jm00530d
35. Kuilla T, Bhadra S, Yao D et al (2010) Recent advances in graphene based polymer composites. Prog Polym Sci 35(11):1350–1375
36. Young RJ, Kinloch IA, Nicolais L (2011) Graphene composites. Wiley encyclopedia of composites. Wiley Inc.
37. Novoselov KS (2004) Electric field effect in atomically thin carbon films. Science (80)306:666–669. https://doi.org/10.1126/science.1102896
38. Gong L, Kinloch IA, Young RJ et al (2010) Interfacial stress transfer in a graphene monolayer nanocomposite. Adv Mater 22:2694–2697. https://doi.org/10.1002/adma.200904264
39. Young RJ, Gong L, Kinloch IA et al (2011) Strain mapping in a graphene monolayer nanocomposite. ACS Nano 5(4):3079–3084. https://doi.org/10.1021/nn2002079
40. Park S, An J, Jung I et al (2009) Colloidal suspensions of highly reduced graphene oxide in a wide variety of organic solvents. Nano Lett 9(4):1593–1597. https://doi.org/10.1021/nl803798y
41. Paredes JI, Villar-Rodil S, Martínez-Alonso A, Tascón JMD (2008) Graphene oxide dispersions in organic solvents. Langmuir 24(19):10560–10564. https://doi.org/10.1021/la801744a
42. Hernandez Y, Nicolosi V, Lotya M et al (2008) High-yield production of graphene by liquid-phase exfoliation of graphite. Nat Nanotechnol 3:563–568. https://doi.org/10.1038/nnano.2008.215
43. Blake P, Brimicombe PD, Nair RR et al (2008) Graphene-based liquid crystal device. Nano Lett 8:1704–1708. https://doi.org/10.1021/nl080649i
44. Bourlinos AB, Georgakilas V, Zboril R et al (2009) Liquid-phase exfoliation of graphite towards solubilized graphenes. Small 5(16):1841–1845. https://doi.org/10.1002/smll.200900242
45. Lotya M, Hernandez Y, King PJ et al (2009) Liquid phase production of graphene by exfoliation of graphite in surfactant/water solutions. J Am Chem Soc 131(10):3611–3620. https://doi.org/10.1021/ja807449u
46. Niu L, Li M, Tao X et al (2013) Salt-assisted direct exfoliation of graphite into high-quality, large-size, few-layer graphene sheets. Nanoscale 5(16):7202–7208. https://doi.org/10.1039/c3nr02173d
47. Xiang J, Drzal LT (2011) Thermal conductivity of exfoliated graphite nanoplatelet paper. Carbon NY 49:773–778. https://doi.org/10.1016/j.carbon.2010.10.003
48. McAllister MJ, Li JL, Adamson DH et al (2007) Single sheet functionalized graphene by oxidation and thermal expansion of graphite. Chem Mater 19(18):4396–4404. https://doi.org/10.1021/cm0630800
49. Yaragalla S, Anilkumar G, Vineeshkumar TV et al (2015) Preparation of epoxy graphene and its structural and optical properties. Adv Mater Lett 6:848–852. https://doi.org/10.5185/amlett.2015.5914
50. Zhang H-B, Wang J-W, Yan Q et al (2011) Vacuum-assisted synthesis of graphene from thermal exfoliation and reduction of graphite oxide. J Mater Chem 21:5392. https://doi.org/10.1039/c1jm10099h
51. Yu Q, Lian J, Siriponglert S et al (2008) Graphene segregated on Ni surfaces and transferred to insulators. Appl Phys Lett 93:113103. https://doi.org/10.1063/1.2982585
52. Zhang Y, Zhang L, Zhou C (2013) Review of chemical vapor deposition of graphene and related applications. Acc Chem Res 46:2329–2339. https://doi.org/10.1021/ar300203n
53. Suk JW, Kitt A, Magnuson CW et al (2011) Transfer of CVD-grown monolayer graphene onto arbitrary substrates. ACS Nano 5:6916–6924. https://doi.org/10.1021/nn201207c

54. Li X, Magnuson CW, Venugopal A et al (2011) Large-area graphene single crystals grown by low-pressure chemical vapor deposition of methane on copper. J Am Chem Soc 133:2816–2819. https://doi.org/10.1021/ja109793s
55. Ruan G, Sun Z, Peng Z, Tour JM (2011) Growth of graphene from food, insects, and waste. ACS Nano 5:7601–7607. https://doi.org/10.1021/nn202625c
56. Li X, Cai W, An J et al (2009) Large-area synthesis of high-quality and uniform graphene films on copper foils. Science (80)324:1312–1314. https://doi.org/10.1126/science.1171245
57. Herron CR, Coleman KS, Edwards RS, Mendis BG (2011) Simple and scalable route for the 'bottom-up' synthesis of few-layer graphene platelets and thin films. J Mater Chem 21:3378. https://doi.org/10.1039/c0jm03437a
58. Edwards RS, Coleman KS (2013) Graphene synthesis: relationship to applications. Nanoscale 5:38–51. https://doi.org/10.1039/C2NR32629A
59. Kim KS, Zhao Y, Jang H et al (2009) Large-scale pattern growth of graphene films for stretchable transparent electrodes. Nature 457:706–710. https://doi.org/10.1038/nature07719
60. Červenka J, Katsnelson MI, Flipse CFJ (2009) Room-temperature ferromagnetism in graphite driven by two-dimensional networks of point defects. Nat Phys 5:840–844. https://doi.org/10.1038/nphys1399
61. Yazyev OV, Louie SG (2010a) Electronic transport in polycrystalline graphene. Nat Mater 9:806–809. https://doi.org/10.1038/nmat2830
62. Yazyev OV, Louie SG (2010b) Topological defects in graphene: Dislocations and grain boundaries. Phys Rev B 81:195420. https://doi.org/10.1103/PhysRevB.81.195420
63. Meyer JC, Geim AK, Katsnelson MI et al (2007) The structure of suspended graphene sheets. Nature 446:60–63. https://doi.org/10.1038/nature05545
64. Tung VC, Allen MJ, Yang Y, Kaner RB (2009) High-throughput solution processing of large-scale graphene. Nat Nanotechnol 4:25–29. https://doi.org/10.1038/nnano.2008.329
65. Xu K, Cao P, Heath JR (2009) Scanning tunneling microscopy characterization of the electrical properties of wrinkles in exfoliated graphene monolayers. Nano Lett 9(12):4446–4451. https://doi.org/10.1021/nl902729p
66. Decker R, Wang Y, Brar VW et al (2011) Local electronic properties of graphene on a bn substrate via scanning tunneling microscopy. Nano Lett 11:2291–2295. https://doi.org/10.1021/nl2005115
67. Bao W, Miao F, Chen Z et al (2009) Controlled ripple texturing of suspended graphene and ultrathin graphite membranes. Nat Nanotechnol 4:562–566. https://doi.org/10.1038/nnano.2009.191
68. Xu B, Yue S, Sui Z et al (2011) What is the choice for supercapacitors: graphene or graphene oxide? Energy Environ Sci 4:2826. https://doi.org/10.1039/c1ee01198g
69. Khanra P, Kuila T, Bae SH et al (2012) Electrochemically exfoliated graphene using 9-anthracene carboxylic acid for supercapacitor application. J Mater Chem 22(46):24403–24410. https://doi.org/10.1039/c2jm34838a
70. Graf D, Molitor F, Ensslin K et al (2007) Spatially resolved raman spectroscopy of single- and few-layer graphene. Nano Lett 7(2):238–242. https://doi.org/10.1021/nl061702a
71. Ferrari AC, Meyer JC, Scardaci V et al (2006) Raman spectrum of graphene and graphene layers. Phys Rev Lett 97:187401. https://doi.org/10.1103/PhysRevLett.97.187401
72. Young RJ, Kinloch IA (2013) Graphene and graphene-based nanocomposites. In: Nanoscience: volume 1: nanostructures through chemistry, The royal society of chemistry, pp 145–179
73. Malard LM, Pimenta MA, Dresselhaus G, Dresselhaus MS (2009) Raman spectroscopy in graphene. Phys Rep 473:51–87. https://doi.org/10.1016/j.physrep.2009.02.003
74. Eckmann A, Felten A, Verzhbitskiy I et al (2013) Raman study on defective graphene: effect of the excitation energy, type, and amount of defects. Phys Rev B 88:035426. https://doi.org/10.1103/PhysRevB.88.035426
75. Kim H, Miura Y, Macosko CW (2010a) Graphene/polyurethane nanocomposites for improved gas barrier and electrical conductivity. Chem Mater 22:3441–3450. https://doi.org/10.1021/cm100477v

76. Araby S, Meng Q, Zhang L et al (2014) Electrically and thermally conductive elastomer/graphene nanocomposites by solution mixing. Polym (Guildf) 55:201–210. https://doi.org/10.1016/j.polymer.2013.11.032
77. Sadasivuni KK, Saiter A, Gautier N et al (2013) Effect of molecular interactions on the performance of poly(isobutylene-co- isoprene)/graphene and clay nanocomposites. Colloid Polym Sci 291(7):1729–1740. https://doi.org/10.1007/s00396-013-2908-y
78. Chen B, Ma N, Bai X et al (2012) Effects of graphene oxide on surface energy, mechanical, damping and thermal properties of ethylene-propylene-diene rubber/petroleum resin blends. RSC Adv 2(11):4683–4689. https://doi.org/10.1039/c2ra01212j
79. Wu S, Tang Z, Guo B et al (2013) Effects of interfacial interaction on chain dynamics of rubber/graphene oxide hybrids: a dielectric relaxation spectroscopy study. RSC Adv 3(34):14549–14559. https://doi.org/10.1039/c3ra41998c
80. Zhan Y, Lavorgna M, Buonocore G, Xia H (2012) Enhancing electrical conductivity of rubber composites by constructing interconnected network of self-assembled graphene with latex mixing. J Mater Chem 22:10464. https://doi.org/10.1039/c2jm31293j
81. Tang Z, Wu X, Guo B et al (2012) Preparation of butadiene-styrene-vinyl pyridine rubber-graphene oxide hybrids through co-coagulation process and in situ interface tailoring. J Mater Chem 22(15):7492–7501. https://doi.org/10.1039/c2jm00084a
82. Dao TD, Lee H, Jeong HM (2014) Alumina-coated graphene nanosheet and its composite of acrylic rubber. J Colloid Interf Sci 416:38–43. https://doi.org/10.1016/j.jcis.2013.10.033
83. Wei J, Qiu J (2014) Allyl-Functionalization enhanced thermally stable graphene/fluoroelastomer nanocomposites. Polym (Guildf) 55(16):3818–3824. https://doi.org/10.1016/j.polymer.2014.06.063
84. Allahbakhsh A, Mazinani S, Kalaee MR, Sharif F (2013) Cure kinetics and chemorheology of EPDM/graphene oxide nanocomposites. Thermochim Acta 563:22–32. https://doi.org/10.1016/j.tca.2013.04.010
85. Kim H, Miura Y, Macosko CW (2010b) Graphene/polyurethane nanocomposites for improved gas barrier and electrical conductivity. Chem Mater 22:3441–3450. https://doi.org/10.1021/cm100477v
86. Shioyama H (2000) The interactions of two chemical species in the interlayer spacing of graphite. Synth Met 114:1–15. https://doi.org/10.1016/S0379-6779(00)00222-8
87. Potts JR, Shankar O, Du L, Ruoff RS (2012a) Processing–morphology–property relationships and composite theory analysis of reduced graphene oxide/natural rubber nanocomposites. Macromolecules 45:6045–6055. https://doi.org/10.1021/ma300706k
88. Lan T, Pinnavaia TJ (1994) Clay-reinforced epoxy nanocomposites. Chem Mater 6:2216–2219. https://doi.org/10.1021/cm00048a006
89. Favier V, Chanzy H, Cavaillé JY (1995) Polymer nanocomposites reinforced by cellulose whiskers. Macromolecules 28(18):6365–6367. https://doi.org/10.1021/ma00122a053
90. Prud'homme RK, Ozbas B, Aksay IA, Register RA, Adamson DH (2009) Functionalized graphene sheets having high carbon to oxygen ratios. WIPO 2009/134492 A2
91. Potts JR, Shankar O, Du L, Ruoff RS (2012b) Processing–morphology–property relationships and composite theory analysis of reduced graphene oxide/natural rubber nanocomposites. Macromolecules 45:6045–6055. https://doi.org/10.1021/ma300706k
92. Potts JR, Shankar O, Murali S et al (2013) Latex and two-roll mill processing of thermally-exfoliated graphite oxide/natural rubber nanocomposites. Compos Sci Technol 74:166–172. https://doi.org/10.1016/j.compscitech.2012.11.008
93. Dong B, Liu C, Zhang L, Wu Y (2015) Preparation, fracture, and fatigue of exfoliated graphene oxide/natural rubber composites. RSC Adv 5:17140–17148. https://doi.org/10.1039/C4RA17051B
94. Beckert F, Trenkle S, Thomann R, Mülhaupt R (2014) Mechanochemical route to functionalized graphene and carbon nanofillers for graphene/sbr nanocomposites. Macromol Mater Eng 299(12):1513–1520. https://doi.org/10.1002/mame.201400205
95. Schopp S, Thomann R, Ratzsch KF et al (2014) Functionalized graphene and carbon materials as components of styrene-butadiene rubber nanocomposites prepared by aqueous dispersion blending. Macromol Mater Eng 299(3):319–329. https://doi.org/10.1002/mame.201300127

96. Zhan Y, Wu J, Xia H et al (2011) Dispersion and exfoliation of graphene in rubber by an ultrasonically-assisted latex mixing and in situ reduction process. Macromol Mater Eng 296:590–602. https://doi.org/10.1002/mame.201000358
97. Xing W, Wu J, Huang G et al (2014) Enhanced mechanical properties of graphene/natural rubber nanocomposites at low content. Polym Int 63:1674–1681. https://doi.org/10.1002/pi.4689
98. Cho D, Lee S, Yang G et al (2005) Dynamic mechanical and thermal properties of phenylethynyl-terminated polyimide composites reinforced with expanded graphite nanoplatelets. Macromol Mater Eng 290:179–187. https://doi.org/10.1002/mame.200400281

Graphene Reinforced Biopolymer Nanocomposites for Water Filtration Applications

Irene S. Fahim

Abstract The world is encountering outrageous water deficiency due to unequal water dispersion, misuse of water and inefficient water system strategies. Rising populations and quick monetary advancement in the nations, contamination and ecological corruption are diminishing water accessibility. The water shortage is causing a real threat. There must be a sustainable solution to this current problem which lies in reusing the water after ensuring its treatment from any type of waste that could be a threat to human beings. The consistent access of fresh water is a humanitarian objective in the twenty-first century. There are several traditions for treating classical water pollutants such as heavy metals, insecticides, poly aromatic hydrocarbons (PAHs), polychlorinated biphenyls (PCBs) but there is less research on nanocomposites filtration membranes. The essential obligations for water purification are suitable materials with elevated separation ability, low cost, and reuse. Nanotechnology offers innovative materials with large surface area thus improving properties like tensile strength, hydrophobicity and porosity. The main disadvantage of nanoparticles is agglomeration; nevertheless, it is minimized by the addition of reinforcements to form nanocomposites. Polymer nanocomposites (PNCs) are used for water purification due to their desired properties such as high mechanical strength.

Keywords Fabrication · Graphene · Nanotechnology · Polymer · Nanoparticles · Chitosan

1 Introduction

People use water filter systems for several reasons: to remove chlorine and bacterial contaminants, protect human body from diseases such as gastrointestinal disease and develop strong immune system. There are several types of water filters such as activated carbon filter; it uses activated carbon pellets that collect chemical impurities by absorption process. Activated carbon is a porous form of the carbon.

I. S. Fahim (✉)
Industrial Engineering Department, SESC Research Center, Nile University, Cairo, Egypt
e-mail: isamy@nu.edu.eg

B. Sharma and P. Jain (eds.), *Graphene Based Biopolymer Nanocomposites*,
Composites Science and Technology,
https://doi.org/10.1007/978-981-15-9180-8_12

Ion Exchange filters are filters used for softening water and make it digestible by removing limescale. Distillation is another method that involves boiling the water to kill the bacteria then cooled in a separate container. Water boils at a lower temperature than other contaminants such as heavy metals. Most of the current filtration systems used now is based on mechanical filtration processes. They remove particles by a mechanical process based on size. However, there are three most common heavy metal contaminants that cannot be removed by the mentioned filtration methods; they are cadmium, lead, and arsenic. The recovery of purified water depends on membrane pore size, temperature, pressure, and membrane surface area. One of the major problems with membranes is the sediment deposition that damages the membranes [1].

Membranes are defined as a barrier separating between two phases and restricting the transport of different matter from one phase to the other preferably in a selective method. A membrane is either a homogenous or heterogeneous structure [2] and holds a positive or negative charge. Diffusion of individual molecules affects the transport through membranes, either by (electric field-concentration-pressure or temperature gradient). The thickness of the membrane ranges from 10 μm to few hundred micrometres. Membrane technology is a promising technology that can be used for water filtration. The transport of water through the membranes happens when a force is applied; the dynamic force is either a change in pressure or concentration difference [3]. The current membranes are categorized according to their pore dimension into microfiltration (MF) and ultrafiltration (UF). MF has the major pore dimension between 0.1 and 3 μm, while UF membranes have pore sizes ranging between 0.01 and 0.1 μm as shown in Fig. 1. Furthermore, nanofiltration (NF) membranes have pore sizes ranging between 1 and 10 nm. Microfiltration membranes remove all bacteria. They are also used in separation of oiler emulsions. Additionally, they are applied in the pre-treatment of water for nano filtration. Meanwhile, the UF process removes all viruses, which are smaller than bacteria. The NF technique is used for the removal of heavy metals [4, 5].

Membrane technology is one of the most promising solutions to air and water pollution. There are different categories of synthesized membranes (also referred to as thin films) manufactured by either synthetic or natural polymers. The preparation of synthetic membranes and their utilization on a large industrial scale is a recent development that gained substantial importance due to the large number of

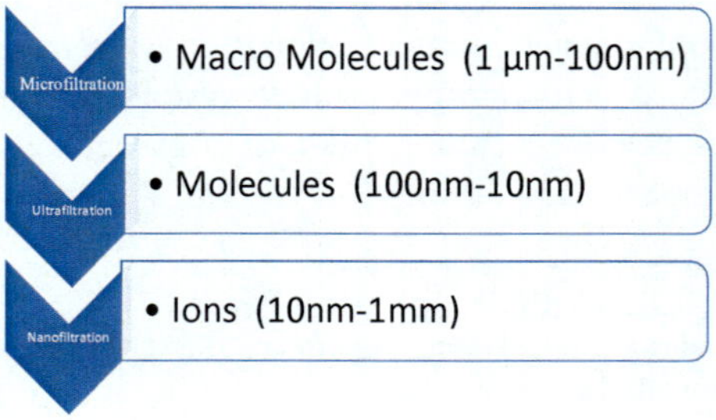

Fig. 1 Different types of membranes

their practical applications including water treatment, filtration, food and drug packaging, gas and vapour separation. The straightforward technology behind membrane filtration requires using a semi-permeable membrane to divide a liquid into two different streams. Injecting this liquid through the membrane surface creates a pressure that push any components less significant than the porosity of the membrane to pass across, forming the permeate. Components larger than the pore size simply cannot pass through and stay behind which is called the retentate. The surface of the membrane is kept free of blockages by the force of the liquid flow moving parallel to the membrane surface.

2 Polymer Nano Composites

Polymer based layers are for the most used membrane material in water filtration, however they are hydrophobic. Thus, polymeric layers tend to foul and block the pores. This is a considered a major disadvantage in the current membranes and needs several cleaning techniques [6, 7]. The combination of nanomaterials such as carbon nanotubes, fullerene and graphene improve the hydrophilic properties of polymers and increase the mechanical strength of filtration membranes, decrease fouling, increase thermal stability [8, 9]. Graphene is extensively used as reinforcement due to its exceptional high surface area, enhanced stiffness, mechanical, thermal and electrical properties. Graphene is composed of closely packed carbon atoms that have high affinity to polymers due to the existence of pores [10, 11].

Polymeric films can be fabricated permeable because of flaws which leave pores inside the polymer lattice. Isotropic layers either need pores or contain little pores. These films are manufactured by solution casting followed by evaporation [12]. Anisotropic membranes (Loeb-Sourirajan layers) are produced by phase separation technique [13]. They comprise of a thick layer of polymer outwardly of the permeable layer [13]. Membrane composites are made from a permeable substrate covered with a polymer layer. They are manufactured through phase separation, track etching. Phase separation produce polymer membranes for separation processes. They are formed by casting a polymer. This arrangement is isolated into two stages: a polymer that frames the strong structure of the layer, and a poor stage that will be evacuated for pores creation. Track-etched layers are set up by lighting a polymer film with charged particles to attack the chains. This procedure abandons harmed atoms. The film is exposed to a scratching arrangement. The harmed atoms separate bringing about pores [13]. Another strategy is the expanded film membranes; they are crystalline polymers with voids. They are made by extrusion; the material is extruded with a quick rate at the melting point, cooled, tempered, and extended up to 300%. The procedure makes pores running in size from 20 to 250 nm [14].

3 Synthetic Polymers Versus Natural Polymers

There are numerous synthetic polymeric membranes materials commercially available such as low-density polyethylene (LDPE), poly vinylidene fluoride (PVDF), polysulfone (PSU), polyacrylonitrile (PAN) and polyvinyl chloride (PVC) copolymers, cellulose acetate (CA), nylons, polytetrafluoroethylene (PTFE) and polyacrylonitrile (PAN) [15, 16]. Each membrane material has a certain disadvantage. PTFE has high thermal stability and exceptional electrical properties but low tensile strength. Nylon membranes are hydrophilic, flexible, durable, and withstand high temperatures but the pore size is not suitable for filtration [17].

Examples of natural polymers include cellulose, it is a natural polymer with an array of hydroxyl groups found in plants, and it tends to form strongly hydrogen bonds. Cellulose acetate membranes are used as filtration membranes. Starch is another natural polymer used for membrane preparation. It has been used to produce biodegradable films with low-cost and good processability. An attempt to overcome the weak mechanical properties of starch is the addition of plasticizers to overcome the brittleness of starch membranes. Chitin is the third natural polymer used to produce membranes. Chitin is the original source of chitosan. Chitin is produced from the shells of crustaceans such as crabs and shrimps. Chitosan membranes have the ability of film forming; hydrophilicity, and biocompatibility but week mechanical properties, thus reinforcing chitosan with graphene is a solution to the strength problem [18, 19].

4 Preparation of Composites Membranes

There are several fabrication techniques for composite membranes. Poly (vinylidene) fluoride (PVDF)-ZrO_2 composite membranes were prepared by casting and immersion into a water bath. Other methods include poly (ethersulfone) (PES)/titanium oxide (TiO_2) composite fabrication. The membranes were prepared by phase inversion method with nano-TiO_2 as an additive. Several testing techniques were applied to investigate the mechanical, thermal and physical properties of the membranes [20]. Additionally, TiO_2 films membranes were synthesized fabricating using dip-coating [21]. This method of fabrication enhanced porosity, homogeneity and increased the surface area of the composite membrane. The applications of these composite membranes include separation, fuel cells, batteries, solar cells and water filtration membranes.

4.1 Preparation of Chitosan Nano Composites Membranes

2 g of Chitosan is dissolved in acetic acid and casted in a container to form a membrane. The containers are dried in air for 48 h. The membranes are immersed in sodium hydroxide (NaOH) to remove impurities. This step is followed by washing in deionized water till the pH is neutral [22]. Additionally, chitosan membranes with big pores are prepared by incorporating silica particles with different particle size [22]. Chitosan can be crosslinked to form chitosan nanoparticles, which display exceptional properties at the nanoscale in comparison with non-cross-linked chitosan particles. Cross linking can be performed either by physical or chemical procedure. An example of cross-linking is the treatment of chitosan by adding glutaraldehyde in a solution form. The glutaraldehyde reacts with the amino groups in the chitosan chains [22]. Furthermore, adding negatively charged ions such as sodium tripolyphosphate (TPP) to chitosan forms chitosan nanoparticles, Different studies show that TPP has the highest mechanical resistance and chemical stability together with membrane flexibility compared to other cross-linking agents [22]. The physical cross-linking of chitosan differs according to the type, time and composition of cross-linking agent. However, crosslinking causes a decline in the crystallinity, growth in pore size, decrease in tensile strength and shrinkage of crystal size. The worsening in the tensile strength is due to the reduction in surface area of the membrane [22]. The growth in length of the molecular chains between bonds declined the pore volume and surface area thus increasing the pore size. Chitosan solution was mixed with the two nanofillers, graphene and fullerene by solvent mixing for 60 min with a VWR® Standard Analog Shaker to form a clear homogeneous solution as shown in Fig. 2.

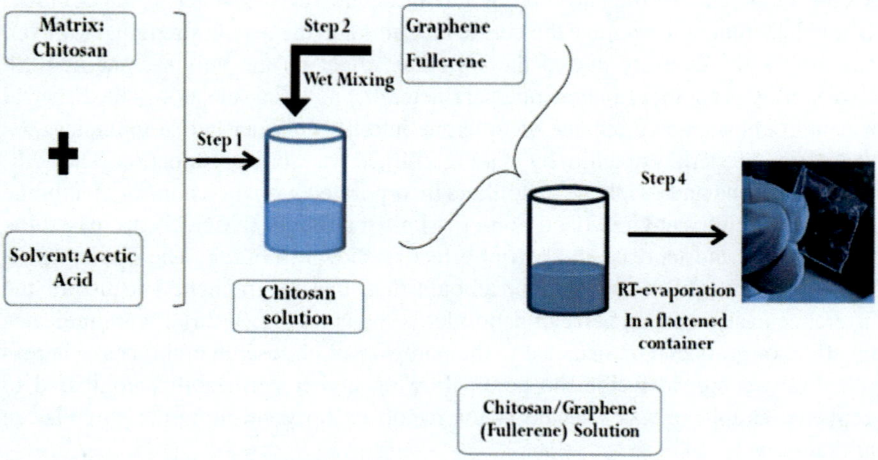

Fig. 2 Manufacture of chitosan nano composite membranes

4.2 Material Selection for Membrane Materials

Selecting the optimum combination of material and process cannot be performed at one certain stage; it should gradually evolve during the different stages of product development. The function of each component is identified; the primary design and material requirements are identified followed by the secondary requirements identification. The initial screening of materials occurs by finding the critical requirements of each part to define the performance requirements of the material. All materials available are studied, and then a narrow down selection occurs based on the rigid requirements. The material performance requirements are divided into functional requirements, processability requirements, cost, reliability requirements, and resistance to service conditions. Having specified the material requirements, the rest of the selection process involves the search for the material that would best meet those requirements. The starting point is the entire range of engineering materials. At this stage, it is essential to open up channels in different directions. Steel may be the best material for one design concept while a plastic is best for a different concept, even though the two designs provide similar functions. The importance of this phase is that it creates alternatives without much regard to their feasibility.

In this work, the mechanical properties of polymeric membranes are critical properties for water filtration applications. They mechanical properties include polymer crystallinity, and addition of plasticizers. The degree of crystallinity affects the Young's modulus, tensile strength and ductility. On the other hand, plasticizers are little- measured particles. They are embedded between chains and debilitate the intermolecular strengths between bonds. They upgrade the processability and mechanical properties of the polymers [23, 24]. Additionally, mechanical properties are vulnerable to the surrounding environment, such as water or temperature. Rising the temperature reduce the elastic modulus and the tensile strength. However, it improves the ductility due to the softening effect on the polymer chains [25]. Permeability is another fundamental parameter that material selection. The dispersal of nanofillers such as graphene or fullerene into the polymer lattice influences the obstruction properties of a film by making a turned way for dissemination. The addition of fullerene and graphene nanofillers incorporated with the chitosan membrane (crosslinked chitosan CLCS and non-crosslinked chitosan (NCLCS) decreased the diffusion rate and increase the barrier effect as shown in Fig. 3. The barrier effect of fullerene nanofiller was smaller amount than that of graphene because of the difference in structure and size of nanofiller is shown in Fig. 3 [26]. Figure indicates the effect of graphene on decreasing the pore size of chitosan membranes, whereas fullerene was spread inside the pores allowing higher permeability compared to graphene. Graphene was selected as the nanofiller for controlling the pore size in this study.

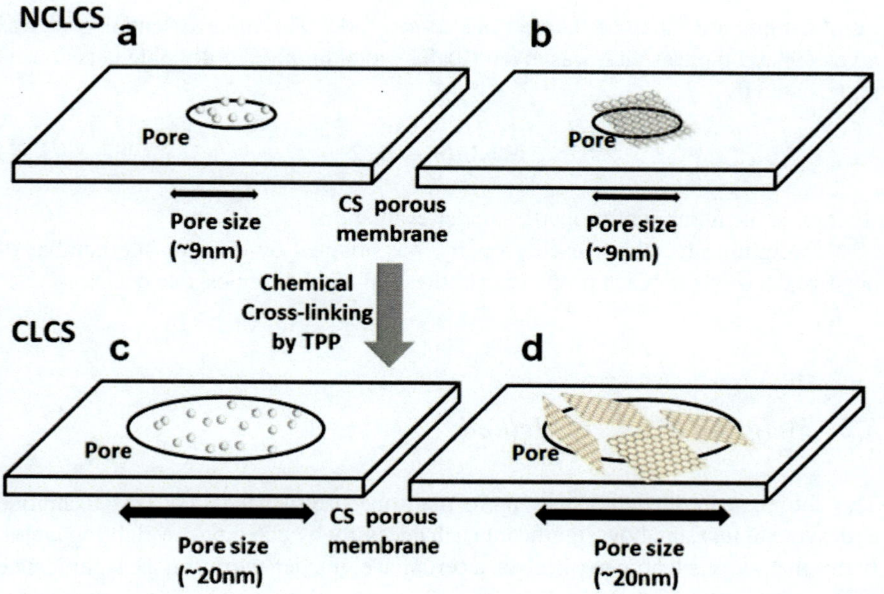

Fig. 3 A schematic illustration of chitosan membranes and their pore sizes before (NCLC) and after physical cross-linking (CLCS) by TTP and after addition of fullerene (**a**, **c**) and graphene (**b**, **d**)

5 Process Selection Procedure

The general goal of process selection is performed to demonstrate the optimum materials for water filtration application. In this work the manufacture of membranes needs a good knowledge of materials properties. The selection job is very challenging due to the enormous number of materials commercially available. The main properties in this work are the mechanical strength and the porosity of three synthetic polymers; Nylon, LDPE, PTFE and one natural polymer which is chitosan reinforced with graphene nanofiller.

5.1 Digital Logic Method

The purpose of using the digital logic (DL) method is the determination of the importance of the selected properties. The selected properties for the water filtration membranes were porosity, ductility biodegradability, biocompatibility, antimicrobial effect, toxicity, carcinogenic effect and tensile strength. Two properties were selected at a time. Each mix of properties is looked at and several decisions are required, just a yes or no choice for every assessment. To decide the general significance of every property a table is built, the properties are recorded in the left-hand section, and examinations are made in the sections to one side. The more essential goal was given (1) and the less vital was given (0). The total number of possible decisions is represented by

$$N = n(n - 1)/2 \tag{1}$$

where n is the number of properties under evaluation.

"A weighting factor for each property was attained by dividing the number of positive decisions for each property over the total number of possible decisions (N)" [27].

5.2 Weighted Property Method

The chosen material guarantees more than one useful necessity. This technique endeavors to measure how significant each necessity by deciding a weighting factor. In this method, each property is given a certain weight, depending on the significance [27].

5.3 Scaled Properties

The weighted-properties method has the drawback of having to combine unlike units, which could yield irrational results. This is particularly true when different mechanical, physical, and chemical properties with widely different numerical values are combined. The property with higher numerical value will have more influence than is warranted by its weighting factor. This drawback is overcome by introducing scaling factors. For scaling material properties, each property is scaled with a highest numerical value equals 100. It depends whether the value specified is a maximum or a minimum value. For a given property, the scaled value, B is the numerical value of property × 100 divided by the maximum value in the list. For properties where a lesser value is required. B is calculated as: minimum value in the list × 100 divided by the numerical value of the property [27].

5.4 Performance Index

For material properties that can be addressed to by numerical qualities, application of the above technique is basic. Numerical qualities are occasionally given, and materials are generally evaluated as excellent, great, reasonable, poor, and so forth. After scaling the properties, the performance is calculated [27].

$$\gamma = \Sigma Bi\alpha i \tag{2}$$

where, γ is Performance index, B_i is Scaled property and α_i is Weighing Factor.

6 Results

Porosity and tensile strength are the main selected factors. Table 1. demonstrate the two properties with highest weighing factor. Table 2 shows the pore size of nylon, LDPE, PTFE and chitosan nano composite membrane and their tensile strength respectively. The pore size and tensile strength are scaled in Table 3. The pore size of chitosan nanocomposite membrane is 20 nm. They of are rated 100, since small pore size represent higher barrier properties in membranes. For, the tensile strength, LDPE membranes were rated 100, since they had the highest tensile strength. The performance indices for LDPE, Nylon, PTFE and chitosan nano composite were calculated as shown in Table 4.

Table 1 Positive decisions for the selected properties and weighing factor

Goals	Positive decisions	Weighing factor
Porosity	*7*	*0.28*
Ductility	4	0.16
Biodegradability	4	0.16
Biocompatibility	0	0
Antimicrobial effect	1	0.04
Tensile strength	*5*	*0.2*
Carcinogenic effect	1	0.04
Toxicity	3	0.12
Total	25	1

Table 2 Pore size and tensile strength

	Pore size (nm)	Tensile strength (MPa)
LDPE	200	95
Nylon	125	65
PTFE	60	57
Chitosan, graphene nanocomposite	20	73

Table 3 Scaled properties

	Pore size (nm)	Tensile strength (MPa)
LDPE	10	100
Nylon	16	68
PTFE	33	60
Chitosan, graphene nanocomposite	100	76

Table 4 Performance indices

	Pore size (nm)	Tensile strength (MPa)	$\gamma = \Sigma\beta_i\alpha_i$
LDPE	2.8	20	22.8
Nylon	4.48	13.6	18.08
PTFE	9.24	12	21.24
Chitosan, graphene nanocomposite	28	15.2	*43.2*

7 Conclusions

Oil-based plastic materials have been generally utilized since mid-twentieth century as a result of their minimal effort, simple processability, quality, and fantastic physiochemical properties. However, these synthetic plastic materials have a genuine downside, they are not biodegradable thus increasing the plastic waste. Chitosan nano composites membranes were selected for the fabrication for water filtration membranes since chitosan is a natural material, biodegradable, biocompatible and has relatively low toxicity. The novelty of this work lies in producing environmentally friendly substitute for water filtration membranes if performance, processing and cost are controlled.

8 Future Prospects

The fabrication of membranes using electrospinning technique can be used in variety of applications like filtration, tissue engineering, and drug delivery. The membrane properties are controlled by their physical properties and pore arrangement and their morphology. A comparison between the fabricated polymer nano composite membranes and electro spun membranes will assess the performance of the fabricated PNC membranes. The reinforcement of membranes with nanofibers will be compared to nanofillers in this study. Nanofibers guarantee uniform size and distribution, thus enhance the membrane filtration properties. The nanofibers technology offers excellent adhesion and structural stability of the membranes [28].

References

1. Leo CP et al (2013) Potential of nanofiltration and low-pressure reverse osmosis in the removal of phosphorus for aquaculture. Water Sci Technol 67(4):831–837
2. Yu B, Jiang X (2013) A fractional anomalous diffusion model and numerical simulation for sodium ion transport in the intestinal wall. Adv Math Phys 1–8
3. Li N, Fane A, Ho W, Matsuura T (eds) (2008) Advanced membrane technology and applications. Wiley, Hoboken
4. Erkanl M, Yilmaz L, Çulfaz-Emecen PZ, Yetis U (2017) Brackish water recovery from reactive dyeing wastewater via ultrafiltration. J Clean Produ 165:1204–1214
5. Marino T, Blasi E, Tornaghi S, Di Nicolò E, Figoli A (2018) Polyethersulfone membranes prepared with Rhodiasolv® Polarclean as water soluble green solvent. J Membr Sci 549:192–204
6. Ahmed F, Lalia BS, Kochkodan V, Hilal N, Hashaikeh R (2016) Electrically conductive polymeric membranes for fouling prevention and detection: a review. Desalination 391:1–15
7. Laohaprapanon S, Vanderlipe AD, Doma BT, You SJ (2017) Self-cleaning and antifouling properties of plasma-grafted poly (vinylidene fluoride) membrane coated with ZnO for water treatment. J Taiwan Inst Chem Eng 70:15–22
8. Pandey RP, Shukla G, Manohar M, Shahi VK (2016) Graphene oxide based nanohybrid proton exchange membranes for fuel cell applications: an overview. Adv Coll Interface Sci 240:15–30
9. Chang Y, Shen Y, Kong D, Ning J, Xiao Z, Liang J, Zhi L (2017) Fabrication of the reduced preoxidized graphene-based nanofiltration membranes with tunable porosity and good performance. RSC Adv 7(5):2544–2549
10. Rao C, Sood A, Voggu R, Subrahmanyam K (2010) Some Novel Attributes of Graphene. J Phys Chem Lett 1:572–580
11. Wujcik E, Monty C (2013) Nanotechnology for implantable sensors: Carbon nanotubes and graphene in medicine. Wiley Interdiscipl Rev Nanomed Nanobiotechnol 233–249
12. Pinnau I, Freemanm BD (2009) Formation and modification of polymeric membranes. American Chemical Society, Washington
13. Daubresse C, Sergent-Engelen T, Ferain E, Schneider Y, Legras R (1995) Characterization of energetic heavy ion track in PVDF: production of PVDF track-etched membrane and application. Nuclear Instrum Methods Phys Res Sect B Beam Interact Mater Atoms 105:126–129
14. Gowin P, Faibish R, El-Dessouky H, Ettouney H (2002) Evaluating the economics of desalination. Chem Eng Prog 98(12):32–39
15. Baker R (2004) Membrane technology and applications, 2nd edn. Wiley, Chichester
16. Nunes S, Peinemann K (2001) Membrane technology in the chemical industry. Wiley-VCH, Weinheim
17. Wu K, Qiu P, Wang Z, Wang J (2002) Large pore size polyacrylonitrile membrane for ultrafiltration. J Environ Sci 12(2):241–251
18. Srinivasa P, Ramesh M, Kumar K, Tharanathan R (2004) Properties of chitosan films prepared under different drying conditions. J Food Eng 63(1):79–85
19. Sebti I, Chollet E, Degraeve P, Noel C, Peyrol E (2017) Water sensitivity, antimicrobial, and physicochemical analyses of edible films based on HPMC and/or chitosan. J Agricult Food Chem 55(3):693–699
20. Guiping W, Shuying G, Longzhe C, Youyi X (2008) Preparation and characterization of PES/TiO_2 composite membranes. Appl Surf Sci 254:7080–7086
21. Hyeok C, Elias S, Dionysios D (2006) Sol–gel preparation of mesoporous photocatalytic TiO_2 films and TiO_2/Al_2O_3 composite membranes for environmental applications. Appl Catal B 63(1):60–67
22. Fahim I, Marei N, Salem H, Mamdouh W (2015) Effect of graphene and fullerene nanofillers on controlling the pore size and physicochemical properties of chitosan nanocomposite mesoporous membranes. J Nanomater 1–10

23. Alberti G, Narducci R, Sganappa M (2008) Effects of hydrothermal/thermal treatments on the water-uptake of Nafion membranes and relations with changes of conformation, counter-elastic force and tensile modulus of the matrix. J Power Sources 178:575–583
24. Satterfield M, Benziger J (2009) Viscoelastic properties of Nafion at elevated temperature and humidity. J Polym Sci Part B Polym Phys 47:11–24
25. Majsztrik P, Bocarsly A, Benziger J (2008) Viscoelastic response of nafion. effects of temperature and hydration on tensile creep. Macromolecules 41:9849–9862
26. Nguyen Q, Baird D (2006) Preparation of polymer–clay nanocomposites and their properties. Adv Polym Technol 25:270–285
27. Farag M. (1979). Materials and process selection for engineering design. London Applied Science Publisher
28. Advantages of A New and Advanced Nanofiber Coating Technology for Filtration Media Compared to The Electrospinning Process (n.d.) Retrieved March 21, 2020 from https://www.hollingsworth-vose.com

Graphene Reinforced Biopolymer Nanocomposites in Energy Storage Applications

Yagyadatta Goswami and Shreya Sharma

Abstract Graphene reinforced biopolymer has lately empowered the dramatic enhancement of portable electronics and electric vehicles by proffering better means for storing electricity. To fulfill the consistently expanding demand for lightweight, compact electronic items, electric vehicles, brilliant lattices with sustainable power source combinations, the hybridization of graphene with several functions have been manifested. Being versatile and powerful procedure to essentially upgrade the performance of different energy frameworks in light of the fact that such hybridization can bring about synergistic impacts that consolidate finest qualities of incorporated segments and profer new capacities and properties. Consequently improving the charge/release efficiencies and abilities, vitality/power densities, and cycle life of these vitality stockpiling frameworks. The proposed chapter discusses the implication of graphene reinforced biopolymer nanocomposites for their utilization in supercapacitors, photovoltaic cells and lithium ion batteries. The modeling of Photovoltaic (PV) cell has been discussed in the later part of the chapter to examine the PV attributes and characteristics of PV cell. The future possibilities and bearings on the investigation of graphene reinforced biopolymer nanocomposites toward the design and development of practical, high-class, and even newly featured vitality stockpiling materials are discussed.

Keywords Biopolymer · Nanocomposites · Battery · Photovoltaic cell · Graphene

1 Introduction

The province of nanotechnology comprises prodigious global consumption of energy and significant discharge of carbon dioxide have prompted another wilderness for

Y. Goswami (✉)
Department of Instrumentation and Control Engineering, Netaji Subhas University of Technology, New Delhi, India
e-mail: yagya1992@gmail.com

S. Sharma
Department of Chemistry, Netaji Subhas University of Technology, New Delhi, India

B. Sharma and P. Jain (eds.), *Graphene Based Biopolymer Nanocomposites*,
Composites Science and Technology,
https://doi.org/10.1007/978-981-15-9180-8_13

research [1–3]. Global energy demand is projected to reach around 778 Etta Joule (18 million tonnes of oil equivalent) by 2035 and carbon-dioxide emissions are anticipated to escalate from 29 Giga-tonne (Gt) per year to 36–43 Gt per year [4, 5]. To mitigate and to adapt to current changes, endeavors have been ventured up to design and build energy frameworks to foster efficient development and exploitation of several intensively distributed clean energy sources and direct the burgeoning of development focus from conventional fossil fuels to clean energy [6].

Carbon-based polymer nanocomposites are perceived as a perfect contender for their use in advanced energy storage devices owing to their remarkable properties including configuration adaptability, lightweight, the comfort of processing, and flexibility to requisites [7]. The wonder material, Graphene is sp^2 hybridized, where the inplane σ_{c-c} bonds are the most grounded bonds and *pi* bonds lie out of the plane, which brings about a delocalized system of electrons, favor the electronic conduction of graphene and proffer a frail communication among graphene layers, or among graphene and a substrate. Contrarily, graphene can be modified by simple arrangement techniques that empower its simple use for different disciplines, Interestingly, the substitution of various sorts of molecules (N, B, S and P) with carbon in graphene further enhance the conductivity and electrochemical properties of graphene. The synergistic impacts of graphene composited with different nanostructured materials like metals, metal oxides, sulfides, alloys and polymeric materials make it propitious for energy storage as depicted in Fig. 1 [8, 9]. Henceforth, graphene, among variety of its potential counterparts, has gained tremendous popularity and has found broad applications in solar cells, fuel cells, supercapacitor, and lithium batteries [10–12].

Biopolymers on its own natural properties are not conductive, but rather procure conductive characteristics through the ascendancy of added conductive metals. Biopolymers are reinforced with various nanoparticle material to combine both the characteristics of the polymer and the metal. The integrated characteristics of biobased nanocomposites being biodegradability, renewability biostability, biocompatibility, carbon-neutrality, and electrical conductivity proffer biopolymer-based composites and nanocomposites a naturally agreeable material [13, 14].

2 Why Biopolymers?

The level of research has considerably enhanced the utilization of biopolymers obtained from products or natural renewable assets from the living organisms including chitosan, soy protein, starch, collagen, hemicellulose, cellulose, and lignin or synthetic biopolymers. An enormous number of research work has been attributed to the utilization of biopolymer-based composites for the creation of electromagnetic shielding materials, actuators, microwave safeguards, sensors, supercapacitors, etc. Biopolymers have progressively discovered their direction into this space because of the environmental threats posed by non-biodegradable, non-renewable, and non-environment friendly petroleum-based polymers. They are broadly utilized for several electrical components and devices. A schematic graph of measured heating

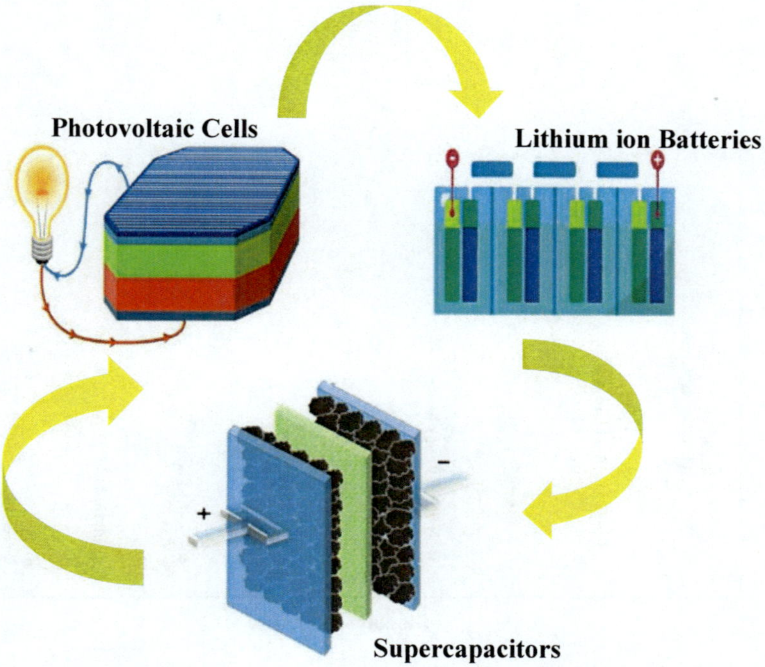

Fig. 1 Graphene reinforced biopolymer nanocomposites used for energy storage

values of different biopolymers is depicted in Fig. 2. Biopolymer frameworks have been fortified with various conductive miniaturized scale and nanostructured metals to deliver composite and nanocomposite materials for different practical implications. Such polymers commingle both the characteristics of the metals and the polymers to shape a conductive polymer.

3 Will the World Get Its Next Energy Fix?

In spite of developing consideration on clean energy, non-renewable energy sources still account for 80% of worldwide vitality utilization and 75% of greenhouse gas emission. Carbon nanostructured materials based biopolymer nanocomposites can possibly eliminate the risks while bestowing scope of economic benefits for organizations and communities to flourish.

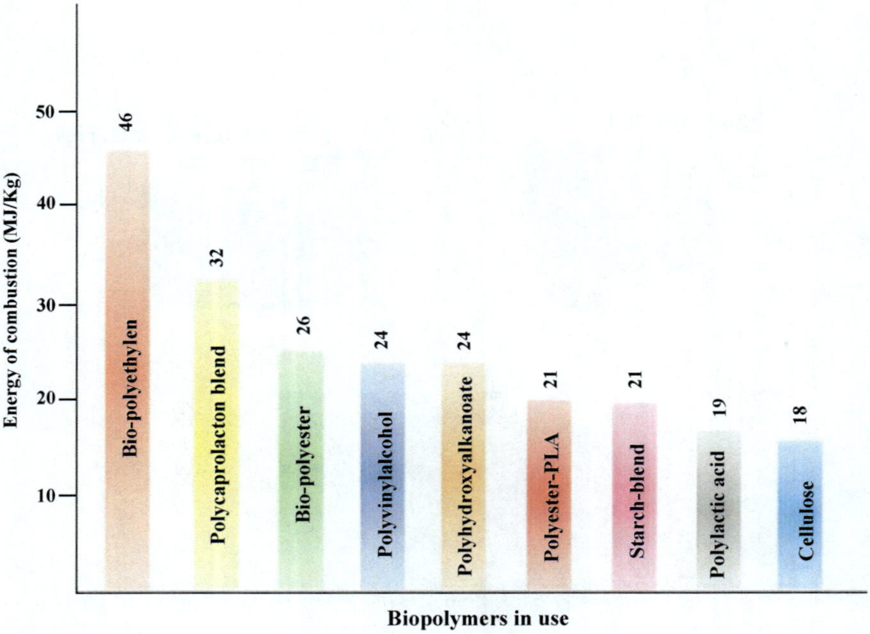

Fig. 2 Measured heating values of different biopolymers will aid in their utilization in enrgy storage applications

3.1 *Graphene Reinforced Biopolymer Nanocomposites in Supercapacitors*

Advancement of profoundly efficient supercapacitors has been one of the most encouraging researches in the scholarly community. Traditional supercapacitors comprise of two electrodes, which stay in contact with a separator and an electrolyte. Generally, for a supercapacitor, it is opined that the positively charged ions move towards the cathode and the negatively charged ions move toward the anode, framing two layers of capacitive stockpiling where the capacitance of the cell is dependent on the individual capacitance value of the two electrodes [15–17]:

$$(C_{cell})^{-1} = C_1^{-1} + C_2^{-1} \tag{1}$$

The stored electrical energy is given by:

$$E = CV^2/2 = QV/2 \tag{2}$$

where E, Q, V represents the amassed electrical vitality, stored charge, voltage, and capacitance of the cell respectively.

Despite the fact that supercapacitors have high force inferable from the nearness of the double layer system, the resistance in series results low output power density. The energy density of a supercapacitor is given by:

$$E = C_{cell}V^2/2m \tag{3}$$

The power density is introduced as:

$$P = E/\Delta t \tag{4}$$

The maximum power density is introduced as:

$$P_{max} = V^2/4mR_s \tag{5}$$

where P_{max}, m, Δt, and R_s represent the maximum power yield, mass of the active material, release time, and equivalent resistance in series, respectively. The leakage current and self-discharge rate decides the overall performance of the supercapacitor, while its cyclic strength is dependent on the nature and chemistry of electrode materials. The essential difference between batteries and supercapacitors is the mechanism of energy storage. For batteries, the energy is stored in the bulk of chemical reactants with the ability to produce charge, while for supercapacitors, energy is stored legitimately as surface charge. Battery release rates will constrain the power execution because of the mass transport and the reaction kinetics, while supercapacitors have high force capacity during both release and charge [18, 19]. There exist three kinds of materials that can be used for manufacturing the terminals-AC materials with high surface region, oxides of progress metals, and electroactive polymers [1].

Biopolymers are reinforced with various nanoparticles to combine both the characteristics of the polymer and the metal. The integrated characteristics of biobased nanocomposites being biodegradability, renewability biostability, biocompatibility, carbon-neutrality and electrical conductivity proffer biopolymer-based composites and nanocomposites a naturally agreeable material. Lately, graphene has attained a great deal of enthusiasm for electrochemical two-fold layer capacitor applications. Graphene reinforced material manifests great specific capacitance, yet its esteem diminishes as the number of cycles increments. Also, their restricted limit and rate limit incarcerate the energy and power density of battery. However, biopolymers have stable capacitance maintenance with somewhat lower explicit capacitance in comparison with graphene-based material [20]. Hence, a constructive combination of biopolymer with graphene is of gigantic importance for high specific capacitance and functional rate capabilities. Shekhar et al. investigated the electrochemical and capacitive properties of chitosan/rGO bionanohybrid. The material demonstrated diffusion controlled in alkaline medium with good electrical conductivity and low resistance [21].

Zhanghua Zou et al. designed a flexible all solid-state supercapacitor by free-standing cellulose/rGO/Ag nanocomposite film. The three-electrode setup demonstrated marvelous supercapacitive performance and performed better than many existing graphene-based symmetric and asymmetric flexible supercapacitors with maximum areal specific capacitance of 683.8 mF cm^{-2} at the current density of 5 mA cm^{-2}, alluring rate ability with the areal specific capacitance up to 347.9 mF cm^{-2} at the current density of 40 mA cm^{-2}, demonstrated no considerable difference in electrochemical properties under bending, and the highest areal energy density of 95.0 μ Wh cm^{-2} at an areal power density of 2.5 mW cm^{-2} [22]. Sevilla et al. established astounding capacitative performance in both solid and liquid electrolytes for the designed graphene-cellulose composite film [23]. Graphene doped lignin/polyacrylonitrile based carbon nanofibre reported in a recent publication demonstrated outstanding potential for electrochemical performance [24]. A supercapacitor device designed from free standing Graphene-PEDOT/PSS films exemplified the functioning of a green LED light. High areal capacitance of 448 mF cm^{-2} accomplished at an output pace of 10 mV s^{-1} with a high mass stacking (8.49 mg cm^{-2}), revealed its potential to be utilized in practical applications [25]. Dependent on three-dimensional nanostructured graphene froth/PVA/formaldehyde and activated carbon terminals, asymmetrical supercapacitors have been designed for high-performance energy storage applications [26].

3.2 Graphene Reinforced Biopolymer Nanocomposites in Lithium Ion Batteries

So as to keep pace with the headway of computers, the battery innovation has additionally enhanced. A great deal from nickel–cadmium (NiCd) and nickel-metal hydride (NiMH) have utilized for remote correspondences, mobile computing and ultimately to lithium-ion batteries with high limit [27]. Progression of electrolyte and the electrode which can hold on against several charge–discharge cycles with minimum capacity fading becomes a challenge for the development of the batteries. High volumetric vitality, insignificant self-release, high power density, and long cycle life have led to its increased demand in the industry. However, it has been discovered that traditional cathodes (including $LiFePO_4$, $LiCoO_2$, $LiMn_2O_4$, and $Li_3V_2(PO_4)_3$), and anodes (including silicon, graphite, and tin) are based and even oxides have individual downsides. Metallic oxides have great cyclic solidness with high reversible limit misfortune. Moreover, carbon-based materials display high cycle execution; be that as it may, they need efficiency. Electrodes comprised of conventional materials experience resulting extension and compression, prompting harm of the anodes. Carbon nanomaterials with high conductivity, high electron versatility, and superconductivity, alongside less resistive heating, are studied for their utilization in lithium-ion batteries [1].

Wang et al. designed lithium-ion phosphate, cellulose nanofibre reinforced graphene electrode for implementation in lithium-ion cells with initial discharge capacity of 151 mAh/g (0.1 C) and a sound capacity retention rate with only 5% loss after 60 cycles due to satisfactory electrolyte wettability at the interface [28]. Luing

Yu et al. focused on an approach to tackle the issue of silicon (Si) reciprocating volume change and poor electrical conductivity during charge/discharge by the arrangement of *N*-doped reduced graphene oxide/Chitosan@Silicon anode material. Biomass amorphous carbon layer and reduced graphene oxide eased the pressure keeping up anode uprightness during lithiation/delithiation and made up for the poor conductivity of silicon. The wettability of electrolyte and electrode interface was further enhanced on doping with nitrogen. The structured *N*-doped rGO/C@Si anode material was reported to exhibit superior cycling performance (1115.8 mAh g^{-1} after 150 cycles at 420 mA g^{-1}) and better rate ability (1077.4 mAh g^{-1} at 4200 mA g^{-1}) [29]. Reduced graphene oxide reinforced chitosan nanocomposite was investigated for amplification of capacitative performance of Li–S batteries [30]. Poly (vinyl alcohol) reinforced graphene demonstrated remarkable electrochemical and cycling performance. The bionanocomposite suggested potential applications for their utilization in lithium ion batteries [31].

3.3 Graphene Reinforced Biopolymer Nanocomposites in Solar Cells

Biopolymers have been utilized as perfect contenders for the Photovoltaic (PV) cell. Biopolymer composite for PV cells can be utilized as photoelectrode layer, substrate, electrolyte, color, etc. An organic electronic material sandwiched between two metal terminals of various work functions constitutes the single-layer organic PV cell. Whereas, two layers of organic electronic materials comprising of electron donor and acceptor molecules sandwiched between the conductive electrodes constitute the bilayer PV cell. As opposed to single and bilayer organic PV cells, mass heterojunction PV cells have higher vitality transformation proficiency because of an intermixed composite of contributor and acceptor as they guarantee a bigger interface region [26]. However, the main challenges put forward by the biopolymer-based solar cells including device stability, reliability, and energy conversion efficiency are dependent on the choice of nanostructured materials, electrolyte, and the preparation strategies.

Relative to the customary battery research, the study on graphene-based biopolymers of sun-powered cells have become a favored research area because of the energy conversion efficiency of graphene and capacity to go about as a PV character. Following the procedures of coating, layering, reduction, and temperature toughening, the graphene in the electrode turn into a natural/inorganic half breed. Regardless of the high work capacity and conductivity exhibited by graphene, graphene polymer-based transparent electrodes have a constraint of 65% of light transmittance. The flexibility of graphene layers permits the sunlight based cell to twist up to 78 °C more than unadulterated indium trioxide electrodes. Electron transporter and acceptor based graphene polymer sun oriented cells rely upon a high electron affinity so as to isolate the electron gap sets into discrete charges. The availability of several acceptor/donor sites and a constant pathway for successful electron transfer is corroborated because of the presence of a larger surface area. A hole transport layer is requisite in sun oriented cells so as to put an end to spillage of the current flow and charge recombination. 2-nm graphene film is considered as the ideal thickness contrasted with the thick film, which can forestall the transmittance of electrons and increments electrical obstruction. High electrical conductivity coupled with 98% transparency of the monolayers of graphene proffers graphene-based materials conceivably appropriate to deliver issues identified with photoexcitation and exciton portability/dissemination as transparent conducting electrodes [1, 32, 33]. Cellulose nanofibre graphene nanoplatelets are contemplated to be excellent candidates. For instance, Valentini et al. designed an optically transparent film of cellulose nanofibre reinforced graphene nanoplatelet and accomplished a higher short circuit current density in comparison to the indium trioxide substrate. The results further inferred that the optimization of mechanical properties of cellulose nanofibre would result into a fully recyclable solar cell technology [34]. Weijie Ye et al. designed cellulose/graphene composite and tested them for electrochemical as well as photothermal properties. The nanocomposite exhibited remarkable photothermal properties for a temperature as high as 75 °C after 1 min of irradiation [35]. As of now sunlight based PV cells/modules have a generally low effectiveness level contrasted not just with the effectiveness of customary petroleum products yet in addition to the proficiency of other sustainable power sources, for example, wind or hydro. Improving the effectiveness of the PV module framework by following the worldwide most extreme force from the PV framework with and without fractional concealing state is contemplated as a hot innovative field.

3.3.1 Modeling Studies

Of the various comparable model circuits for depicting the identical electrical network of PV cell, Single Diode Model (additionally, known as the single diode Rp-model), the Double Diode Model and the Triple Diode Model have been acknowledged and utilized in a large portion of the detailed literary works [36–45].

Single Diode Model (SDM) of a PV Cell

This segment introduces the SDM of a PV cell along with PV module model. In this modeling, different parameters have been identified to entirely delineate the nonlinear $I-V$ characteristics of a PV cell/module. In SDM, a photovoltaic cell is contemplated as a current source (I_{ph}) which is in parallel with a rectifying diode (D) along with a shunt resistance (R_{sh}) as shown in Fig. 3. Material resistivity and ohmic loss at contacts are represented by the series resistance, R_s [46].

When illuminated output current is given by [47]:

$$I = I_{ph} - I_d - I_{sh} \tag{6}$$

Fig. 3 Single diode model of a photovoltaic cell

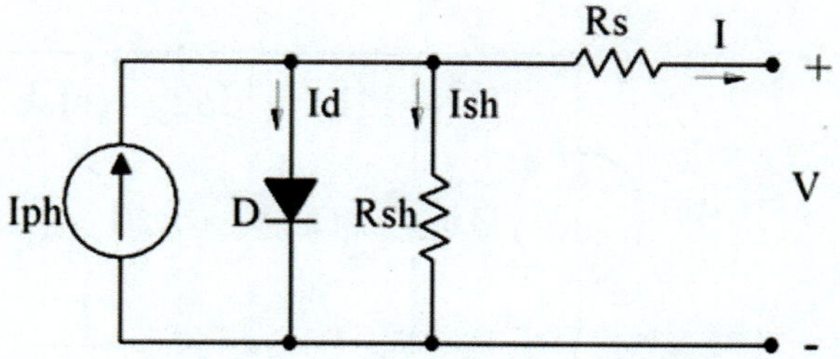

In deficiency of sunlight, cell performs as a p–n junction diode, which represents diode current as:

$$I_d = I_o\left[e^{\frac{(V+IR_s)}{aV_t}} - 1\right] \tag{7}$$

where, V_t is the thermal voltage of the diode and is given by:

$$V_t = \frac{kT}{q} \tag{8}$$

I_{sh} is obtained as:

$$I_{sh} = \left[\frac{(V + IR_s)}{R_{sh}}\right] \tag{9}$$

By using Eqs. (7) and (9), Eq. (6) is modified as:

$$I = \left[I_{ph} - I_o\left[e^{\frac{(V+IR_s)}{aV_t}} - 1\right] - \frac{(V + IR_s)}{R_{sh}}\right] \tag{10}$$

The non-linear $I-V$ characteristic is obtained by Eq. (10). The values of five parameters, namely I_o, I_{ph},R_s,R_{sh} and a are required to be estimated in SDM of PV cell.

Single Diode Model (SDM) of a PV Module

A SDM based PV module constitutes an amalgamation of series and parallel connections of SDM based PV cells. The $I-V$ characteristics of a PV module utilizing the use of number of parallel PV cell strings (N_p) coupled with number of series PV cells string (N_s) is given by Eq. (11) [48].

$$I = \left[I_{ph}N_p - I_oN_p\left[e^{\frac{q\left(\frac{V}{N_s} + \frac{IR_s}{N_p}\right)}{akT}} - 1\right] - \frac{\frac{VN_p}{N_s} + IR_s}{R_{sh}}\right] \tag{11}$$

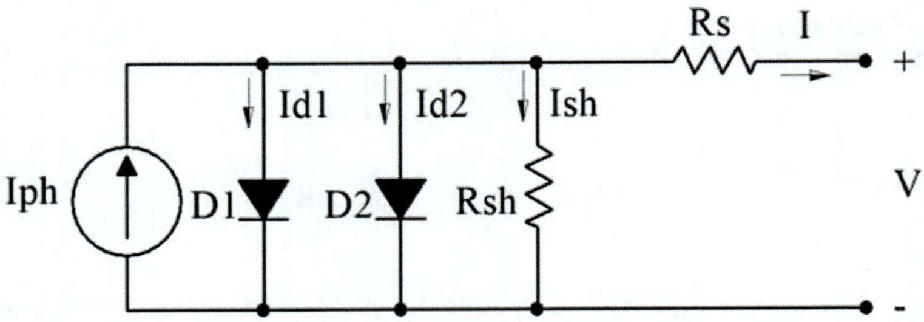

Fig. 4 Double diode model of a photovoltaic cell

Double Diode Model (DDM) of a PV Cell

This segment gives the DDM of a PV cell alongside PV module model. The parameters have been distinguished to totally portray the nonlinear $I-V$ attributes of a photovoltaic cell/module. The above depicted SDM has certain natural downsides, essentially because of the suspicion that the ideality factor of diode (a) remains consistent all through the variety of yield voltage. Be that as it may, diode ideality factor is a function of voltage across the device. SDM additionally disregards the impact of recombination at intersection at lower voltages (close to open circuit voltage Voc). Hence, ideality factor is near 2 at lower voltages, [46]. Considering the aforementioned issues and to produce the model progressively sensible, the current because of recombination at intersection is demonstrated by including a second diode in corresponding with the primary dispersion diode prompting DDM of a PV cell as appeared in Fig. 4 [44].

When illuminated conditions, the output current is given by [44]:

$$I = I_{ph} - I_{d1} - I_{d2} - I_{sh} \tag{12}$$

As described for SDM, each cell behaves as a p–n junction diode and gives the diode currents as:

$$I_{d1} = I_{o1}\left[e^{\frac{(V+IR_s)}{a_1 V_t}} - 1\right] \tag{13}$$

and

$$I_{d2} = I_{o2}\left[e^{\frac{(V+IR_s)}{a_2 V_t}} - 1\right] \tag{14}$$

Using KVL, I_{sh} is obtained, as:

$$I_{sh} = \left[\frac{(V + IR_s)}{R_{sh}}\right] \tag{15}$$

By using Eqs. (13), (14) and (15), Eq. (12) is modified as:

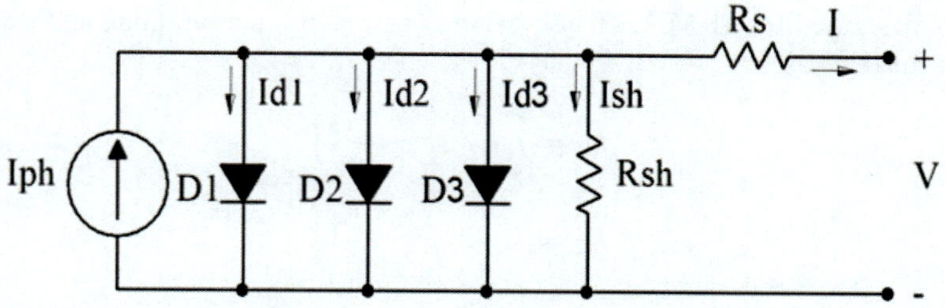

Fig. 5 Three diode model of a PV cell

$$I = \left[I_{ph} - I_{o1}\left[e^{\frac{(V+IR_s)}{a_1 V_t}} - 1 \right] - I_{o2}\left[e^{\frac{(V+IR_s)}{a_2 V_t}} - 1 \right] - \frac{(V+IR_s)}{R_{sh}} \right] \tag{16}$$

As seen from Eq. (16), seven parameters are required to be estimated which are $I_{o1}, I_{o2}, I_{ph}, R_s, R_{sh}, a_1, a_2$ for DDM of PV cell.

Double Diode Model (DDM) of a PV Module

A DDM based PV module constitutes an amalgamation of series and parallel connections of DDM based PV cells. The $I-V$ characteristics of a PV module utilizing number of parallel (N_p) strings along with number of cells in series (N_s) in a string is given by Eq. (17) [49].

$$I = \left[I_{ph} N_p - I_{o1} N_p \left[e^{\frac{q\left(\frac{V}{N_s} + \frac{IR_s}{N_p}\right)}{a_1 kT}} - 1 \right] - I_{o2} N_p \left[e^{\frac{q\left(\frac{V}{N_s} + \frac{IR_s}{N_p}\right)}{a_2 kT}} - 1 \right] - \frac{\frac{VN_p}{N_s} + IR_s}{R_{sh}} \right] \tag{17}$$

Triple Diode Model (TDM) of a PV Cell

In this model, to overcome the limitation posed by DDM of different current components due to the variation of ideality factor, a three diode-based model is presented. In this model, a third diode is added in parallel with the two diodes of DDM. The three diode current as shown in Fig. 5, the diode current (I_{d1}) which is due to diffusion and recombination in the quasi-neutral regions of the PN junction, second diode current (I_{d2}) due to recombination in the space charge region and the third diode current (I_{d3}) is modeled due to the recombination in the defect regions, grain sites, etc.

Under illuminated conditions, the output current is given by [45]:

$$I = I_{ph} - I_{d1} - I_{d2} - I_{d3} - I_{sh} \tag{18}$$

As described for SDM, each cell behaves as a p–n junction diode and gives the diode currents as:

$$I_{d1} = I_{o1}\left[e^{\frac{(V+IR_s)}{a_1 V_t}} - 1\right] \tag{19}$$

$$I_{d2} = I_{o2}\left[e^{\frac{(V+IR_s)}{a_2 V_t}} - 1\right] \tag{20}$$

and

$$I_{d3} = I_{o3}\left[e^{\frac{(V+IR_s)}{a_3 V_t}} - 1\right] \tag{21}$$

I_{sh} is obtained as:

$$I_{sh} = \left[\frac{(V + IR_s)}{R_{sh}}\right] \tag{22}$$

Substituting Eqs. (19), (20), (21) and (22) in (18), results in Eq. (23)

$$I = \left[I_{ph} - I_{o1}\left[e^{\frac{(V+IR_s)}{a_1 V_t}} - 1\right] - I_{o2}\left[e^{\frac{(V+IR_s)}{a_2 V_t}} - 1\right] - I_{o3}\left[e^{\frac{(V+IR_s)}{a_3 V_t}} - 1\right] - \frac{(V + IR_s)}{R_{sh}}\right] \tag{23}$$

As seen from Eq. (23), seven parameters, namely, I_{o1},I_{o2}, $I_{o3}I_{ph}$, R_s, R_{sh}, a_3 are required to be estimated while a_1, a_2 are taken as 1 and 2, respectively for TDM of PV cell.

In this chapter, a solar panel (KC200GT by M/S Kyocera) has been used for investigations. The $I-V$ and $P-V$ non-linear characteristics are shown in Figs. 6 and 7 along with the parameters of the module in Table 1 at Standard Test Condition (STC) [50]. Further, Fig. 8a, b show the $I-V$ and $P-V$ characteristic for different solar insolation at constant temperature (25 °C), respectively. Additionally, the $I-V$ and $P-V$ characteristic for different temperature of 25, 50 and 75 °C at constant solar insolation (1000 W/m^2) are presented in Fig. 9a, b, respectively.

4 Limitations and Challenges

There are still many challenges to be addressed to realize the practical applications of graphene reinforced biopolymer-based energy storage devices. To enhance the characteristics, the regulation of the distribution, density, kinds of chemical bonds, as well as three-dimensional arrangements of the composites is a requisite. Several assembly methodologies have been developed for the constant demand of property optimization; the particular efforts are directed towards the design and formation of specially constructed hybrid architectures rather than random mixtures.

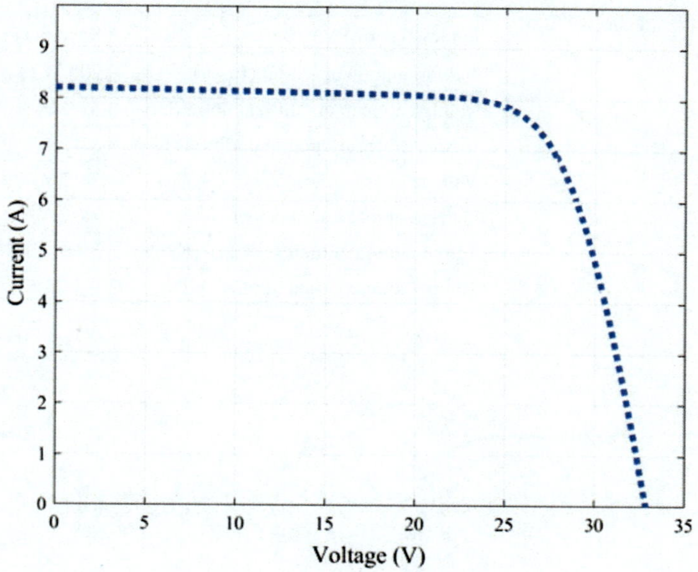

Fig. 6 $I-V$ characteristics of solar cell

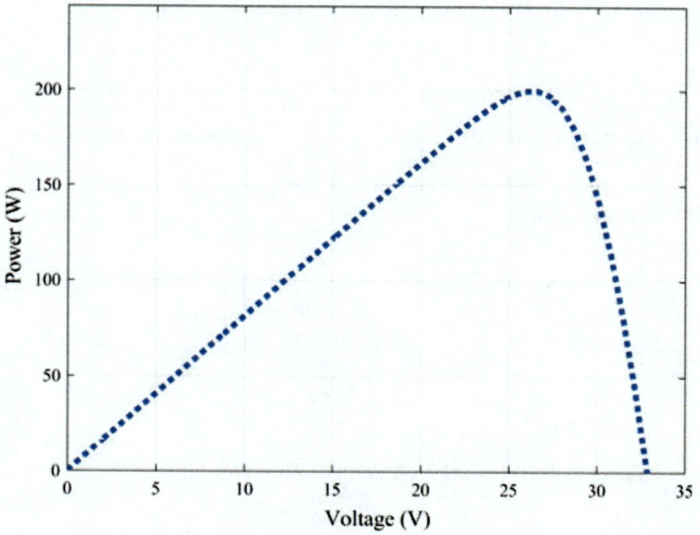

Fig. 7 $P-V$ characteristics of solar cell

Table 1 Parameters of KC200 GT module [51]

Parameters of PV cell	Specification
Maximum power output (P_{MPP})	200 W (+10%/−5%)
Voltage at open circuit condition (V_{OC})	32.9 V
Current at short circuit condition (I_{SC})	8.21 A
Voltage at maximum power (V_{MPP})	26.3 V
Current at maximum power (I_{MPP})	7.61 A
Temperature coefficient of I_{SC} (K_I)	3.18×10^{-3} A/°C
Temperature coefficient of V_{OC} (K_V)	-1.23×10^{-1} V/°C

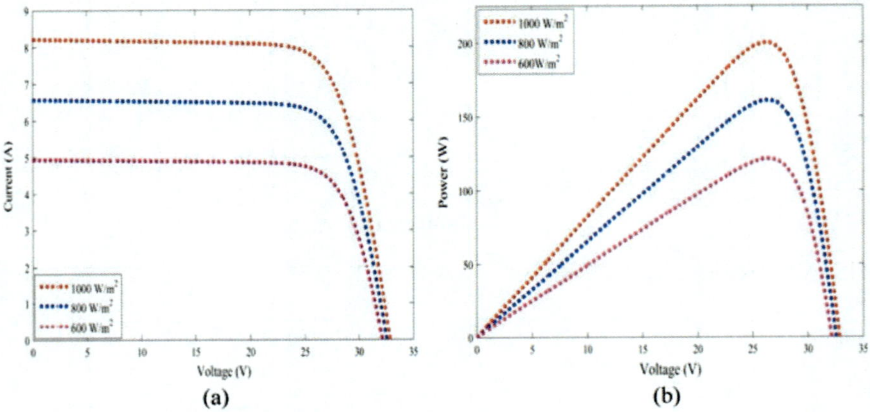

Fig. 8 (**a**) $I-V$ characteristics (**b**) $P-V$ characteristics at different irradiation conditions

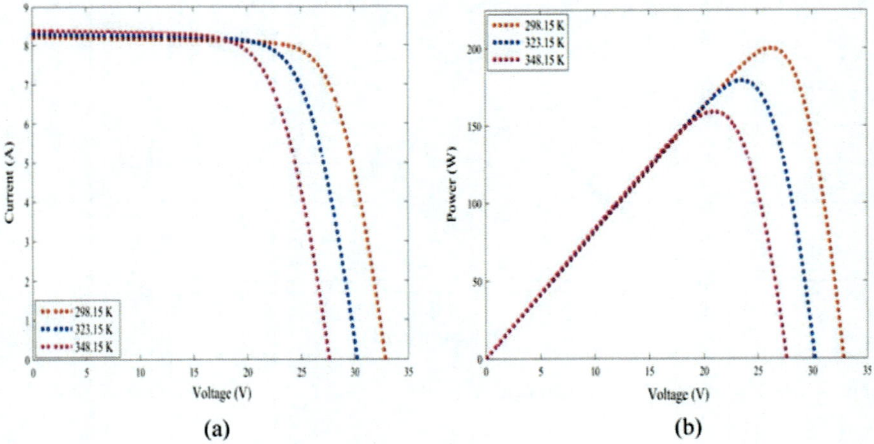

Fig. 9 (**a**) $I-V$ characteristics (**b**) $P-V$ characteristics at different temperature conditions

- Although researchers have depicted graphene-based biopolymer batteries and supercapacitors with high-performance characteristics far surpassing those of commercially accessible ones, the absence of achievable procedures for the large scale manufacturing of top-notch graphene limits their latent capacity.
- Significant effort must be orchestrated for beneficial enlargement of the specific surface area, porosity, reduction of the Oxygen/Carbon ratio and to enhance the conductivity of the electrode. During the composite formation, graphene layers restacked and the actual surface area of the graphene is not exposed.
- Estimates for the expense of the creation of graphene differ contingent upon the nature of the material from tens to thousands of dollars for every kilogram, but it is still not competitive with state-of-the-art materials. For instance, the extreme ease of initiated carbon release right now utilized in supercapacitors (US$15 per kilogram) presents a troublesome boundary to the passage of different materials.
- Undoubtedly graphene-based biopolymer nanocomposites have changed the landscape of energy storage because of its remarkable electrochemical properties and unique combinations, the overall prospect is yet to be realized. Many challenges remain, particularly concerning feasible techniques for the low-cost mass production of graphene with controlled microstructure and low residual oxygen content.
- More research efforts should be made for resolving the existing challenges using theoretical calculations together with experimental investigations. Furthermore, the knowledge of the interaction of graphene sheets at the nanoscale to form self-assembled structures of different shapes and dimensionalities will drive more promising implications of graphene and the nanocomposites. Graphene research is expected to continue to expand rapidly during the next decade with the promise that it will make a real impact on our lives.

5 Conclusion and Future Trends

The upsurge in the stipulation of clean, sustainable power sources for the age of economical society prompted the design of propelled vitality capacity gadgets, as supercapacitors, energy units, sun oriented cells, and Li-ion batteries. Graphene and leading biopolymers have the option to tie down a noteworthy situation so as to encounter several implications in the field of vitality stockpiling gadgets. Be that as it may, there subsist lot problems to scale up and commercialize these stockpiling sources and capacity gadgets with the consideration of nanostructured materials. Consequently, ongoing examination has been engaged to improve productivity and execution. The biopolymer composites are adaptable and can be effortlessly designed. However, the acceptor–donor issue on account of directing polymers is as of now limiting the implication in photovoltaic cells. The effective photon-engrossing limit, capacity to produce photocarriers, great photovoltaic characteristics, and their capacity to isolate charge transporters to frame heterojunctions mark their potential for implication in photovoltaic cells. Furthermore, modeling of PV cell/module necessarily provides with the electrical behavior of the practical cell. Supercapacitors have high force densities; nonetheless, their vitality densities are lower contrasted with traditional batteries. The capacitance of a supercapacitor not just rely on on the particular surface area of graphene yet additionally on a few different variables, for example, the amalgamation of graphene, pore size, pore size appropriation, etc.

Carbon nanomaterials assume a huge job in creating anodes for Li-ion batteries. High electron conductivity and high surface area of nanostructured materials lead to improved rate capacity and charge stockpiling. Be that as it may, the significant expense of the materials is as yet constraining their commercial implication.

Acknowledgements Authors would like to acknowledge Netaji Subhas University of Technology for providing facilities and financial support.

References

1. Banerjee J, Dutta K, Rana D (2019) Carbon nanomaterials in renewable energy production and storage applications. In: Emerging nanostructured materials for energy and environmental science. Springer, Cham, pp 51–104
2. Huang Y, Zhu Y, Bao S, Kang S, Lu J (2019) Interfacial polymerisation of polyaniline/$Fe(OH)_3$/graphene nanocomposites and its energy storage properties. Mater Technol 35:1–12
3. Zhu J, Deng D (2018) Design and synthesis of structured particles for next-generation lithium-ion batteries. In: Anisotropic particle assemblies. Elsevier, pp 261–278
4. Wang H, Liang X, Wang J, Jiao S, Xue D (2020) Multifunctional inorganic nanomaterials for energy applications. Nanoscale 12(1):14–42
5. Rafiee A, Khalilpour KR (2019) Renewable hybridization of oil and gas supply chains. In: Polygeneration with polystorage for chemical and energy hubs. Academic Press, pp 331–372
6. Liu Z (2015) Global energy interconnection. Academic Press
7. Siwal SS, Zhang Q, Devi N, Thakur VK (2020) Carbon-based polymer nanocomposite for high-performance energy storage applications. Polymers 12(3):505
8. Mahmood N, Zhang C, Yin H, Hou Y (2014) Graphene-based nanocomposites for energy storage and conversion in lithium batteries, supercapacitors and fuel cells. J Mater Chem A 2(1):15–32
9. Pollak E, Geng B, Jeon KJ, Lucas IT, Richardson TJ, Wang F, Kostecki R (2010) The interaction of Li+ with single-layer and few-layer graphene. Nano Lett 10(9):3386–3388
10. Young RJ, Kinloch IA, Gong L, Novoselov KS (2012) The mechanics of graphene nanocomposites: a review. Compos Sci Technol 72(12):1459–1476
11. Kim H, Abdala AA, Macosko CW (2010) Graphene/polymer nanocomposites. Macromolecules 43(16):6515–6530
12. Cai D, Song M (2010) Recent advance in functionalized graphene/polymer nanocomposites. J Mater Chem 20(37):7906–7915
13. Ibrahim ID, Sadiku ER, Jamiru T, Hamam Y, Alayli Y, Eze AA, Kupolati WK (2019) Biopolymer composites and bionanocomposites for energy applications. In: Green biopolymers and their nanocomposites. Springer, Singapore, pp 313–341
14. Jian M, Zhang Y, Liu Z (2020) Natural biopolymers for flexible sensing and energy devices. Chin J Polym Sci 38(5):1–32
15. Pandolfo AG, Hollenkamp AF (2006) Carbon properties and their role in supercapacitors. J Power Sources 157(1):11–27

16. Wang G, Zhang L, Zhang J (2012) A review of electrode materials for electrochemical supercapacitors. Chem Soc Rev 41(2):797–828
17. Frackowiak E, Beguin F (2001) Carbon materials for the electrochemical storage of energy in capacitors. Carbon 39(6):937–950
18. Balli B, Şavk A, Şen F (2019) Graphene and polymer composites for supercapacitor applications. In: Nanocarbon and its composites. Woodhead Publishing, pp 123–151
19. Arico AS, Bruce P, Scrosati B, Tarascon JM, Van Schalkwijk W (2011) Nanostructured materials for advanced energy conversion and storage devices. In: Materials for sustainable energy: a collection of peer-reviewed research and review articles from Nature Publishing Group, pp 148–159
20. Selvaraj T, Perumal V, Khor SF, Anthony LS, Gopinath SC, Mohamed NM (2020) The recent development of polysaccharides biomaterials and their performance for supercapacitor applications. Mater Res Bull 126:110839
21. Shekhar S, Sharma R, Sharma S, Sharma B, Sarkar A, Jain P (2020) An exploration of electrocatalytic analysis and antibacterial efficacy of electrically conductive poly (D-glucosamine)/graphene oxide bionanohybrid. Carbohydr Polym 240:116242
22. Zou Z, Zhou W, Zhang Y, Yu H, Hu C, Xiao W (2019) High-performance flexible all-solid-state supercapacitor constructed by free-standing cellulose/reduced graphene oxide/silver nanoparticles composite film. Chem Eng J 357:45–55
23. Sevilla M, Ferrero GA, Fuertes AB (2016) Graphene-cellulose tissue composites for high power supercapacitors. Energy Storage Mater 5:33–42
24. Dai Z, Ren PG, Jin YL, Zhang H, Ren F, Zhang Q (2019) Nitrogen-sulphur Co-doped graphenes modified electrospun lignin/polyacrylonitrile-based carbon nanofiber as high performance supercapacitor. J Power Sources 437:226937
25. Liu Y, Weng B, Razal JM, Xu Q, Zhao C, Hou Y, Seyedin S, Jalili R, Wallace GG, Chen J (2015) High-performance flexible all-solid-state supercapacitor from large free-standing graphene-PEDOT/PSS films. Sci Rep 5:17045
26. Bello A, Barzegar F, Momodu D, Dangbegnon J, Taghizadeh F, Fabiane M, Manyala N (2015) Asymmetric supercapacitor based on nanostructured graphene foam/polyvinyl alcohol/formaldehyde and activated carbon electrodes. J Power Sources 273:305–311
27. Banerjee J, Dutta K (2017) Materials for electrodes of Li-ion batteries: issues related to stress development. Crit Rev Solid State Mater Sci 42(3):218–238
28. Mohiuddin M, Kumar B, Haque S (2017) Biopolymer composites in photovoltaics and photodetectors. In: Biopolym Compos Electron. Elsevier, pp 459–486
29. Wang Y, He ZY, Wang YX, Fan C, Peng QL, Chen JJ, Feng ZS (2018) Preparation and characterization of flexible lithium iron phosphate/graphene/cellulose electrode for lithium ion batteries. J Colloid Interface Sci 512:398–403
30. Yu L, Liu J, He S, Huang C, Gong Z, Gan L, Long M (2020) N-doped rGO/C@Si composites using sustainable chitosan as the carbon source for lithium-ion batteries. Appl Surf Sci 501:144136
31. Kim S, Cho M, Lee Y (2020) Multifunctional Chitosan–rGO network binder for enhancing the cycle stability of Li–S batteries. Adv Func Mater 30(10):1907680
32. Hu M, Zhou Y, Nie W, Chen P (2018) Functionalized graphene nanosheets with fewer defects prepared via sodium alginate assisted direct exfoliation of graphite in aqueous media for lithium-ion batteries. ACS Appl Nano Mater 1(4):1985–1994
33. Omar G, Salim MA, Mizah BR, Kamarolzaman AA, Nadlene R (2019) Electronic applications of functionalized graphene nanocomposites. In: Functionalized graphene nanocomposites and their derivatives. Elsevier, pp 245–263
34. Tsang CHA, Huang H, Xuan J, Wang H, Leung DYC (2020) Graphene materials in green energy applications: recent development and future perspective. Renew Sustain Energy Rev 120:109656
35. Ye W, Li X, Zhu H, Wang X, Wang S, Wang H, Sun R (2016) Green fabrication of cellulose/graphene composite in ionic liquid and its electrochemical and photothermal properties. Chem Eng J 299:45–55

36. Zhang Y, Hao N, Lin X, Nie S (2020) Emerging challenges in the thermal management of cellulose nanofibril-based supercapacitors, lithium-ion batteries and solar cells: a review. Carbohydr Polym, 115888
37. Green MA (1987) High efficiency silicon solar cells. In: Seventh EC photovoltaic solar energy conference. Springer, The Netherlands
38. Chin VJ, Salam Z, Ishaque K (2015) Cell modelling and model parameters estimation techniques for photovoltaic simulator application: a review. Appl Energy 154:500–519
39. Jervase JA, Bourdoucen H, Al-Lawati A (2001) Solar cell parameter extraction using genetic algorithms. Measur Sci Technol 12(11):1922
40. Zagrouba M, Sellami A, Bouaïcha M, Ksouri M (2010) Identification of PV solar cells and modules parameters using the genetic algorithms: application to maximum power extraction. Sol Energy 84(5):860–866
41. Ye M, Wang X, Yousheng Xu (2009) Parameter extraction of solar cells using particle swarm optimization. J Appl Phys 105(9):094502
42. Wei H, Cong J, Lingyun X (2011) Extracting solar cell model parameters based on chaos particle swarm algorithm. In: 2011 international conference on electric information and control engineering (ICEICE). IEEE
43. Ishaque K, Salam Z (2011) An improved modeling method to determine the model parameters of photovoltaic (PV) modules using differential evolution (DE). Sol Energy 85(9):2349–2359
44. Gong W, Cai Z (2013) Parameter extraction of solar cell models using repaired adaptive differential evolution. Solar Energy 94:209–220
45. El-Naggar KM, AlRashidi MR, AlHajri MF, Al-Othman AK (2012) Simulated annealing algorithm for photovoltaic parameters identification. Solar Energy 1(86):266–274
46. Khanna V, Das BK, Bisht D, Vandana SPK (2015) A three diode model for industrial solar cells and estimation of solar cell parameters using PSO algorithm. Renew Energy 78:105–113
47. El Tayyan AA (2015) An approach to extract the parameters of solar cells from their illuminated IV curves using the Lambert W function. Turkish J Phys 39(1):1–5
48. Easwarakhanthan T, Bottin J, Bouhouch I, Boutrit C (1986) Nonlinear minimization algorithm for determining the solar cell parameters with microcomputers. Int J Sol Energy 4(1):1–12
49. Hamid NFA (2016) Nasrudin Abd Rahim, Jeyraj Selvaraj, Solar cell parameters identification using hybrid Nelder-Mead and modified particle swarm optimization. J Renew Sustain Energy 8(1):015502
50. Ram JP, Babu TS, Dragicevic T, Rajasekar N (2017) A new hybrid bee pollinator flower pollination algorithm for solar PV parameter estimation. Energy Convers Manage 135:463–476
51. KC200GT high efficiency multicrystal photovoltaic module datasheet. https://www.energymatters.com.au/images/kyocera/KC200GT.pdf. Accessed on 20 Nov 2019

Functionalization of Graphene Based Biopolymer Nanocomposites for Packaging and Building Applications

Prakash Chander Thapliyal and Neeraj Kumar

Abstract Graphene has attracted interests in both academic and industrial areas in recent times because it can produce improvement in material properties even with low quantity. In fact, utilization of graphene-based materials in the fabrication of various nanocomposites with different polymer matrixes has been continuously explored all over the world. There are, however, several types of graphene containing powder materials known such as graphene oxide, graphene nanoplatelets, graphene nanoribbons, graphene quantum dots and graphene enabled products such as graphene ink or graphene master batches. Flexible packaging is one of important areas where incorporating graphene in to polymer matrix in most cases among other improves the thermal, mechanical and barrier properties to a larger extent. In addition, well known antibacterial and/or antimicrobial activity of most of the graphene based materials can also be utilized for packaging purpose. Another interesting and important possible newer application of graphene-based biopolymers is in the thermal management in buildings for cooling and heating purposes. This chapter presents and discusses the development of graphene-based biopolymer nanocomposites and their applications in packaging and building applications with special focus on protective coatings and phase change materials.

Keywords Graphene · Packaging · Building applications · Biopolymers · Composites

P. C. Thapliyal (✉)
Advanced Structural Composites and Durability Group, CSIR-Central Building Research Institute, Roorkee 247667, India
e-mail: pcthapliyal@cbri.res.in; pct866@yahoo.com

N. Kumar
Building Energy Efficiency Group, CSIR-Central Building Research Institute, Roorkee 247667, India
e-mail: kumar.nrj.neeraj@gmail.com

B. Sharma and P. Jain (eds.), *Graphene Based Biopolymer Nanocomposites*, Composites Science and Technology,
https://doi.org/10.1007/978-981-15-9180-8_14

1 Introduction

Graphene based biopolymer nanocomposites can act as barrier materials. Barrier is a basic requirement for providing shelf life to the various products such as packaging materials and some building materials. In general, products require barrier all the time on account of their inherent complex chemical formulations and associated degradation with time. Most common barrier properties are:

- Moisture vapour barrier
- Gas barrier–Oxygen, Carbon dioxide
- UV light barrier.

One can have single, dual or multiple barriers for packaging and protective coatings as per requirements and uses. All materials have different level of barrier properties. Permeability properties determine the packaging and coating characteristic of the materials. Packaging requirements depend on product's shelf life which in turn is co-related with climatic zones. The ICH guideline classifies them into following zones:

- Zone I: Temperate
- Zone II: Mediterranean/Subtropical
- Zone III: Hot Dry
- Zone IVA: Hot Humid/Tropical
- Zone IVB: Hot/Higher Humidity.

Materials with higher packaging barrier strength will have higher shelf life based on climatic conditions. It is possible to have a material having higher gas barrier but lower moisture vapour permeability and vice versa. Higher the barrier property, higher is the coat of a packaging material. Typically shelf life requirements govern the packaging requirements depending on climatic zone conditions. Graphene was first isolated by Dr. A. Geim and Dr. K. Novoselov at the University of Manchester in 2004 and for this work both researchers received the Nobel Prize in Physics in 2010. Graphene based biopolymer nanocomposites show enhanced mechanical, thermal, gas barrier, electrical and flame-retardant properties, when compared to the other biopolymers [1–3]. It was also reported by several researchers from time to time that the improvement in mechanical and electrical properties of graphene-based biopolymer nanocomposites are much better in comparison to use of the clay or other carbon filler-based nanocomposites [4–6].

2 Packaging Applications

In recent few decades, the packaging materials are big challenge in term of disposal and to protect the environment. The waste disposal of packaging materials is harming the environment due to use of non-biodegradable or plastics materials as food packaging and increasing significantly. These materials are produced from petroleum products and create environmental problems. So, to save and protect our environment, alternate biodegradable materials have been developed for packaging applications. The biopolymer leaves behind the organic bio-products, which are biodegradable and environmental friendly [7, 8]. Today, biopolymers have gained attention between researchers to replace the commercial used plastic packaging materials due to increase the importance of sustainable development [9]. The biopolymers are renewable in nature, biodegradable and leave less carbon footprint so they can be used to develop better biodegradable materials required for packaging and may offer a better solution to protect the environment [10]. Therefore, biopolymers can be applied to resolve the environmental issue and to develop the biodegradable packaging materials [7, 11]. The main challenges associated with the current biopolymers are to achieve the good stability and mechanical properties, water vapour barrier and weak mechanical properties, which is limiting the applications of packaging materials on industrial scale [12]. Beside of their mechanical properties, some problems also associate with bio-based materials such as their cost and difficult to process in compare to plastic or commercial available non-biodegradable materials in packaging [12, 13]. The non-biodegradable packaging material has become more challenging for environment. Therefore, currently more and more research is focused on the biodegradable materials for food packaging from renewable source [12]. The biodegradable polymers made from renewable resources can help us to reduce the dependence on petroleum-based polymers used in packaging and reduce the waste material, while still requiring a product that having the similar properties with commercially available plastics [14]. So, improvement of the properties of the biopolymer can be made possible by using filler like carbon nanotubes, nanoclays, graphene and their related materials. Graphene got much more attention as compare to carbon nanotubes and nanoclays due to their outstanding electronic and mechanical properties, low density, high aspect ratio and low cost [15]. Graphene has good compatibility with biopolymers due to π-conjugation [11, 16]. In recent years, graphene and their related materials gained enormous attention in various fields. Graphene has honeycomb like structure with sp^2 hybridized carbon and compatible with biopolymers [11]. It has been used to design the several functionalized materials with biopolymers. Graphene oxide (GO) is derivative of graphene with abundance of oxygen functional groups like epoxy, hydroxyl and carboxyl groups. In polymeric structures, GO enhances the properties of host polymers by generating some new properties due to presence of strong hydrogen bonding [17]. The biopolymer nanocomposite is playing significant role to improve the chemical, mechanical and physical properties [18]. In biopolymer nanocomposites, the biopolymer acts as matrix, while the dispersing materials (nanofillers) improve the functional properties. Biopolymer nanocomposites have a set of improve properties like physical and mechanical properties due to matrix-filler interface. These properties in biopolymers do not come naturally but can be gained by addition of nanocomposite or nanofiller [19]. On reviewing the literature, it was observed that biodegradable biopolymers such as starch, cellulose, polyamides, chitosan, protein, polysaccharides and chemically synthesized polymer derive from natural monomers (polylactic acid) have been applied to synthesis the biopolymer-based nanocomposites [9–11, 18, 20, 21]. The graphene related materials can form more heat resistant and higher barrier nanocomposite which is promising for the food packaging [9]. The nanocomposite is promising a better method to increase the biopolymer properties. The composites of these biopolymers have been prepared with graphene related materials including reduced graphene oxide, graphene oxide, graphene nanoplatelets etc. [9, 15, 18–23].

2.1 Starch

Starch is known biologically produced materials and second most abundant after cellulose. It is also known as one of the significant biopolymer to remove the fossil resources. Starch is a biocompatible, renewable and biodegradable naturally occurring polymer, which is frequently applied to develop the environmentally friendly materials and compatible with different polymers. It has been used to prepare the biodegradable film due to cost effective, renewable, good processability and easy to chemical and physical modifications. However, because of low barrier properties, poor mechanical properties, high sensitivity to moisture, these properties limited the development of starch-based materials [24]. Numerous polymers and different nanofillers have been used to improve the water barrier properties, mechanical properties and other properties of the starch [14]. The graphene related materials were also used to improve up on the properties of the starch-based nanocomposites. The starch and graphene based nanocomposite have good compatibility, high thermal properties and resistance to moisture uptake [15, 22, 24, 25]. Jose et al. prepared the poly (vinyl alcohol)/starch-based blend with graphene-based polymer nanocomposite. The maximum tensile strength was noticed at 5% wt. of graphene used in poly (vinyl alcohol)/starch blend nanocomposite. The increment in the thermal properties of the nanocomposites material was noticed while the crystallinity of the poly (vinyl alcohol) decreases due to presence of graphene and starch [15]. Biopolymers were used to minimize the plastic wastes for environmental sustainability. Over the last few decades poly (vinyl alcohol) has been considered one of the main synthetic biopolymers. The poly (vinyl alcohol) on blending with glycerol and starch improves the compatibility and biodegradability even at reduction in material cost. The poly (vinyl alcohol)/starch/glycerol composite films had poor water resistance due to these belonging to hydrophilic materials. The addition of halloysite nanotubes improves the water resistance of nanocomposite film due to hydrogen bonding between polymers and nanofillers [26]. Zhou et al. proposed the pea starch and poly (lactic acid) bilayer and aloe vera banana starch based packaging materials respectively. The interaction between pea starch and poly (lactic acid) layers, water resistance, morphology, barrier properties, mechanical and thermal properties of bilayer were investigated. The barrier and mechanical properties of bilayer films were improved due to synergistic effect between them, and this has played a significant role in food packaging [27]. Suaza et al. synthesized the aloe vera gel, banana starch and curcumin loaded acetylated and native starch nanoparticles composite films and characterized. The water vapour properties improved due hydrophobic nature of curcumin. The properties of the starch have been improved by using the chemical methods or formation of composites with more hydrophobic compounds such as lipids and chitosan compounds [28].

2.1.1 Starch and Graphene-Based Nanocomposites

The polylactide (PLA) and starch both are represent the commercially and environmental friendly bio-based polymers. These are among those bio-based plastic polymer which are used for short-term and packaging applications. Several studies have been found that on addition of starch in PLA, reduce the cost of PLA/starch composites and also maintain the biodegradable nature of the polymer. The poor interfacial adhesion was observed in between hydrophilic starch and hydrophobic PLA in PLA/starch. However, the interfacial adhesion was improved by using nano-size graphene oxide (nGO). While in composite, the amount of nGO was very low but nGO increase the barrier and mechanical properties and thermal stability of the PLA/starch composite. The nano-size graphene oxide (nGO) was capable in crystallization of PLA in composite materials by providing number of nucleation sites. The compatibility of PLA and starch has been achieved by using nGO in PLA/starch composite. The nGO induced the reinforced interfacial adhesion and crystallization between starch and PLA, and also the mechanical and thermal properties of the PLA/starch composite [17].

Ashori et al. synthesised the starch and chitosan-based composites film with graphene. Biopolymers containing starch and chitosan composite film with graphene have potential applications in food, agriculture and pharmacy. While the film only made of chitosan or starch have lack of water resistance, less stability and hold low mechanical properties. So, to obtain the biodegradable, miscible films of starch and chitosan nano-particles is a good substitute. The properties of the starch and chitosan were improved by using graphene nano-particles. The composite films have improvement in tensile moduli and tensile strength, and showed decreasing elongation at breaks with graphene amount. By increasing the graphene content in biopolymers composite, it was demonstrated the significant reduction in elongation values. Irrespective of biopolymers, the obtained results showed the increment in the graphene content, decreased the water vapour transmission rate and moisture uptake and important in food packaging. The high content of graphene (1.5–3%) was easily agglomerated. This behaviour decreased the mechanical properties of the composite films [22].

Aqlil et al. proposed the starch/lignin and GO based nanocomposite material and possibility in packaging application. The GO and starch/lignin blend based biopolymer nanocomposite has been prepared by using solution casting evaporation method. The films of the biopolymer nanocomposite were characterized by using FTIR, thermal gravimetric analysis (TGA), SEM, Ultraviolet–visible (UV–Vis), and tensile tests. The result showed that GO significantly influenced the mechanical properties of the produced film and also compatibility of the biopolymer with GO. The water permeability and moisture uptake of starch/lignin blend film reduced by using GO. The thermal stability of the nanocomposite film was found stable than starch film. The starch was blended with lignin by hydrogen bond and analyzed by FTIR. The hydroxyl group of starch and the oxygenated group on GO have an interconnection network with each other. These interactions improved the resulting properties of the bio-nanocomposites like moisture uptake, water swelling, thermal stability, water permeability, hydrolytic degradation and mechanical properties [29].

Ge et al. developed an effective method to prepare the reduced graphene oxide and starch composites. The reduced graphene oxide (r-GO) was functionalized with sodium dodecyl benzene sulfate. The film of the r-GO/oxidized starch (OS) was obtained by incorporating the r-GO based filler into the oxidized starch matrix. The sodium dodecyl benzene sulfate increased the hydrophilic behaviour of r-GO in OS matrix. The results obtained in the FTIR studies demonstrated that the sodium dodecyl benzene sulfate adhere to the surface of r-GO. The r-GO/OS composite films were characterized by using SEM, TEM, XRD, TGA and UV-spectroscopy. The tensile strength of r-GO/OS composites film was improved by using the r-GO filler, which was three times more than OS film. Some other properties of the r-GO/OS based composite film were also improved such as oxygen barrier properties, moisture barrier properties, thermal stability and light performing performance in term of protective against UV light. It was concluded that r-GO/OS based composites films can play significant role in packaging and other fields [24].

The starch and poly (vinyl alcohol) composite based packaging material have great importance over the petroleum based packaging due to the biodegradability, abundant and compatibility. The biopolymer nanocomposite was prepared by using melt and solution-mix method. The film of PVOH/starch/laponite were prepared with GO to improve the hygroscopic properties of composite material. Laponite was used to improve the mechanical barrier of the starch/PVOH matrix. However, the GO coating has been applied on PVOH/starch/laponite RD to enhance the water barrier of the nanocomposite. GO and laponite level have significant effects on the bionanocomposites film at $P < 0.05$ and $P < 0.01$, respectively. While water absorption was found minimum on 10% laponite RD and 25:75 graphene: lipid ratio. The nanocomposites have high potential for industrial applications and specifically in packaging [30].

Mohanty et al. proposed a nano silver decorated starch-based nanocomposite biodegradable antimicrobial packaging material. The nano silver decorated starch hybrid poly (ethymethacrylate) (PEMA) and GO based nanocomposite was prepared by in situ polymerization method. The synthesized PEMA-co-starch/GO/Ag nanocomposites were characterized by using the FTIR, SEM, XRD, UV–visible and TEM. The thermal studies of nanocomposite were performed by using the TGA. The intermolecular hydrogen bonding in PEMA-co-starch is confirmed by FTIR. The thermal stability, biodegradability, oxygen barrier properties, antimicrobial and chemical resistance properties of nanobiocomposites were improved due to good dispersion of AgNPs and GO in PEMA-co-starch. The degradation temperature and oxygen barrier properties of nanobiocomposites material increased in this copolymer. A significant improvement in tensile strength of material was found in between 0.5 to 1 wt% of AgNPs, while the oxygen barrier properties increase about 40% in 2 wt% AgNPa decorated nanocomposites material. The biodegradability of PEMA-co-starch/GO/Ag nanocomposites was observed 4.5% after 180 days. The nanobiocomposites material was found suitable for packing industry due to antimicrobial, thermal stability, high barrier properties, and biodegradability properties [31]. Afshar et al. synthesized the starch/gelatin and graphene oxide based biodegradable nanocomposite films. Experiments were executed by using Design Expert software via RSM. The gelatin/starch weight ratio and GO wt% affect the properties of the nanocomposite films. The optimized film was achieved by investigation of various properties. The nanocomposite films were characterized by SEM, FTIR, TGA and XRD. The observed results indicated that uniform dispersion of GO into the polymer matrix and as a result proper interaction was achieved in between nanoparticles and polymer. Intercalation structure was obtained via solution casting method. The presence of GO improved the thermal stability, mechanical strength, water vapour permeability and surface hydrophilicity of the starch/gelatin/GO nanocomposite films. The starch/gelatin/GO composite films can be used in food packaging application [32].

Amri et al. prepared a cassava starch and GO composites based bioplastic successfully shown improvement in the properties of the cassava starch based bioplastic by addition of GO filler. The composite of starch and GO was prepared by starch intercalation method applying glycerol plasticizer with filler GO (5–15% v/v) and at mixing time of 30 and 60 min. The addition of GO with various content improved the mechanical properties of composite. Higher the concentration of GO and more the mixing time, more was the Young's modulus and tensile strength and lower the elongation. The homogeneity of the GO in the bioplastic composite was characterized by SEM and FTIR. While, GO was characterized by using XRD. The water uptake was increased by addition of GO and decreased with increasing the mixing time demonstrated the completeness of interactions. The more GO added; more was the biodegradation rate, this attribute to hydrophilic properties of the GO. Shorter mixing time indicated the higher biodegradable rate because of empty space in bioplastic–composites as conductive position for biodegradation activities [33].

Nuona et al. prepared PLA and starch based composite materials with GO. The high cost of polylactide effectively reduced using the starch, but it has also limited mechanical properties composites. GO has been used as compatibilizer with polylactide and starch. Firstly, the natural starch was cationized and then GO encapsulated by electrostatic force between the positive charged starch and negatively charge GO. The encapsulated GO was reduced by ammonium ion on the positively charged cationic starch, which help to achieve the starch from hydrophilic to hydrophobic. The composite starch-GO/polylactide has high thermal and mechanical properties. The polylactide composite supported by GO encapsulated starch have much higher yield strength from pure polylactide. The morphology and structure of the starch-GO/polylactide composites were characterized by FTIR, SEM, XPS and Zeta potential test. The hydrophobicity of the starch was elucidated by contact angles found for starch, cationic starch, cationic-starch and polylactide 33°, 19°, 64° and 72° respectively. These advantages in composite hold the improvement in mechanical properties as shown by tensile strength testing and good thermal stability in TGA study [34]. Chen et al. prepared biodegradable composites with graphene quantum dots (GQDs) and thermoplastic starch (TPS) based materials by melt-extrusion combine with hot pressing. The composite films of GQDs/TPS were characterized as high-performance, novel, and ecofriendly to replacing the traditional commercial available plastics. The different concentrations of GQDs were added into the TPS matrix to examine the solid-state fluorescent properties and conductive properties of the GQDs/TPS composite films. The composite films of GQDs/TPS were characterized by using UV-visible spectroscopy, FTIR, surface resistance measurement, SEM, tensile testing and XRD. FTIR studies demonstrated the hydrogen bonding between –OH group of TPS and oxygen containing functional groups on GQDs. The XRD and SEM shows the uniform distribution of graphene in the TPS and mechanical studies showed the optimum loading of graphene about to 10.9 wt%. Further, increasing the GQDs loading $\leq$10.9 wt% led to agglomeration. High loading of the GQDs in the TPS reduced the transparency of the film due GQDs themselves brown and the loaded film hold greater scattering effect. The results demonstrated the ecofriendly composite film of GQDs/TPS prepared by the melt extrusion contain advantages of easy processing, low materials and production cost, biodegradable and suitability of larger production. GQDs were prepared by pyrolysis of citric acid. The high photoluminescence intensity was observed at $\leq$0.9 wt% loaded GQDs in the TPS. The developed GQDs/TPS composite have potential application in optoelectronic packaging materials [35].

Ferreira et al. developed the reduced graphene oxide (rGO) incorporated poly(lactic acid), plasticizer cornstarch and their blend. The cornstarch/poly(lactic acid) were mixed at 70:30 ratio, the incorporation of rGO was maximum 5.0 mass%, while the best dispersion was found with 1.0 mass% and 3.0 mass% in filler and starch. At high content 5.0 mass% of rGO percolated into the related network and realm of correlated sheet. These properties of the composite provided the high viscosities, high conductivity, hindered shear flow, enhanced the storage moduli, and biodegradability. Small-angle X-ray scattering was applied to analysis the dispersion; these results reflected the electric conductivity, biodegradability, capillary rheometry and dynamic mechanical analysis. The GO and rGO were characterized by using Raman spectroscopy. The rGO composites and hybrids were synthesed by melt extrusion. The hydrogen bonding in between –OH groups on plasticizer cornstarch and oxygenated groups on partially rGO could stabilize the fatal structure [36].

2.2 Chitosan

Chitosan is a polysaccharide normally obtained by alkaline deacetylation of chitin and composed of randomly distributed N-acetyl-D-glucosamine and β-linked D-glucosamine, which is known as a most abundant natural polymer on earth. It is a biodegradable biopolymer contains some unique properties like antibacterial, moisture permeability, film forming properties, biocompatible and multifunctional groups [23, 37]. The amino functional groups of chitosan can be protonated to NH_3 upon dissolution at pH 6 or below, increase the electrostatic interaction with anionic group, which help to attain the polycationic polymer and bio-composite films. The chitosan has been used in various applications like biosensors, membrane separation, biological engineering and packaging materials. But due the poor mechanical properties of chitosan hold limited applications. Therefore, to achieve the better mechanical properties of chitosan, different composites have been prepared by using nanofillers such as carbon nanotubes, graphene to improve the mechanical properties of chitosan [22, 23, 38]. In biopolymers, chitosan has gained much attention in between the researcher to develop the antimicrobial packaging materials. Several attempts have been made to achieve the biopolymer based antimicrobial packaging materials to attain the food quality and shelf life. Chitosan has antimicrobial properties and potential biopolymer for antimicrobial applications in food industry, contains biodegradable, biocompatible and nontoxic properties, as well as abundant in nature and low cost [38, 39]. The poor mechanical properties of the chitosan are a big challenge for scientific community to improve the mechanical properties of chitosan. So, to improve the mechanical properties several nanocomposites have been proposed [40].

2.2.1 Chitosan and Graphene-Based Nanocomposites

Terzioglu et al. proposed the chitosan/GO/zinc oxide composites film through the casting method. The surface morphology and properties of chitosan/GO/zinc oxide and chitosan were investigated by FTIR, UV-visible, mechanical and thermal gravimetric analysis. The observed results showed that the incorporation of GO and zinc oxide into the chitosan matrix and improvements in thermal stability and mechanical properties of chitosan biocomposite films. The zinc oxide and GO both increased the tensile strength and Young's modulus in compare to neat chitosan. The improvement in the mechanical properties of composites film was indicated by Young's modulus (1549.43–2527.72) compared to pure chitosan. However, incorporation of GO into the composites decreased the transmittance, due to the brown colour of GO. Thus, developing these composite films can protect food against UV light [40]. Barra et al. prepared biodegradable electric conductive nanocomposite materials by using r-GO and chitosan. The r-GO was prepared with hydrothermal method in presence of natural compound caffeic acid and dispersed in chitosan. The hydrothermal method is a green method and preserved the oxygen groups at GO, necessary for interaction with amino groups of chitosan. These interactions between the r-GO and chitosan increased the reinforcement of biocomposite materials, achieve the better mechanical properties and water resistance materials. The r-GO enhanced the mechanical properties and antioxidant activity of chitosan matrix, decreasing the water solubility and increases the tensile strength. The antioxidant activity, electric conductivity and mechanical properties of nanocomposite can be reflected through the filler content. The ecofriendly behaviour of bio-nanocomposite materials and with their mechanical properties, antioxidant activity and electric conductivity make this material suitable for body sensors, food packaging materials and electro-responsive biocompatible devices [41].

Lyn et al. prepared GO and chitosan composite film by solution casting method. The GO of different oxidation degree has been prepared by varying amount of potassium permanganate ($KMnO_4$) by modified and improved Hummer's method. The increment in oxygen groups on GO with increasing the $KMnO_4$ were confirmed by FTIR and Raman spectroscopy. The composite film of GO and chitosan was characterized by SEM, mechanical test, FTIR and contact angle analysis. The composite GO and chitosan were prepared by incorporation of highly oxidized GO_4 (Graphite: $KMnO_4$, 1:8) into chitosan matrix. The high degree of oxidation in the GO increased the tensile strength of the composite due to increment in the interaction between GO and chitosan. The GO_4 enhanced the mechanical properties, water vapour permeability and lower the UV transmission in compare to pure chitosan. For, chitosan/GO_4 composite film resulted in three folds improvement in Young's Modulus and tensile strength in compared to pure chitosan film. The composite film with GO_4 also had denser structure and all films were found biodegradable within 20 days in soil. The present work has been considered that the chitosan/GO4 composite films as important materials for food packaging [42].

Ahmed et al. developed the chitosan and GO hydrogel based blend material and discuss their applications in for 3-D printing and food packaging. The lactic acid and acetic acid were found suitable to prepare the chitosan solutions as observed in dielectric and rheometry measurements. The dielectric and rheological properties were significantly influenced by the temperature blew or above the gel point. The chitosan was capable to form the gel without using any cross-linking agent at below the gel temperature. On dispersion of GO in chitosan, it did not influence the gel point, however, it improved the dielectric properties, while the both G' and G" values of the solution increased in between 0.5 to 1% GO and marginally dropped at 2% GO observed with AFM study. The dispersibility, miscibility and functionality of GO into chitosan matrix were characterized by XRD, FTIR, SEM, TGA and AFM. The chitosan/GO based inks have been successfully applied to 3D print in various objects, which could significantly be applied for pharmaceutical solutions and medical right at the point need. This can be play significant role to develop the bioprinting and 3D printing in biomedical applications [43].

2.3 Cellulose

Cellulose is a most abundant biodegradable biopolymer on earth and frequently found in cell walls of the green plants. It is a main component in ligno cellulosic biomass which primarily present around 35–40% in plant cell wall. Cellulose is a polysaccharide consisted from several linear chain of β (1–4) linked D-glucose unit. The pure cellulose film contains good mechanical performance and transparency, but due the poor properties against the Ultraviolet (UV) radiation limiting their applications [44–46]. The barrier properties towards the UV radiation have significant impact on polymer films when used in biomedical products and food packaging. The barrier properties towards the UV radiation have significant impact on polymer films when used in biomedical products and food packaging. The cellulose fiber contains excellent mechanical and physical properties due to several strong hydrogen bonding network and hydroxyl groups [45, 46]. Different types of cellulose have been used as additive for preparing the composites films: nanofiber cellulose, nanowhiskers cellulose, nanocrystalline and bacterial nanocellulose. Their nanocrystals can be prepared via acidic hydrolysis of cellulose fibers and flocculation of cellulose and depends upon type of hydrolysis. The nanocellulose fiber contains several micrometer lengths and diameter less than 100 nm [44, 46, 47]. Cellulose nanowhiskers are commonly produced from the natural cellulose based substance such as bacteria, sea creature and plants through the hydrolysis methods. Cellulose nanowhiskers contains larger aspect ratios, high elastic modulus and high surface areas, and normally considered as reinforcing filler in polymer composites [45]. The biodegradable nature and other outstanding properties of nanocellulose have been used in various applications like surface modified materials, nanocomposite materials and transparent paper [48]. Cellulose is very interesting material and having various applications. But some deficiencies like poor properties against UV radiation, hydrophobic matrices and agglomeration, limiting its applications. The graphene oxide improves the tensile strength and barrier properties against UV radiation of cellulose [44, 49–51]. The UV radiation causes the some serious problem on human skin and also breaks the covalent bond in organic compounds. Beside of this, UV radiation is also known as inducer of DNA mutation and carcinogenesis. Therefore, UV shielding coating and UV protected films which can block full UV spectrum but can behave good in transmittance of visible light become a hot topic of research [44].

2.3.1 Cellulose and Graphene-Based Nanocomposite

Zhang et al. prepared the graphene oxide and cellulose based transparent film. The composite film of GO/cellulose contains high barrier properties against the UV radiation. The UV shielding properties of GO/cellulose composite films has been achieved with GO nanosheets and regenerated cellulose. The regenerated cellulose was obtained from $ZnCl_2/CaCl_2$/GO solution. The homogeneous dispersion of GO was found in composites film. The UV protective properties has been achieved with addition of GO. The addition of GO reached 2%, the UVB and UVA shielding percentage were found 55.2 and 66.7%, respectively. The developed composite films also had outstanding visible light transmittance. The GO/cellulose composite films would be suitable as the materials for packaging and protective materials [44]. Liu et al. prepared the cellulose nanowhiskers/graphene nanoplatelets (CNW/GN) composite films through solution casting method. The developed CNW/GN films showed improvement in the electrical, thermal and mechanical properties. The microstructure and morphologies of CNW/GN films were investigated by using

XRD, SEM and Raman spectroscopy. The SEM images indicated the strong interaction between CNWs and GN when CNWs randomly attached to graphene sheet. The electric conductance of CNW/GN composite films was increased with GN content 0.5 wt%. Thermal study of CNWs indicated that the initial thermal decomposition temperature and melting temperature increased 70.2, 88.4, 87.8 °C and 23.2, 29.3, 26.3 °C with addition of 0.1, 0.25 and 0.5 wt% GNs, respectively. The 33.7% improvements in the tensile strength of composites were also noticed with GN content up to 0.25 wt%. These composite films can be significantly applied for electronic and green anti-static packaging applications [45].

Ramezani et al. prepared biodegradable nanocomposite films for packaging. The nanocomposite films were prepared by using thermoplastic starch (TPS), cellulose nanofibers (CNF) and GO nanoplatelets. The GO nanoplatelets were characterized by using AFM, XRD and FTIR. Thermal properties of the composite films were characterized by TGA and DTA. The observed results demonstrate that the synergistic effect of GO nanoplatelets and cellulose nanofibers has played significant role to improve the mechanical properties of the TPS/CNF/GO nanocomposite films. The mechanical properties of samples showed that with increasing the loading of CNF up to 15 wt% in TPS/CNF composites, the elastic modulus and tensile strength was improved 527 and 66%, respectively, comparison to pure polymer film. The elastic modulus and tensile strength of starch film were increased with addition of GO nanoplatelets and cellulose nanofibers from 32 and 3 to 436 and 13 MPa, which related to 1435 and 438% enhancement respectively. The water barrier properties and oxygen permeability resistance for films containing 3 wt% GO (3GO) nanoplatelets were decreased by 30 and 78% respectively, in comparison with thermoplastic starch film. The permeability coefficient of carbon dioxide, nitrogen and oxygen for TPS/CNF/3GO film were found 0.047, 0.054 and 0.051 bar, which indicated that composite films can be used as packaging films [52].

2.4 *Other Biopolymers*

In addition to above, Babu et al. and Bhasha et al. reported in their reviews that there were few others biopolymers modified and synthesized in lab and their use for preparing nanocomposites, such as poly hydroxy alkanoates, poly (vinyl alcohol), alginate and polycaprolactone [53, 54].

3 Building Applications

In addition to potential applications in different widely distributed areas such as advanced batteries, transistors, solar cells, displays and sensors; there is keen interest in the application of graphene by itself as a coating material or as a functional additive for improving the performance of various coating technologies. In particular, several approaches have been used to develop graphene-based corrosion protection coatings where superhydrophobic nature of graphene has also been exploited. There are other many applications for conductive graphene coatings too. Some important examples of current research on graphene based coatings are highlighted in following discussions. In addition to the intrinsic mechanical properties of graphene, its low mass and low loading requirements (i.e. that you can get a big change with relatively little material) make graphene stand out as a reinforcing agent in composites and possible future applications as phase change materials. Coating objects and reinforcing composites with graphene can serve different purposes such as:

- Lightweight composites for body structures
- Lubricants with enhanced anti-wear properties
- Nanolaminates as ultra-high permeation barrier films
- Corrosion protection
- Transparent conductive coatings for photonic devices
- Radiation shielding and lightening strike protections
- Superhydrophobic coatings
- Transparent, flexible and conductive thin films
- High performance sheets for EMI shielding.

3.1 Protective Coating

Protective coating is one of the most growing area where graphene-based polymer nanocomposites have been utilized, because graphene has excellent chemical and solvent resistance, gas impermeability, antibacterial and antimicrobial properties, mechanical strength, lubricity and thermal stability. Hence, graphene can be modified and mixed with various types of polymers for synthesizing self cleaning and corrosion barrier coatings. Numerous different properties of graphene can also be used in the production of composite coatings and to obtain multifunctional layers. For homogeneous dispersion of graphene in a polymer matrix, its sheets are oxidised and properly functionalised with different organoalkoxysilanes [55]. Moreover, graphene is widely used in organic coatings [56, 57]. If added to a polymeric matrix, the graphene sheets improve the corrosion protection properties of the coating owing to their barrier effect towards water molecules and aggressive ions [58–60].

Ramanathan et al. [2] demonstrated the ability of FGS to disperse well in solution and interact intimately with polar polymers such as PMMA, thus generating interphase polymer that dramatically affected thermal and mechanical properties at loadings of 0.05 wt% [61]. Nzenguet et al. [62] prepared novel bio-nanocomposite films from wheat starch (ST) by blending with high molecular weight polyacrylamide and filled with graphene oxide nanosheets. Strong hydrogen bonding between ST and GO nanosheets was revealed from FTIR which resulted in a significant improvement of the physicochemical, thermal and mechanical properties of bio-nanocomposites. Moreover incorporation of only 1 wt% of GO into polymer blend led to a significant enhancement of tensile strength (more than 167.17%) and water vapour permeability and moisture uptake of the films significantly reduced by increasing amount of GO as nanofiller [62]. Liu et al. [63] fabricated a series of NSC/GO-Zn composite films with different GO concentrations having the layered structure confirmed by SEM. These Zinc ion-enhanced hierarchical NSC/GO-Zn composite films exhibited remarkable mechanical properties (strength of 212.6 ± 10.1 MPa and toughness of 2.51 ± 0.08 MJ m^{-3}. Furthermore, a remarkable broad spectrum bactericidal property was also observed against E. coli and S. aureus exposure. From these experimental observations it is clear that NSC/GO-Zn composite film has inherent properties and can be used in corrosion protection, aerospace, and as biological materials [63]. Poly (butylene succinate) (PBS) and poly(l-lactide) (PLLA) have found applications in bio-based thermoplastics. To improve the properties of such types of polymers, Sisti et al. [64] devised a novel multicomponent reinforcing system where nanofibrous electrospun mats of PLLA containing graphene were incorporated into the PBS matrix. The results on final biocomposite showed PBS matrix does not improve upon thermal behaviour in the presence of PLLA and graphene. However, the tensile strength of these composites showed gradual enhancement with addition of PLLA mats, and the resulting improvement found to be proportional to the number of PLLA layers and amount of graphene added. These multicomponent reinforcing systems have shown 21% increment in tensile strength [64].

Biopolymers have shown a broader prospect as an effective alternative to petroleum-based materials. Jin et al. [65] developed a novel versatile mussel-inspired modification design for tough and high-performing graphene oxide (GO)/soy protein isolate nanocomposite films. The resulting nanocomposite films exhibited favourable vapour barrier behaviour and water-resistance and can be used as protective coatings [65]. Ebrahimzadeh et al. [66] prepared carboxymethyl cellulose based bio-nanocomposite and graphene nano-platelets (GNPs) and studied their morphological, thermal, light barrier, water vapor permeability, contact angle, moisture absorption and mechanical properties. It was found that by adding 0.5% w/w GNPs to polymer matrix, ultimate tensile strength came down from 7.74 MPa (in the pure film) to 5.69 MPa with strain at break increasing from 12% to more than 19% and thus reducing light transmission and increasing the water repelling nature of nano-biocomposites [66]. Rissa et al. [67] showed that Kaolin based barrier coatings can replace fluorocarbons in extruded polymer barrier coatings. Arora and Padua [9] evaluated changes in water resistance of zein coated paper as a result of added kaolinite. Water absorption and oil permeation rate reduced and water vapour permeation decreased by 50% due to the formation of a tiled nanoplate structure [67, 68]. Chiu and Huang prepared polyamide GO nanocomposites by solution mixing technique. The morphological observations showed the compatibility between PA46 and GO. Small amounts of GO facilitate the nucleation and overall crystallization of PA46. The stiffness increased with GO and hence wear life of coatings increased in comparison to the neat coating [69–71]. Cai et al. [72] reported development of highly stiff PU graphene nanocomposites. Young's modulus of modified PU has been improved by nearly nine times with the addition of 4.4 wt% FG due to the strong covalent interface resulting from the reaction between the hydroxyl groups on FG and terminal isocyanate group of PU chains. Nanoscratch test carried out on these samples reflected their better protective ability as surface coatings [72]. Later Cai et al. in 2016 prepared polyaniline/graphene (PANI/RGO) composite anti-corrosion coatings with improved performance by in situ polymerization using PANI/RGO as anti-corrosive filler and waterborne polyurethane as matrix and results showed that the composite coatings reinforced by 0.75 wt% of PANI/RGO composites resulted in superior anti-corrosive performance with the 4 wt% graphene content [73].

Water sorption of sodium alginate significantly reduced by the inclusion of graphene oxide sheets due to crosslinking of alginate with divalent metal ions and in the presence of reduced graphene oxide improved the swelling resistance due to the formation of strong widespread hydrogen bonding network between oxygenated groups and hence resulting in the strong interactions between metal ions, alginate, and filler sheets making them attractive for moisture barrier coating applications [74]. Graphene oxide (GO) nanoplatelets have been chemically functionalized with starch biopolymer to proceed with the epoxy ring opening and resulting in high-performance epoxy nanocomposite for coating applications where starch molecules on the surface of GO efficiently contributed to the crosslinking through exfoliation of GO interlayers leading to a significant enhancement in the heat of cure of epoxy/amine system from 448.3 mJ/mg for neat epoxy to 515.6 mJ/mg for epoxy/GO and 468.1 mJ/mg for epoxy/GO-St. Some of these epoxy nanocomposites showed acceptable curing potential and fracture resistance and appropriate for coating applications [75].

Fayyad et al. [76] reported an anticorrosion coating film consisting of chitosan (green matrix), oleic acid and graphene oxide. Compared to the pure chitosan coating, the corrosion resistance of oleic acid-modified chitosan/graphene oxide film rose by more than 100 times. Well-dispersed smart grafted nanolayers of these coatings delayed the penetration rate of corrosive species and thus maintain long term anticorrosive stability correlated with better hydrophobicity and permeability [76].

3.2 Thermal Management

The phase change materials (PCMs) are an effective alternative for thermal energy management to store the latent heat during phase transition [77]. Although the small temperature deviation and high heat storage density of the PCMs, leakage and low thermal conductivity in the liquid phase are some of the major challenges with these materials. So, to overcome these problems microencapsulation of the PCMs were carried out. Several types of inorganic or organic materials have been used to coat the PCMs cores. The low thermal conductivities led to limit the heat transfer through the microencapsulated PCM walls. The addition of thermal conductive promoter materials into the PCMs was a confirmed route to increase the rate of heat transfer [78–80]. Carbon allotropes were playing significant role enhance the PCMs performances, due to their thermal stability, good compatibility and high thermal conductivity [81, 82]. In building applications reduction of flammability is a major concern. Ming et al. reported increase in flame retardance properties by incorporating montmorillonite clay in PVA graphene nanocomposites. They have showed by cone calorimeter experiment that flame retardance property increased with graphite content [83].

3.3 *Phase Change Materials*

High performance composite phase change materials (PCMs) for use as energy storage materials have been developed in recent years owing to the progress achieved in multifunctional 3D structural materials such as porous scaffolds, graphene aerogels, metallic and carbon foams [84]. Polyethylene glycol (PEG) is a well established organic solid–liquid phase change material (PCM) offering a wide range of enthalpies and phase transition temperatures as a function of its molecular weight and these PCMs are known for their high-energy absorbance. However, these PCMs have two main drawbacks of leakage and enthalpy reduction during melting.

Properties of phase change materials depend on the loading of particles in matrix. Use of nanoparticles has advantages as tiny structure cause large surface area. Mohamad et al. reported use of inorganic salt hydrates in these nano-enhanced phase change materials having better thermal physical properties and concluded that enhancement in thermal conductivity and supercooling degree reduction of PCM can be obtained by employment of nanoparticles [85]. Recently Feng et al. developed phase change materials with modified graphene oxide as fillers for thermal interface applications in silicone rubber. In their work graphene oxide was chemically modified by 3-aminopropyltriethoxysilane resulting in spherical paraffin modified graphene oxide particles with a core–shell structure [86].

3.3.1 Graphene Based PCMs

Very few papers have appeared on this topic of research on graphene based PCMs till date. Sabagh et al. [87] used polyethylene glycol as a phase change material with graphene oxide (GO), expanded graphene (EG) and epoxy resin (EP) as shape stabilizing materials and showed that the epoxy resin with graphene oxide caused a major reduction in molten PEG leakage (less than 5 wt%) because of hydrogen bonding and subsequent trapping of PEG among GO plates and the barrier effect. Moreover, the expanded graphene due to heterogeneous nucleation of molten PEG in a cooling cycle caused a dramatic increment in crystallinity and enthalpy of melting (8%) [87]. Irani et al. prepared the microencapsulated phase change materials (PCM) by using n-heptadecane with self-assemble graphene/starch composite shell. The microencapsulated composites were characterized by FTIR. The microcapsules were observed through the adsorption of amylose–lipid complex on the n-heptadecane core. The microcapsules of graphene nanoparticles were characterized by using SEM, optical microscopy images and Raman spectroscopy. The results explained the step by step mechanism of microencapsulation. The obtained results of TGA indicated that with increasing the graphene content the thermal stability increased. The graphene also facilitated the heat transfer and temperature control performance by 54% with 11.35 wt%. The thermal reliability and phase change properties of microcapsules were examined by DSC. The graphene/starch encapsulated n-heptadecane composite shell hold properties to apply in thermal storage applications like fluids, surface coating for buildings [88].

4 Conclusions

Graphene-based biopolymer nanocomposites are a relatively newer research area. As per the demand of chapter, we have tried to include all the relevant literature especially for the applications of graphene-based biopolymer nanocomposites in packaging and buildings. From the discussion in this chapter it is clear that graphene-based biopolymer nanocomposites have found significant application in packaging but for buildings there is need for more research. At our institute CSIR-Central Building Research Institute, Roorkee work is in initial stage of research and also to find newer applications of graphene-based biopolymer nanocomposites in buildings especially protective coatings and PCM.

Acknowledgements Authors wish to acknowledge the kind support of Director, CSIR-Central Building Research Institute, Roorkee for initiating the work on graphene-based biopolymer nanocomposites.

References

1. Stankovich S, Dikin DA, Dommett GHB, Kohlhaas KM, Zimney EJ, Stach EA et al (2006) Graphene-based composite materials. Nature 442:282–286
2. Ramanathan T, Abdala AA, Stankovich S, Dikin DA, Alonso MH, Piner RD et al (2008) Functionalized graphene sheets for polymer nanocomposites. Nat Nanotechnol 3:327–331
3. Lee YR, Raghu AV, Jeong HM, Kim BK (2009) Properties of waterborne polyurethane functionalized graphene sheet nanocomposites prepared by an in situ method. Macromol Chem Phys 210:1247–1254
4. Xu Y, Wang Y, Jiajie L, Huang Y, Ma Y, Wan X et al (2009) A hybrid material of graphene and poly (3,4-ethyldioxythiophene) with high conductivity, flexibility, and transparency. Nano Res 2:343–348
5. Quan H, Zhang B, Zhao Q, Yuen RKK, Li RKY (2009) Facile preparation and thermal degradation studies of graphite nanoplatelets (GNPs) filled thermoplastic polyurethane (TPU) nanocomposites. Composites a 40:1506–1513
6. Das TK, Prusty S (2013) Graphene based Polymer Composites and their Applications. Polym Plast Technol Eng 52:319–331
7. Petersen K, Nielsen PV, Bertelsen G, Lawther M, Olsen MB, Nilsson NH, Mortensen G (1999) Potential of biobased materials for food packaging. Trends Food Sci Technol 10:52–68
8. Othman SH (2014) Bio-nanocomposite materials for food packaging applications: types of biopolymer and nano-sized filler. Agric Agric Sci Procedia 2:296–303

9. Arora A, Padua GW (2010a) Review: nanocomposites in food packaging. J Food Sci 75:R43–R49
10. Zubair M, Ullah A (2020) Recent advances in protein derived bionanocomposites for food packaging applications. Food Sci Nutr 60:406–434
11. Ferreira WH, Andrade CT (2018) Preparation and properties of Poly(lactic Acid)/thermoplastic starch/reduced graphene oxide composites. Macromol Symp 381:1800113
12. Tang XZ, Kumar P, Alavi S, Sandeep KP (2012) Recent advances in biopolymers and biopolymer-based nanocomposites for food packaging materials. Crit Rev Food Sci Nutr 52:426–442
13. Pandey JK, Reddy KR, Kumar AP, Singh RP (2005) An overview on the degradability of polymer nanocomposites. Polym Degrad Stab 88:234–250
14. Gadhave RV, Das A, Mahanwar PA, Gadekar PT (2018) Starch based bio-plastics: the future of sustainable packaging. Open J Polym Chem 8:21–33
15. Jose J, Al-Harthi MA, AlMa'adeed MAA, Dakua JB, De SK (2015) Effect of graphene loading on thermomechanical properties of poly(vinyl alcohol)/starch blend. J Appl Polym Sci 41827
16. Cheng C, Li S, Nie S, Zhao W, Yang H, Sun S, Zhao C (2012) General and biomimetic approach to biopolymer-functionalized graphene oxide nanosheet through adhesive dopamine. Biomacromol 13:4236–4246
17. Wu D, Xu H, Hakkarainen M (2016) From starch to polylactide and nano-graphene oxide: fully starch derived high performance composites. RSC Adv 6:54336
18. Rouf TB, Kokini JL (2016) Biodegradable biopolymer–graphene nanocomposites. J Mater Sci. https://doi.org/10.1007/s10853-016-0238-4
19. Jamróz E, Kulawik P, Kopel P (2019) The effect of nanofillers on the functional properties of biopolymer-based films: a review. Polymers 11:675
20. Norazlina H, Kamal Y (2015) Graphene modifications in polylactic acid nanocomposites: a review. Polym Bull 72:931–961
21. Belgacem MN, Gandini A (2008) Monomers, polymers and composites from renewable resources, 1st edn. Elsevier, UK
22. Ashori A (2014) Effects of graphene on the behavior of chitosan and starch nanocomposite films. Polym Eng Sci 2258–2263
23. Ryu HJ, Mahapatra SS, Yadav SK, Cho JW (2013) Synthesis of click-coupled graphene sheet with chitosan: effective exfoliation and enhanced properties of their nanocomposites. Eur Polymer J 49:2627–2634
24. Ge X, Li H, Wu L, Li P, Mu X, Jiang Y (2017) Improved mechanical and barrier properties of starch film with reduced graphene oxide modified by SDBS. J Appl Polym Sci 44910
25. Ma T, Chang PR, Zheng P, Ma X (2013) The composites based on plasticized starch and graphene oxide/reduced graphene oxide. Carbohyd Polym 94:63–70
26. Abdullah ZW, Dong Y (2019) Biodegradable and water resistant poly(vinyl) alcohol (PVA)/starch (ST)/glycerol (GL)/halloysite nanotube (HNT) nanocomposite films for sustainable food packaging. Front Mater. https://doi.org/10.3389/fmats.2019.00058
27. Zhou X, Yang R, Wang B, Chen K (2019) Development and characterization of bilayer films based on pea starch/polylactic acid and use in the cherry tomatoes packaging. Carbohyd Polym 222:114912
28. Suaza LN, Guevara LA, Sanchez LT, Pinzon MI, Villa CC (2019) Characterization of Aloe vera banana starch composite films reinforced with curcumin-loaded starch nanoparticles. Food Struct 22:100131
29. Aqlil M, Nzenguet AM, Essamlali Y, Snik A, Larzek M, Zahouily M (2017) Graphene oxide filled lignin/starch polymer bionanocomposite: structural, physical and mechanical studies. J Agric Food Chem 65:10571–10581
30. Sharma C, Manepalli PH, Thatte A, Thomas S, Kalarikkal N, Alavi S (2017) Biodegradable starch/PVOH/laponite RD-based bionanocomposite films coated with graphene oxide: preparation and performance characterization for food packaging applications. Colloid Polym Sci

31. Mohanty F, Swain SK (2019) Nano silver embedded starch hybrid graphene oxide sandwiched poly(ethylmethacrylate) for packaging application. Nano-Struct Nano-Obj 18:100300
32. Afshar S, Banisadi H (2018) Investigation the effect of graphene oxide and gelatin/starch weight ratio on the properties of starch/gelatin/GO nanocomposite films: the RSM study. Int J Biol Macromol 109:1019–1028
33. Amri A, Ekawati L, Herman S, Yenti SR, Zultiniar AY, Utami SP, Bahruddin (2018) Properties enhancement of cassava starch based bioplastics with addition of graphene oxide. IOP Conf Ser: Mater Sci Eng 345:012025
34. Nuona A, Li X, Zhu X, Xiao Y, Che J (2015) Starch/polylactide sustainable composites: Interface tailoring with graphene oxide. Composites: Part A 69:247–254
35. Chen J, Long Z, Wang S, Meng Y, Zhang G (2019) Biodegradable blends of graphene quantum dots and thermoplastic starch with solid-state photoluminescent and conductive properties. Int J Biol Macromol 139:367–376
36. Ferreia WH, Dahmouche K, Andrade CT (2019) Dispersion of reduced graphene oxide within thermoplastic starch/poly(lactic acid) blends investigated by small-angle X-ray scattering. Carbohyd Polym 208:124–132
37. Caner C, Vergano PJ, Wiles JL (1998) Chitosan Film Mechanical and Permeation Properties as Affected by Acid, Plasticizer, and Storage. J Food Sci 63:1049–1053
38. Kumar S, Mukherjee A, Dutta J (2020) Chitosan based nanocomposite films and coatings: emerging antimicrobial food packaging alternatives. Trends Food Sci Technol 97:196–209
39. Wu Z, Huang X, Li YC, Xiao H, Wang X (2018) Novel chitosan films with laponite immobilized Ag nanoparticles for active food packaging. Carbohyd Polym 199:210–218
40. Terzioglu P, Altin Y, Kalemtas A and Bedeloglu AC (2020) Graphene oxide and zinc oxide decorated chitosan nanocomposite biofilms for packaging applications. J Polym Eng 40:152–157. (https://doi.org/https://doi.org/10.1515/polyeng-2019-0240)
41. Barra A, Ferreira NM, Martins MA, Lazar O, Pantazi A, Jderu AA, Neumayer SM, Rodriguez BJ, Enachescu M, Ferreira P, Nunes C (2019) Eco-friendly preparation of electrically conductive chitosan—reduced graphene oxide flexible bionanocomposites for food packaging and biological applications. Compos Sci Technol 173:53–60
42. Lyn FH, Peng TC, Ruzniza MZ, Hanani ZAN (2019) Effect of oxidation degrees of graphene oxide (GO) on the structure and physical properties of chitosan/GO composite films. Food Packag Shelf Life 21:100373
43. Ahmed J, Mulla M, Maniruzzaman M (2020) Rheological and dielectric behaviour of 3D-printable chitosan/graphene oxide hydrogels. ACS Biomater Sci Eng 6:88–99
44. Zhang XF, Song L, Wang Z, Wang Y, Wan L, Yao J (2020) Highly transparent graphene oxide/cellulose composite film bearing ultraviolet shielding property. Int J Biol Macromol 145:663–667
45. Liu D, Dong Y, Liu Y, Ma N and Sui G (2019) Cellulose nanowhisker (CNW)/Graphene nanoplatelet (GN) composite films with simultaneously enhanced thermal, electrical and mechanical properties 6:235
46. Phanthong P, Reubroycharoen P, Hao X, Xu G, Abudula A, Guan G (2018) Nanocellulose: extraction and application. Carbon Resour Convers 1:32–43
47. Yadav M, Behera K, Chang YH, Chiu FC (2020) Cellulose nanocrystal reinforced chitosan based uv barrier composite films for sustainable packaging. Polymers 12:202
48. Abitbol T, Rivkin A, Cao Y, Nevo Y, Abraham E, Ben-Shalom T, Lapidot S, Shoseyov O (2016) Nanocellulose, a tiny fiber with huge applications. Curr Opin Biotechnol 39:76–88
49. Jamroz E, Kulawik P, Ko P (2019) The effect of nanofillers on the functional properties of biopolymer-based films: a review. Polymers 11675
50. Khosravi A, Fereidoon A, Khorasani MM, Naderi G, Ganjali MR, Zarrintaj P, Saeb MR, Gutierrez TJ (2020) Soft and hard sections from cellulose-reinforced poly(lactic acid)-based food packaging films: A critical review. Food Packaging Shelf Life 23:100429
51. Idumah CI, Zurina M, Ogbu J, Ndem JU, Igba EC (2020) A review on innovations in polymeric nanocomposite packaging materials and electrical sensors for food and agriculture. Compos Interf 27:1–72

52. Ramezani H, Behzad T, Bagheri R (2020) Synergistic effect of graphene oxide nanoplatelets and cellulose nanofibers on mechanical, thermal, and barrier properties of thermoplastic starch. Polym Adv Technol 31:553–565
53. Babu RP, O'Connor K, Seeram R (2013) Current progress on bio-based polymers and their future trends. Progr Biomater 2:8
54. Sharma B, Malik P, Jain P (2018) Biopolymer reinforced nanocomposites: a comprehensive review. Mater Today Commun 16:353–363
55. Calovi M, Dirè S, Ceccato R, Deflorian F, Rossi S (2019) Corrosion protection properties of functionalised graphene–acrylate coatings produced via cataphoretic deposition. Prog Org Coatings 136:105261
56. Dong R, Liu L (2016) Preparation and properties of acrylic resin coating modified by functional graphene oxide. Appl Surf Sci 368:378–387. https://doi.org/10.1016/j.apsusc.2016.01.275
57. Tong Y, Bohm S, Song M (2017) The capability of graphene on improving the electrical conductivity and anti-corrosion properties of Polyurethane coatings. Appl Surf Sci 424:72–81. https://doi.org/10.1016/j.apsusc.2017.02.081
58. Ramezanzadeh B, Ghasemi E, Mahdavian M, Changizi E, Moghadam MHM (2015) Covalently-grafted graphene oxide nanosheets to improve barrier and corrosion protection properties of polyurethane coatings. Carbon 93:555–573. https://doi.org/10.1016/j.carbon.2015.05.094
59. Chang KC, Hsu MH, Lu HI, Lai MC, Liu PJ, Hsu CH, Ji WF, Chuang TL, Wei Y, Yeh JM, Liu WR (2014) Room-temperature cured hydrophobic epoxy/graphene composites as corrosion inhibitor for cold-rolled steel. Carbon 66:144–153. https://doi.org/10.1016/j.carbon.2013.08.052
60. Ramezanzadeh B, Niroumandrad S, Ahmadi A, Mahdavian M, Moghadam MHM (2016) Enhancement of barrier and corrosion protection performance of an epoxy coating through wet transfer of amino functionalized graphene oxide. Corros Sci 103:283–304. https://doi.org/10.1016/j.corsci.2015.11.033
61. Ramanathan T, Abdala AA, Stankovich S, Dikin DA, Herrera-Alonso M, Piner RD, Adamson DH, Schniepp HC, Chen X, Ruoff RS, Nguyen ST, Aksay IA, Prud'homme RK, Brinson LC (2008) Functionalized graphene sheets for polymer nanocomposites. Nat Nanotechnol 3:327–331. https://doi.org/10.1038/nnano.2008.96
62. Nzenguet AM, Aqlil M, Essamlali Y, Amadine O, Snik A, Larzek L, Zahouily M (2018) Novel bionanocomposite films based on graphene oxide filled starch/polyacrylamide polymer blend: structural, mechanical and water barrier properties. J Polym Res 25:86. https://doi.org/10.1007/s10965-018-1469-7
63. Liu S, Yao F, Oderinde O, Li K, Wang H, Zhang Z, Fu G (2017) Zinc ions enhanced nacre-like chitosan/graphene oxide composite film with superior mechanical and shape memory properties. Chem Eng J 321:502–509
64. Sisti L, Belcari J, Mazzocchetti L, Totaro G, Vannini M, Giorgini L, Zucchelli A, Celli A (2016) Multicomponent reinforcing system for poly(butylene succinate): Composites containing poly(l-lactide) electrospun mats loaded with grapheme. Polym Testing 50:283–291
65. Jin S, Li K, Li JZ (2018) A general bio-inspired, novel interface engineering strategy toward strong yet tough protein based composites. Appl Surf Sci 447:452–462
66. Ebrahimzadeh S, Ghanbarzadeh B, Hamishehkar H (2016) Physical properties of carboxymethyl cellulose based nano-biocomposites with Graphene nano-platelets. Int J Biol Macromol 84:16–23
67. Rissa K, Lepisto T, Yrjola K (2006) Effect of Kaolin content on structure and functional properties of water-based coatings. J Prog Org Coatings 55(2):137–141
68. Arora A, Padua GW (2010b) Nanocomposites in food packaging. J Food Sci 75:43–49
69. Pan B, Xu G, Zhang B et al (2012) Preparation and tribological properties of polyamide 11/Graphene coatings. Polym Plast Technol Eng 51:1163–1166
70. Chiu FC, Huang IN (2012) Phase morphology and enhanced thermal/mechanical properties of polyamide 46/graphene oxide nanocomposites. Polym Test 31:953–962

71. Madhad HV and Vasava DV (2019) Review on recent progress in synthesis of graphene–polyamide nanocomposites. J Therm Compos Mater 1–29. doi: https://doi.org/10.1177/0892705719880942
72. Cai D, Yusoh K, Song M (2009) The mechanical properties and morphology of a graphite oxide nanoplatelet/polyurethane composite. Nanotechnology 20:85712
73. Cai K, Zuo S, Luo S, Yao C, Liu W, Ma J, Mao H, Li Z (2016) Preparation of polyaniline/graphene composites with excellent anti-corrosion properties and their application in waterborne polyurethane anticorrosive coatings. RSC Adv 6:95965–95972. https://doi.org/10.1039/C6RA19618G
74. Vilcinskas K, Zlopasa J, Jansen KMB, Mulder FM, Picken SJ, Koper GJM (2016) Water sorption and diffusion in (reduced) graphene oxide-alginate biopolymer nanocomposites. Macromol Mater Eng 301:1049–1063. doi: https://doi.org/10.1002/mame.201600154
75. Yarahmadi E, Didehban K, Sari GM, Saeb MR, Shabanian M, Aryanasab F, Zarrintaje P, Paran SMR, Mozafarig M, Rallinih M, Puglia D (2018) Development and curing potential of epoxy/starch-functionalized graphene oxide nanocomposite coatings. Prog Org Coatings 119:194–202
76. Fayyad EM, Sadasivuni KK, Ponnamma D, Al-Maadeed MAA (2016) Oleic acid-grafted chitosan/graphene oxide composite coating for corrosion protection of carbon steel. Carbohyd Polym 151:871–878
77. Irani F, Ranjbar Z, Jannesari A, Moradian S (2019) Fabrication and characterization of microencapsulated n-heptadecane with graphene/starch composite shell for thermal energy storage. Prog Org Coat 131:203–210
78. Paloma JG, Alkan C, Chimenos JM, Fernandez AI (2017) Comparison of microencapsulated phase change materials prepared at laboratory containing the same core and different shell material. Appl Sci 7:723
79. Jegadheeswaran S, Pohekar SD (2009) Performance enhancement in latent heat thermal storage system: a review. Renew Sustain Energy Rev 13:2225–2244
80. Paloma JGiro, Martínez M, Cabeza LF, Fernández AI, (2016) Types, methods, techniques, and applications for microencapsulated phase change materials (MPCM): a review. Renew Sustain Energy Rev 53:1059–1075
81. Ola O, Chen Y, Zhu Y (2019) Three-dimensional carbon foam nanocomposites for thermal energy storage. Sol Energy Mater Sol Cells 191:297–305
82. Amaral C, Vicente R, Marques PAAP, Barros-Timmons A (2017) Phase change materials and carbon nanostructures for thermal energy storage: a literature review. Renew Sustain Energy Rev 79:1212–1228
83. Huang G, Chen S, Liang H, Wang X, Gao J (2013) Combination of graphene and montmorillonite reduces the flammability of poly(vinyl alcohol) nanocomposites. Appl Clay Sci 80:433–437
84. Yang J, Tang L, Bai L, Bao R, Liu Z, Xie B, Yang M, Yang W (2019) High-performance composite phase change materials for energy conversion based on macroscopically three-dimensional structural materials. Mater Horiz 6:250–273
85. Mohamad AT, Sidik NAC (2019) Nano-enhanced phase change material effects on the supercooling degree improvement: a review. IOP Conf Ser Mater Sci Eng 469:1–11
86. Feng J, Liu Z, Zhang D, He Z, Tao Z, Guo Q (2019) Phase change materials coated with modified grapheme oxide as fillers for silicone rubber used in thermal interface applications. New Carbon Mater 34(2):188–195
87. Sabagh S, Bahramian AR, Madadi MH (2020) Improvement in phase change hybrid nanocomposites material based on polyethylene glycol/epoxy/graphene for thermal protection systems. Iran Polym J 29:161–169
88. Irania F, Ranjbara Z, Moradian S, Jannesari A (2017) Microencapsulation of n-heptadecane phase change material with starch shell. Prog Org Coatings 113:31–38

Graphene Based Biopolymer Nanocomposites in Sensors

Shreya Sharma, Bhasha Sharma, and Purnima Jain

Abstract The marvelous electrical, optical, mechanical and thermal characteristics coupled with high specific surface area of graphene label it as an appealing integrant for high-performance stimuli responsive smart materials. Typical graphene-based smart materials include mechanically exfoliated perfect graphene, chemical vapor deposited high-quality graphene, chemically modified graphene including, graphene oxide and reduced graphene oxide and their macroscopic assemblies or composites. The interaction of graphene based materials with biopolymers to deliver surprisingly beguiling electrical, mechanical, optical, thermal and sensing properties have received considerable attention. Biopolymer-based materials have the advantage of light weight, enhanced transparency, good elasticity, and biodegradability but are abided by drawback of poor mechanical strength, slow response and disappointing environmental stability. Howbeit, graphene fabricated biopolymers proffer propitious characteristics including sensitivity to a range of stimuli including gas molecules or biomolecules, pH value, mechanical flexibility, electrical and thermal conductivity to aid ceaseless promising development in sensor technology. In the chapter, we outline several graphene fabricated biopolymer nanocomposite-based smart sensors and their implication in chemical or strain sensors, actuators, gas sensors, detection of several biomolecules and their delivery. We also shed light on the sensing properties and debate about the challenges facing realization of their practicality.

Keywords Graphene · Sensors · Biosensor · Electrochemical sensors · Biopolymers · Biomedical · Nanotechnology

1 Introduction

Nature has metamorphosed the most modern sensor ideas familiar to mankind and much research progresses to acquire comprehension on how nature's noses, eyes,

S. Sharma (✉) · B. Sharma · P. Jain
Department of Chemistry, Netaji Subhas University of Technology, Delhi, India
e-mail: shreya.k2k@gmail.com

B. Sharma and P. Jain (eds.), *Graphene Based Biopolymer Nanocomposites*,
Composites Science and Technology,
https://doi.org/10.1007/978-981-15-9180-8_15

and ears reach their unparalleled sensitivities and specificities. Among the multitudinous divaricates of science, sensing is leading a reawakening era, on account of the exploitation of cross-cutting advancements including nanotechnology, microfluidics, and lab-on-chip, rational design, printing technology, and internet of things. Particularly, advances in nano-synthesis technologies stimulate entirely novel sensor theories [1, 2]. Carbon-based nanostructured materials have been reported to flaunt parts-per-billion sensitivities, a repercussion of high surface area of these materials. Single-iota thick, sp^2 graphene is an especially fascinating material inferable from its surprising electrical, mechanical, optical, thermal and sensing properties [3–5]. The remarkable attributes of graphene emanate from the 2*p* orbitals, manifesting the *p* state groups to delocalize over the sheet of carbons that comprise graphene [6]. Reduction of the existing cohesive forces between different graphene molecules in several forms through functionalization results in further change in the physiochemical chemistry, thus expanding its applications. The contribution of graphene to design sensors is credited to a portion of the perceptible advantages including huge surface-to-volume proportion, remarkable optical properties, astounding carrier mobility and anomalous thermal and electrical characteristics contrasted with the different allotropes of carbon. These properties are steady for the two-fold and multi-layered graphene structures [7].

The structure of graphene gives an extraordinary surface to the adsorption of atoms. Cavities present in the graphene sheet improve the ingestion proficiency of particles to the surface, permitting them to be identified. Graphene can likewise go about as a p-type semiconductor where they have hole like charge carriers. For certain applications, the exhaustion of holes from the valence band can prompt growth in obstruction and along these lines improve sensitivity and response. The adsorption of particles onto the surface can likewise trigger an adjustment in the electrical conductivity of graphene. The surface adsorbents can either go about as a benefactor or acceptor atom, giving or evacuating electrons individually. The voltage change with graphene gives a quantifiable reaction which shows that a particle has been distinguished. The π-π stacking interaction with different parts of hybrid prompts a charge-transfer over delocalized electrons, following upgraded sensitivity. Notwithstanding the adsorption into the holes on the graphene surface, particles can likewise communicate with the π-bond. This can prompt a more noteworthy number of potential adsorption locales boosting the sensitivity of the material. A higher consolidation of graphene into a composite material has been found to deliver a bigger π-network. Aside from the distinction in the structure, working conditions, the utilization of these focal points in graphene sensors lie principally in their ability to adapt as per the use [7, 8].

The cognizance regarding deleterious environmental effects of engineered, oil-based polymers is consistently progressing. By and by, this concern is increasing rapidly in light of the fact that the utilization of engineered polymers is expanding instead of declining. As of late, enthusiasm for rationing the earth by not just utilizing products produced using sustainable assets and products that debase into environment-friendly constituents has been consistently and quickly augmenting. Bio-based polymers are an unambiguous, multifaceted class of materials that have potential implications in virtually all segments of the economy [9]. Biopolymers are utilized for the reduction of sensor carbon footprint. Moreover, the proposed functionalities provide higher affinity polar target molecules, biocompatibility, biodegradability, and so on [10]. Be that as it may, they for the most part display poor mechanical properties, a short lifespan, low chemical resistance, poor long haul toughness, and restricted processing capability. All together to defeat these inadequacies and create propelled materials for a wide scope of uses, biopolymers are fortified with fillers or nanofillers to contour bio-based composites or bio-based nanocomposites. Specifically, nanostructures can display higher explicit surface area, surface vitality, and thickness, contrasted with regular microfillers, and can prompt materials with as good as ever properties because of synergistic impacts that are superior to those emerging from the ordinary principle of polymers [11].

2 Graphene

A sp^2 hybridized carbon structure, Graphene, displays exceptional characteristics. The high thermal conductivity, and charge versatility with values in the order of 2000–4000 $Wm^{-1} K^{-1}$ and 200,000 $cm^{-1} V^{-1}s^{-1}$at room temperature, respectively has been a debated issue of research among theoretical physicists. Graphene unit cell constitutes two carbon iotas, framed by two sub-lattices. Two of the six carbon atoms located at apex of hexagon comprise the long-extended hexagonal graphene organization [11, 12]. Three σ bonds link the carbon atom. The σ bondings are framed by sp^2 cross breed orbitals, which are made by the superposition out of *s*, *px* and *py* orbitals. Perpendicularity of unhybridized *pz* orbital stays to sp^2 hybrid orbitals frame π bonds. These bonds form a π framework that prompts the delocalization of electrons. The band structure of graphene can be portrayed by a basic tight binding estimation of π groups, where sigma groups have a negligible graphene network. It has been demonstrated that the conductivity of graphene never falls beneath a base worth comparing to the quantum unit of conductance, in any event, when groupings of charge bearers tend to zero [13]. At room temperature, the transporter portability in graphene can reach 15,000 $cm^2V^{-1}s$ [14]. Graphene can be single-layered, twofold layered, multiple layered, or in the graphene oxide/reduced graphene oxide structures. The structure of graphene and chemically modified states of graphene have been illustrated in Fig. 1. Weak *Van der Waal* interactions hold the diverse layers of together with the counterpoise separation of 3.35 Å [15].

Graphene oxide (GO) is synthesized by oxidation of exfoliated graphite with strong oxidizing agents as fumic nitric acid, sulfuric acid (Hummer's method). The higher hydrophobic behavior and interlayer distance exhibited by GO is attributed to higher oxygen content. Also, the presence of several functional groups such as carbonyl, hydroxyl groups, and phenols are useful in terms of biofunctional pathways proteins, antibody, or aptamers/DNA in order to realize biosensors [15]. The exceptional properties of graphene are without a doubt the primary explanation

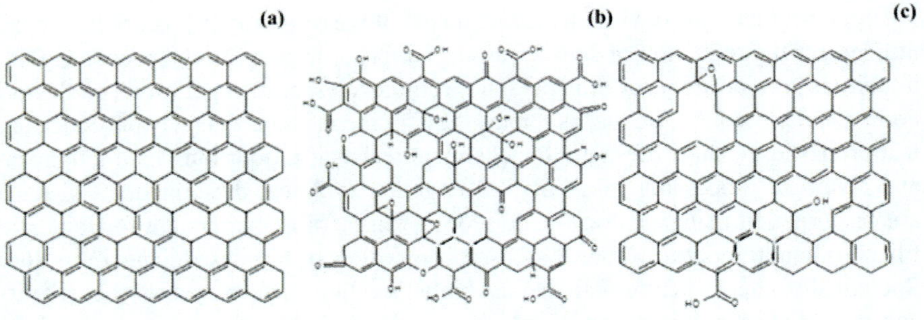

Fig. 1 Structure of **a)** Graphene. **b)** Graphene Oxide and **c)** Reduced Graphene Oxide [17]

behind the material's subjectivity to prodigious enthusiasm for several disciplines of study.

GO is reduced to reduced Graphene Oxide (rGO), using chemical agents like Hydrazine, electrochemically by directly applying a reduction potential or thermally at elevated temperature. Reduction helps in the removal of chemical and structural defects in GO and thus helps in restoring the semiconductor properties for applications in electronics, especially for FET-based biosensing [16].

3 Sensing Properties

Sensor is a device that detects changes in the physical or chemical stimulus generating a signal that can be measured or recorded i.e. can be electronically analyzed. A sensor consists of two fundamental components; a receptor that interacts with a specific analyte and a transducer that transforms the recognition event which occurred between the analyte and the receptor into an accessible detection signal. The selection of an appropriate sensor is based on the possibility of quick response to an external stimulus, be able to detect an analyte in proportion as low as possible, should provide the result in high accuracy and precision, have minimal recuperation time, distinguish the right analyte among others, and be an easy-to-operate system. Furthermore, eco-friendly and cheap devices are today's needs. Therefore, accelerating research in the field to improve already being used systems and find a promising alternative that allows changing technologies.

In the past decade, nanomaterials have begun to emerge as viable alternatives. Specifically, nanomaterials based on carbon are the most widely discussed, researched, and applied synthetic nanomaterials due to their magnificent characteristics. The typical characteristics of every carbonaceous nanostructure enable it to be utilized in a particular sensor model [18, 19]. Graphene and its derivatives have become the most used raw material for sensors in different applications due to their portion of the perceptible advantages including huge surface-to-volume proportion, remarkable optical properties, astounding carrier mobility, and anomalous thermal and electrical characteristics contrasted with the different allotropes of carbon. However, the optimal no of graphene layers, and the best way to synthesize and measure response is still in need.

4 Graphene-Based Biopolymer Nanocomposites

Biopolymers require strengthened and improved properties for utilization in several fields of study, as they have customarily greater expenses and underlying property profiles contrasted with commercial petroleum-based polymers. Incorporation of an enormous assortment of fillers like layered silicates, nanotubes, in polymers prompts prominent increment in their mechanical, thermal, rheological, and gas-barrier properties [20, 21]. In contrast to all the aforementioned fillers, graphene seems to have a condescending capability to bring about generous upgrades in useful properties of polymers. Graphene-based biopolymer nanocomposites have tremendous applications including utilization in electronic gadgets, photocatalytic materials, fuel cells, lithium-ion batteries, sensing devices, etc. without causing any harm to the environment. Graphene and its subsidiaries may be modified by treatment with various strengthening operators, such change should be possible by thermal and chemical reduction and furthermore treating its surface by the scope of surfactants so its reactivity gets expanded. The reduction process removes oxygen functionalities and can enhance the electronic and mechanical properties. Be that as it may, the performance of the reduced nanostructured materials relies upon the synergetic impact of functional gatherings present at the edges and into the polymer grid. Contrasted with CNTs, graphene-based nanomaterials have pulled in gigantic consideration because of its high area and amazingly high elasticity [22].

Coordination of the extremities between the polymer and filler is critical for ideal interfacial communications between the filler and the lattice. Furthermore, sound geometrical and aspect proportion, an enormous number of functional gatherings are a few characteristics that unequivocally influence the resultant property enhancement. Consequently, high surface area, high aspect ratio, and high strength of graphene empower it to beat a variety of nanofillers used [23]. Uniform dispersion promoted as a result of sufficient interaction and interfacial bonding between the graphene/graphene subordinate and the biopolymer is of most extreme significance. The lower number of functional groups in graphene when contrasted with graphite oxide, (for instance, carboxyl, epoxide, and hydroxyl) brings about low surface vitality and lower affinity to polar polymer frameworks. On account of the higher affinity with the functional gatherings of biopolymers, the oxidized type of graphene, Graphene oxide as received gigantic attention for the fabrication of bio-based polymer nanocomposites in contrast to the different subsidiaries of graphene including reduced graphene oxide, graphene nanoplatelets, etc. A decent dispersion of the nanostructured materials in polymers relies upon the preparation of polymer-based nanocomposites, which eventually can influence the mechanical, thermal, electrical, water vapor, and gas-barrier properties of nanostructured composites. Three main pathways for incorporation of graphitic nanofiller into polymer lattices include: (i) arrangement intercalation, (ii) liquefy intercalation, and (iii) in situ intercalative polymerization [21–24].

5 Sensing Applications of Graphene Based Biopolymer Nanocomposites

5.1 Electrochemical Sensor

Electrochemical sensor utilizes an electrode as a transducing component to convert chemical information generated by the substance in the response of the analyte into a quantifiable electrical sign, in this manner accomplishing the immediate recognition of the analyte in the example. Teodo et al. designed electrochemical sensor detection of mercury(II) utilizing a green hybrid sensor comprising reduced graphene oxide reinforced cellulose nanowhiskers polyamide nanofibres. The electrode exhibited stability, high selectivity, low detection limit and wide dynamic linear range for the detection of the heavy metal in water samples [25]. Shuping Wu et al. designed chitosan-graphene-gold nanocomposite based electrode for the detection of toxic metal ion, Cadmium (II) in water and milk samples. The sensor had sound repeatability, stability, selectivity with good linear response towards Cd(II) in the range of 0.1–0.9 μm (R2 = 0.999) with LOD of 1.62×10^{-4} μm [26]. Reduced graphene oxide-carboxymethyl cellulose nanocomposite was utilized by Arenas et al. for the design of an electrochemical immuno-sensor. The disposable electrode scaffold developed was first of its kind with the detection limit of 61 mg/ml and a substantial stockpiling steadiness for the assurance of the objective analyte in human serum of hypercholesterolemia or diabetic patients as it was shown by contrasting the outcomes presented by commercial ELISA unit. A sigmoidal calibration plot for APN in the 0.1–50 μg/mL range, with a linear portion in the 0.5–10.0 μg/mL adiponectin cytokine concentration ranges was obtained [27].

Determination of nitrobenzene in sewage water by reduced graphene oxide based cellulose microfiber was reported lately. The sensor displayed extraordinary electrocatalytic activity and stability because of the interaction of polar hydroxyl group of cellulose microfiber with that of defective active sites of reduced graphene oxide [28]. A recent paper by Poletti et al. analyzed the enhanced sensitivity for glucose sensing induced by the presence of glucose oxidase and an electrochemical transduction. Covalent binding of chitosan with graphene oxide inferred enhance activity up to eight times, and exhibiter superior reliability and efficiency in comparison with the results inferred by utilization of prisitine graphene oxide. Covalent alteration of GO was accomplished by a standard carboxylic actuation/amidation approach by misusing the copious amino pendants of chitosan. Fruitful functionalization was demonstrated by correlation with a specially appointed blended control test acknowledged by utilizing non-initiated graphene oxide as precursor [29].

5.2 Gas Sensors

Gas sensors with low working temperature, high affectability, astounding unique response recuperation qualities, and great selectivity are in incredible interest for the preservation of the environment and human wellbeing. Ongoing examinations have seen the enormous advancement in the design of gas sensors dependent on several conducting polymer. Zhang et al. designed a graphene oxide-chitosan nanocomposite based sensor to detect amine vapors. The quartz crystal microbalance sensor demonstrated excellent sensitivity to aliphatic amines include methyl, dimethyl, and trimethylamine with 2.7, 2.3, 4.8 Hz/pm sensitivity values respectively. An adsorption-desorption process was involved as a result of the hydrogen bonding of the graphene oxide-chitosan coated sensor [30]. Nanocellulose assisted dispersion of graphene to polyvinyl alcohol was exploited to design humidity sensors with sound repeatability between RH at 50–70%. The results demonstrated the role of nanocellulose to act as a bridge between load transfer and humidity transfer and also as a water absorber [31].

Thangamani et al. [32] designed chemo resistive sensors by utilizing the matrix of graphene reinforced polypyrrole and poly (vinyl alcohol) nanocomposite for the determination of petroleum gases at room temperature. Thangamani et al. designed poly (vinyl alcohol) reinforced graphene nanoplatelets and CuO nanoparticles in another pioneer work for the detection of ethanol and propanol vapors. Incorporation of CuO nanoparticles and graphene nanoplatelets inferred enhancement in thermal and mechanical sensing of the volatile compounds [33]. Lim et al. designed humidity sensor by utilization of poly (vinyl alcohol) reinforced reduced graphene oxide and tannic acid. Long term stability with excellent humidity sensing demonstrated its potentiality for use in humidity sensor systems [34].

5.3 Biosensor

Innovative work on biosensors is turning into the most widely contemplated discipline on the grounds that the simple, fast, ease, profoundly sensitive, and exceptionally selective biosensors add to progress in cutting edge medicines, for instance, personalized medication and ultrasensitive purpose of-care identification of markers for diseases [35]. Biosensors can be grouped based on the type of physicochemical transduction or the kind of biorecognition component. In light of the transducer, biosensors can be characterized as electrochemical, optical, warm, and piezoelectric biosensors. Electrochemical biosensors can be additionally delegated as, amperometric biosensors (measure the current created during oxidation or decrease of electroactive item or reactant), potentiometric biosensors (measure the capability of the biosensor anode concerning a reference terminal), and conductometric biosensors (measure the adjustment in conductance emerging because of the biochemical response). Electrochemical biosensors are the most broadly explored biosensors as they proffer the benefit of low discovery limit, particularity, effortlessness of development, and simplicity of activity. Inferable from late headways in electronic instrumentation, these biosensors can be scaled down as lab-on-chip gadgets for in vivo checking or as a handheld gadget for on-location checking. Optical biosensors depend on the estimation of light consumed or transmitted as a result of a biochemical response. Optical biosensors depend on different optical procedures including fluorescence, absorption, iridescence, surface plasmon reverberation (SPR), and so on. Thermal biosensors depend on the estimation of the thermal changes happening on biochemical acknowledgment.

The vast majority of the biochemical responses include an adjustment in enthalpy furthermore; the warmth changes can be estimated by touchy thermistors. Piezoelectric biosensors include the estimation of mass change happening because of biomolecular connection. Piezoelectric gems are utilized to quantify the mass change by associating with the adjustment in wavering recurrence of the piezo crystal. Given the sort of biorecognition unit utilized, biosensors can be delegated enzymatic, nucleic acid based, aptamer-based, antibody-based, entire cell-based biosensors (Table 1) [36]. Nanocomposite sensor comprised of Co_3O_4 nanorod, chitosan, and graphene was utilized as a biosensor for the detection of Staphylococcus aureus nuc gene sequence with the detection limit of 4.3×10^{-3} M and its polymerase chain reaction products. The DNA sensor demonstrated several benefits including minimal cost production, fast response, sound selectivity, wide linear, and high selectivity [37]. Levofloxacin, a typical orally dispensed third-generation fluoroquinolones anti-bacterial agent was detected with the utilization of poly(L-cysteine)-reduced graphene oxide reinforced gold nanoparticle. The biosensor displayed considerable potential for detection in blood serum [38]. A non-enzymatic biosensor based on chitosan, reduced graphene oxide reinforced chromium oxide was designed by *Ghanbari and* Nejabati et al. for detection of uric acid, dopamine, xanthine and hypoxanthine in fish meat with a detection limit of 0.65 μM, 0.80 μM, 1.20 μM, and 0.85 μM, respectively [39]. DNA based electrochemical sensor for the determination of typhoid was designed by covalent immobilization of *Salmonella typhi* specific 5 -amine labeled single-stranded (ss) DNA probe on graphene oxide/chitosan nanocomposite via glutaraldehyde. The detection range of 10 FM to 50 nM and limit of detection of 10 FM within 60 s hybridization times for complementary sequence [40]. Graphene-chitosan-Bi_2O_3 nano-based composite was studied for the detection of epinephrine, neurotransmitter in the central nervous system with a linear response in the range of 100–500 nM, and sensitivity of 1.3 nA/nM [41]. The infographic illustration of implementation of graphene based nanocomposites in design of nanosensors is depicted in Fig. 2.

Table 1 Applications of graphene based biopolymer discussed in the review and previously reported

S. No	Biopolymer/Biopolymer derivative used	Type of sensor	Application	Reference
1.	Cellulose nanowhiskers	Electrochemical sensor	Detection of heavy metal (Mercury (II))	[25]
2.	Chitosan	Electrochemical sensor	Detection of heavy metal (Mercury (II))	[26]
3.	Carboxymethyl Cellulose	Electrochemical sensor	Immuno-sensor	[27]
4.	Cellulose microfiber	Electrochemical sensor	Detection of nitrobenzene	[28]
5.	Chitosan	Electrochemical biosensor	Glucose sensing	[29]
6.	Chitosan	Gas sensor	Detection of amine vapors	[30]
7.	Nanocellulose, Poly (vinyl alcohol)	Gas sensor	Humidity sensor	[31]
8.	Polypyrrole, Poly (vinyl alcohol)	Gas sensor	Detection of petroleum gases	[32]
9.	Poly (vinyl alcohol)	Gas sensor	Detection of ethanol and propanol vapors	[33]
10.	Chitosan	Biosensor	Detection of Staphylococcus aureus nuc gene sequence	[37]
11.	Poly(L-cysteine)	Biosensor	Detection of Levofloxacin	[38]
12.	Chitosan	Biosensor	Detection of uric acid, dopamine, xanthine, and hypoxanthine	[39]
13.	Chitosan	Biosensor	Determination of typhoid	[40]
14.	Chitosan	Biosensor	Detection of epinephrine	[41]
15.	Chitosan	FRET	Glucose sensing	[42]
16.	Peptide	FRET	Detection of proteases, chymotrypsin and network metalloproteinase-2	[43]
17.	Polypyrrole	Electrochemical	Detection of bisphenol A	[45]
18.	Polypyrrole	Electrochemical	Detection of heavy metal (Cd(II) and Pb(II))	[46]
19.	Starch	Electrochemical	Detection of transferrin	[47]

(continued)

Table 1 (continued)

S. No	Biopolymer/Biopolymer derivative used	Type of sensor	Application	Reference
20.	Starch	Electrochemical	(Immunosensor) Detection of neuron-specific enolase	[48]

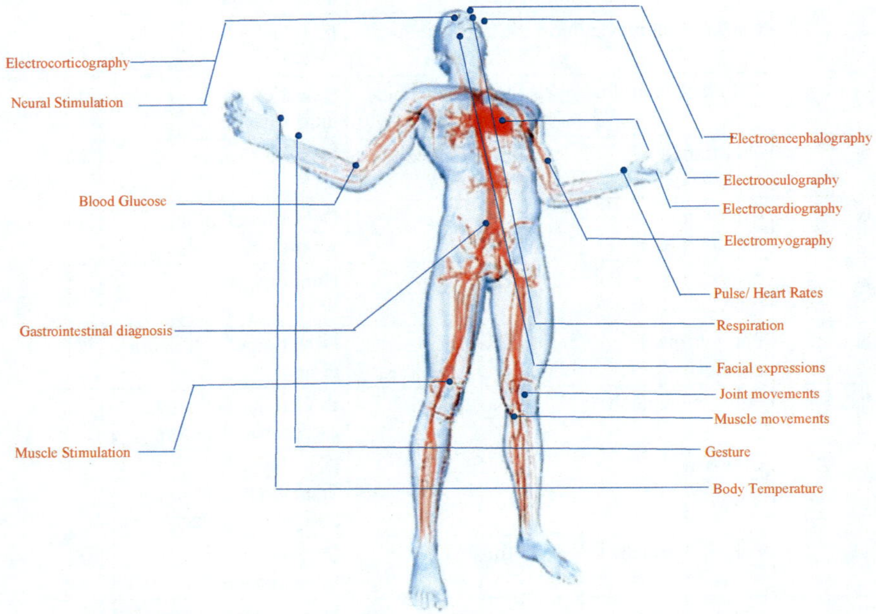

Fig. 2 Application of Graphene based biopolymer nanocomposites for biosensors

5.4 *Fluorescence Resonance Energy Transfer (FRET) Sensors*

Graphene and graphene derivatives are acceptable quenchers for several fluorescent labels, and hence can be used to build numerous sorts of fluorescent biosensors for distinguishing biomolecules including DNA, protein, antigen, antibody, enzyme, cells, etc. Wennjun Li et al. utilized fluorescence quenching for the determination of the concentration of glucose with the creation of chitosan modified graphene oxide nanocomposite. The results inferred the utilization of the designed biosensor for future applications in detection of other biomolecules by changing the maltose-binding protein component [42]. Kwak et al. exhibited the creation of a luminescent GO, peptide quencher-based fluorescent biosensors for straightforward, fast, and touchy discovery of proteases, chymotrypsin, and network metalloproteinase-2 (MMP-2). So as to apply GO as a transmitting benefactor for the manufacture of

FRET-based biosensors, two potential quencher possibilities for PL of GO like metalloprotoporphyrin subordinates (MePPs) and QXL570 were used. It was discovered that the photoluminescence of alkyne-altered GO could be extinguished by restricting a QXL-connected peptide (AAPF and GPLGVRG for chymotrypsin and MMP-2, individually) onto the surface of graphene oxide, but the graphene oxide photoluminescence was recuperated by adding protease to proteolytic separate the peptide chain to expel the Q quencher. The planned GO-peptide-QXL framework indicated high affectability and selectivity for distinguishing chymotrypsin (0.16 mM) and MMP-2 (0.76 mM), and displayed potential application for in situ assurance of MMP-2 (2.0 ng/mL) in living human hepatocytes HepG2 cells [43, 44].

6 Visualizing the Challenges

Graphene-based biopolymer nanocomposites exhibit providential development in technology and implications and are becoming prominent as one of the priority materials, both as a material and as a composite, yet several challenges lie on the way to be acknowledged and resolved for utilization at their fullest potential.

- Firstly, the development of graphene fabricated biopolymer based nanosensors is still deficit of less costly, scalable, and controllable strategies. Furthermore, the size and composition still cannot be precisely controlled.
- Secondly, microstructures designed by reasonable and logical means are required to improve the performance and augmentation of its applications.
- Thirdly, the toxicity and biocompatibility of graphene should be taken care of before their utilization in clinical diagnosis and delivery of drugs. Some of the catalysts used during the growth of graphene increase its toxicity after its production. The investigation, both in vitro and in vivo is necessary.
- The nanocomposites formed alter the intrinsic characteristics. For instance, the electrical conductivity is reduced to 10^3 in contrast to pristine graphene, which could be vital concerning application demanding profoundly conductive materials. Weaker interaction between the graphene and the polymer matrix can also result in reduction of thermal stability.
- The vulnerability of graphene to be oxidized in oxidative conditions is another major hindrance to be managed. This makes it very hard to use graphene in its unadulterated structure. Another inconvenience of getting unadulterated graphene lies in the measure of crude materials required to create it. For instance, when graphene oxide (GO) is utilized to deliver rGO, the substance decrease causes an immense mass misfortune driving to limited quantities of produced tests.
- Finally, the design of multi-stimuli sensors that are operational under harsh conditions is a challenge still facing the future development.

7 Conclusion and Future Prospects

Graphene-and biopolymers-based nanocomposites have unmistakably exhibited a high potential to aid ceaseless propitious development in innovation and applications. However, a couple of key difficulties majorly allude to synthesis strategies, expenses, and implications that need to be addressed to and feasibly conquered to additionally augment their potential. Despite the progress and advances in the design of graphene nanocomposite sensors and their characterization, there are as yet a few moves identified with reasonable execution, including restricted lifetime, operational steadiness, and reproducibility. Future fascinating examination drifts in the field of the present paper will incorporate the improvement of versatile detecting stages in light of graphene-based biopolymer nanocomposites and related materials, just as the concurrent multiplexed discovery of analytes. The fabrication of graphene reinforced biopolymer-based sensors and purpose of care gadgets, could prompt better in situ clinical analysis or as in situ detecting stages for ecological investigation ought to likewise be acted in future exploration.

Acknowledgement Authors would like to thank Netaji Subhas University of Technology for providing facilities and financial support.

References

1. Arduini F, Cinti S, Scognamiglio V, Moscone D (2019) 13 Nanomaterial-based sensors. In: Handbook of nanomaterials in analytical chemistry: modern trends in analysis, 329
2. Arduini F, Cinti S, Scognamiglio V, Moscone D, Palleschi G (2017) How cutting-edge technologies impact the design of electrochemical (bio) sensors for environmental analysis. A review. Anal Chim Acta 959:15–42
3. Kranz C (ed) (2019) Carbon-based nanosensor technology, vol 17. Springer
4. Nehra M, Dilbaghi N, Hassan AA, Kumar S (2019) Carbon-based nanomaterials for the development of sensitive nanosensor platforms. In: Advances in nanosensors for biological and environmental analysis. Elsevier, pp 1–25
5. Mannoor MS, Tao H, Clayton JD, Sengupta A, Kaplan DL, Naik RR, … McAlpine MC (2012) Graphene-based wireless bacteria detection on tooth enamel. Nat Commun 3:763
6. Papageorgiou DG, Kinloch IA, Young RJ (2017) Mechanical properties of graphene and graphene-based nanocomposites. Prog Mater Sci 90:75–127
7. Nag A, Mitra A, Mukhopadhyay SC (2018) Graphene and its sensor-based applications: a review. Sens Actuators, A 270:177–194
8. https://www.cheaptubes.com/graphene-sensors/
9. Abhilash M, Thomas D (2017) Biopolymers for biocomposites and chemical sensor applications. In: Biopolymer composites in electronics. Elsevier, pp 405–435
10. https://www.oxfordscholarship.com/view/10.1093/acprof:oso/9780199581924.001.0001/acprof-9780199581924-chapter-15
11. Kausar A (2019) Applications of polymer/graphene nanocomposite membranes: a review. Mater Res Innovations 23(5):276–287
12. Lawal AT (2019) Graphene-based nano composites and their applications. A review. Biosens Bioelectron 111384

13. Itkis ME, Borondics F, Yu A, Haddon RC (2007) Thermal conductivity measurements of semitransparent single-walled carbon nanotube films by a bolometric technique. Nano Lett 7(4):900–904
14. Cai D, Song M (2010) Recent advance in functionalized graphene/polymer nanocomposites. J Mater Chem 20(37):7906–7915
15. Chaudhry AU, Mittal V (2013) High-density polyethylene nanocomposites using master-batches of chlorinated polyethylene/graphene oxide. Polym Eng Sci 53(1):78–88
16. Moon IK, Lee J, Ruoff RS, Lee H (2010) Reduced graphene oxide by chemical graphitization. Nat commun 1(1):1–6
17. Sharma B, Malik P, Jain P (2018) Biopolymer reinforced nanocomposites: a comprehensive review. Mater Today Commun 16:353–363
18. Sharma S, Shekhar S, Gautam S, Sharma B, Kumar A, Jain P (2020) Carbon-based nanomaterials as novel nanosensors. In: Nanofabrication for smart nanosensor applications (pp 323–347). Elsevier
19. Baptista FR, Belhout SA, Giordani S, Quinn SJ (2015) Recent developments in carbon nanomaterial sensors. Chem Soc Rev 44(13):4433–4453
20. Rouf TB, Kokini JL (2016) Biodegradable biopolymer–graphene nanocomposites. J Mater Sci 51(22):9915–9945
21. Nasir S, Hussein MZ, Zainal Z, Yusof NA (2018) Carbon-based nanomaterials/allotropes: a glimpse of their synthesis, properties and some applications. Materials 11(2):295
22. Terzopoulou Z, Kyzas GZ, Bikiaris DN (2015) Recent advances in nanocomposite materials of graphene derivatives with polysaccharides. Materials 8(2):652–683
23. Roy N, Sengupta R, Bhowmick AK (2012) Modifications of carbon for polymer composites and nanocomposites. Prog Polym Sci 37(6):781–819
24. Young RJ, Kinloch IA, Gong L, Novoselov KS (2012) The mechanics of graphene nanocomposites: a review. Compos Sci Technol 72(12):1459–1476
25. Teodoro KB, Migliorini FL, Facure MH, Correa DS (2019) Conductive electrospun nanofibers containing cellulose nanowhiskers and reduced graphene oxide for the electrochemical detection of mercury (II). Carbohyd Polym 207:747–754
26. Wu S, Li K, Dai X, Zhang Z, Ding F, Li S (2020) An ultrasensitive electrochemical platform based on imprinted chitosan/gold nanoparticles/graphene nanocomposite for sensing cadmium (II) ions. Microchem J 155:104710
27. Arenas, C. B., Sánchez-Tirado, E., Ojeda, I., Gómez-Suárez, C. A., González-Cortés, A., Villalonga, R., … & Pingarrón, J. M. (2016). An electrochemical immunosensor for adiponectin using reduced graphene oxide–carboxymethylcellulose hybrid as electrode scaffold. Sens Actuators B: Chem 223:89–94
28. Balasubramanian P, Balamurugan TST, Chen SM, Chen TW, Tseng TW, Lou BS (2018) A simple architecture of cellulose microfiber/reduced graphene oxide nanocomposite for the electrochemical determination of nitrobenzene in sewage water. Cellulose 25(4):2381–2391
29. Poletti F, Favaretto L, Kovtun A, Treossi E, Corticelli F, Gazzano M, … Melucci M (2020) Electrochemical sensing of glucose by chitosan modified graphene oxide. JPhM 3(1):014011
30. Zhang K, Hu R, Fan G, Li G (2017) Graphene oxide/chitosan nanocomposite coated quartz crystal microbalance sensor for detection of amine vapors. Sens Actuators B: Chem 243:721–730
31. Xu S, Yu W, Yao X, Zhang Q, Fu Q (2016) Nanocellulose-assisted dispersion of graphene to fabricate poly (vinyl alcohol)/graphene nanocomposite for humidity sensing. Compos Sci Technol 131:67–76
32. Thangamani JG, Deshmukh K, Sadasivuni KK, Ponnamma D, Goutham S, Rao KV, … Pasha SK (2017) White graphene reinforced polypyrrole and poly (vinyl alcohol) blend nanocomposites as chemiresistive sensors for room temperature detection of liquid petroleum gases. Microchim Acta 184(10):3977–3987
33. Thangamani GJ, Deshmukh K, Chidambaram K, Ahamed MB, Sadasivuni KK, Ponnamma D, … Pasha SK (2018) Influence of CuO nanoparticles and graphene nanoplatelets on the sensing behaviour of poly (vinyl alcohol) nanocomposites for the detection of ethanol and propanol vapors. J Mater Sci Mater Electron 29(6):5186–5205

34. Lim MY, Shin H, Shin DM, Lee SS, Lee JC (2016) Poly (vinyl alcohol) nanocomposites containing reduced graphene oxide coated with tannic acid for humidity sensor. Polymer 84:89–98
35. Kawamura A, Miyata T (2016) Biosensors. In: Biomaterials nanoarchitectonics. William Andrew Publishing, pp 157–176
36. Sawant SN (2017) Development of biosensors from biopolymer composites. In: Biopolymer composites in electronics. Elsevier, pp 353–383
37. Qi X, Gao H, Zhang Y, Wang X, Chen Y, Sun W (2012) Electrochemical DNA biosensor with chitosan-Co_3O_4 nanorod-graphene composite for the sensitive detection of staphylococcus aureus nuc gene sequence. Bioelectrochemistry 88:42–47
38. Ghanbari MH, Khoshroo A, Sobati H, Ganjali MR, Rahimi-Nasrabadi M, Ahmadi F (2019) An electrochemical sensor based on poly (L-Cysteine)@ AuNPs@ reduced graphene oxide nanocomposite for determination of levofloxacin. Microchem J 147:198–206
39. Ghanbari K, Nejabati F (2020) Ternary nanocomposite based reduced graphene oxide/Chitosan/Cr_2O_3 for simultaneous determination of Dopamine, Uric acid, Xanthine, and Hypoxanthine in fish meat. Anal Methods
40. Singh A, Sinsinbar G, Choudhary M, Kumar V, Pasricha R, Verma HN, … Arora K (2013) Graphene oxide-chitosan nanocomposite based electrochemical DNA biosensor for detection of typhoid. Sens Actuators B: Chem 185:675–684
41. Devnani H, Satsangee SP, Jain R (2016) A novel graphene-chitosan-Bi_2O_3 nanocomposite modified sensor for sensitive and selective electrochemical determination of a monoamine neurotransmitter epinephrine. Ionics 22(6):943–956
42. Li W, Jiang T, Pu Y, Jiao X, Tan W, Qin S (2019) Glucose biosensor using fluorescence quenching with chitosan-modified graphene oxide. Micro Nano Lett 14(3):344–348
43. Kwak SY, Yang JK, Jeon SJ, Kim HI, Yim J, Kang H, … Kim JH (2014) Luminescent graphene oxide with a peptide-quencher complex for optical detection of cell-secreted proteases by a turn-on response. Adv Funct Mater 24(32):5119–5128
44. Wang L, Zhang Y, Wu A, Wei G (2017) Designed graphene-peptide nanocomposites for biosensor applications: A review. Anal Chim Acta 985:24–40
45. Tan F, Cong L, Li X, Zhao Q, Zhao H, Quan X, Chen J (2016) An electrochemical sensor based on molecularly imprinted polypyrrole/graphene quantum dots composite for detection of bisphenol A in water samples. Sens Actuators B: Chem 233:599–606
46. Dai H, Wang N, Wang D, Ma H, Lin M (2016) An electrochemical sensor based on phytic acid functionalized polypyrrole/graphene oxide nanocomposites for simultaneous determination of Cd (II) and Pb (II). Chem Eng J 299:150–155
47. Srivastava J, Kushwaha A, Srivastava M, Srivastava A, Singh M (2019) Glycoprotein imprinted RGO-starch nanocomposite modified EQCM sensor for sensitive and specific detection of transferrin. J Electroanal Chem 835:169–177
48. Fang Y, Li Y, Zhang M, Cui B, Hu Q, Wang L (2019) A novel electrochemical strategy based on porous 3D graphene-starch architecture and silver deposition for ultrasensitive detection of neuron-specific enolase. Analyst 144(6):2186–2194

Graphene Based Biopolymer Nanocomposite Applications in Drug Delivery

Sudip Majumder, Sujata Kumari, and Debasree Ghosh

Abstract In previous years graphene based material have found extensive uses in the field biotechnology and medical sciences including drug delivery. By virtue of their exceptional properties like two-dimensional planar structure, huge surface area, chemical, and mechanical stability, super conductivity and good biocompatibility. Graphene and graphene oxide have been predominantly analyzed as few of the supreme auspicious biomaterials for biomedical applications. These properties consequence in encouraging utilization designed for the proposal of progressive drug delivery systems and delivery of a wide scope of therapeutics. In this review, we commenced an outline of novel development in this area of research. In present study we also elaborate the functionalization i.e. covalent and non-covalent functionalization of graphene oxide. We shortly illustrate recent procedures for the surface amendment of graphene-based nanocarriers, their biocompatibility, and toxicity, monitored by compacted of the most engaging cases established for the distribution of anticancer drugs and genes. Lastly, the review is concise, by a short consequence of upcoming forecasts and tasks in this field.

Keywords Graphene · Graphene oxide · Drug delivery · Anticancer activity · Functionalization

1 Introduction

1.1 Graphene and Graphene Oxide (GO)

The discovery of graphene results in the rise of a novel period in material sciences research and can be the trademark of a hypothetically newfangled material phase.

S. Majumder · S. Kumari · D. Ghosh (✉)
Department of Chemistry, Amity School of Applied Science, Amity University Haryana, Gurugram, India
e-mail: ddebasreeghosh@gmail.com

B. Sharma and P. Jain (eds.), *Graphene Based Biopolymer Nanocomposites*, Composites Science and Technology,
https://doi.org/10.1007/978-981-15-9180-8_16

Graphene is an uniform monolayer of carbon atoms decisively crammed into two-dimensional (2D) honeycomb lattice having sp^2 hybridization [1–4]. It is the main construction block of further significant allotropes; it can be arranged to form 3D graphite, rolled to form 1D nanotubes, and enveloped to form 0D fullerenes. Graphite is remarkably considered earlier, because of its tremendous properties and several applications. It is made up of carbon which is the fourth utmost ample component in the earth by mass [5]. In Graphite, each carbon is sp^2 hybridized and the atomic layers of carbon are stacked over each other by van der Waal's interactions and its one-atom-broad layer of carbon is acknowledged as "Graphene". Over 50 years ago, it was synthesized experimentally for the first time and few of its surface area properties like squeaky carbon sheet configuration were established [6]. However, the name Graphene was first time introduced by Boehm et al. [6]. Graphene is known as a prodigy material because of its extended surface dimension, unique mechanooptical and thermoelectric properties [7–11]. Due to distant π-π conjugation, Graphene shows diverse applications in the zone of materials science, electronics, and photonics [12–14]. Precisely, the huge surface area and optical properties of graphene were treated of enormous attention for biomedical applications, mainly in drug delivery [15] and these have enlarged since 2008 [16]. Graphene has synthesized via diverse methods like micro-mechanical exfoliation of pyrolytic graphite, Chemical Vapour Deposition (CVD), and epitaxial growth [17–22]. As, graphene is exclusive and relatively difficult to synthesis, subsequently, graphene oxide can be used to make graphene due to its cheap and easy synthesis nature [23]. Graphene oxide was produced by the strong oxidation of Graphite in acidic medium, as it is easily dispersible in water and other solvents. Distinct methods have been used to manufacture graphene oxide such's, Staudenmaier, Hofmann, Brodie, and Hummers method [24]. Like graphite, graphene oxide has a layered confirmation which contains various functional groups such as epoxy, carbonyl, hydroxyl, and carboxyl. The oxygen-containing functional groups like epoxy and hydroxyl are attached above the upper and lower surface of graphene layers or a plane and other functional groups like carbonyl and carboxyl are decorated on the edges of graphene layer [25–27].

Once again, biopolymers like polysaccharide [28–31], protein [32–34] and DNA [35] have been used to modify the chemical structure of GO. The unique amphiphilic structure of GO became excessively hydrophilic when GO was adorned by negative or positive-charged polymer [32, 36] to interact with or partition into lipid bilayers later, the strong cell membrane interactions were prominently repressed [37]. Hence to advance the biocompatibility of GO and rGO, the modest blending biopolymers as capping agents are the utmost extensively used process [29, 30, 32, 35]. Hydrophilic polymers [38, 39] chitosan, [31] and proteins [33, 34] are embedding onto the surface or the edge of GO by the covalent bonding. The specific functionalization of GO makes it advisable material for biomedical applications and also boosts its solubility under physiological environments. Additionally the large surface area of GO is responsible for its application in drug delivery. This chapter designates several significant efforts done on the functionalization and application of GO for efficient binding to drugs.

1.2 Drug Delivery

Construction of equipment that enhances the efficiency of a therapeutic material and ensures its secure application via regulating its pharmacokinetics and pharmacodynamics is known as drug delivery system (DDS) [40]. The maximum accepted method of authority of a medication is the pill. Even in the Ebers Papyrus, there is indication of the consumption of pills in ancient Egypt (1500 BC). They comprise of a prescription that includes preservatives like sugars and starches to maintain the construction [41]. The elementary approach is that after swallowing, the pill should directly reach the stomach and absorbed through the intestines into the blood flow. This is not applicable in case of those drugs which undergo decomposition in chemical environment of the stomach. Insulin is one such drug which is inserted directly in the fatty tissue [42] which is then followed by systemic circulation.

2 Synthesis of GO

Earlier graphene oxide was prepared through the oxidation of graphite using H_2SO_4 and HNO_3. In 1859, Brodie et al. improved the Schafhaeutl method comprises of oxidants like $KClO_3$ along with H_2SO_4 and HNO_3 [43]. He found intercalation between graphite and chemical oxidation of graphite forming graphene oxide. Nowadays, the graphene is being enclosed by oxygen atoms in its oxidized form. This results in the contraction in inter-planar forces and thus these graphene oxide layers can be exfoliated by the application of thermal, ultrasonic and other energetic conditions. It's essential to prepare GO for the synthesis of graphene. In 1858, Hummer and Off man et al. recommended different methods for GO oxidation by proceeding to the graphite with potassium permanganate ($KMnO_4$) and concentrated H_2SO_4 [44]. Modified Hummer's method is the maximum adopted scheme for the fabrication of GO. Hummer's method is most extensively used to produce GO, which comprises oxidation of graphite by using potassium permanganate ($KMnO_4$) and sulfuric acid (H_2SO_4) [45]. The graphite salts are suspended in solvents and subject to extensive sonication to undergo exfoliation, so they can work as starting units for GO (Fig. 1). GO can be transformed into graphene by various methods that includes thermochemical reduction [46]. Chemical vapor deposition (CVD) produced a crystalline and electronic quality of graphene sheet as compared to the graphene sheet developed by GO. In few applications GO is favored due to its smooth transfer route (surface coating on virtually any surface is extremely easy by spin coating or usual spraying methods), manageable scaling-up, and cost-effective production. The distinctive chemical structure of GO correspondingly permits diverse chemical alteration or functionalization beneficial for a wide array of applications that covers from electrochemistry to biomedical sciences.

Initially graphene-like structures were described by Boehm et al. in 1962 [47]. They had incorporated graphene or reduced graphene oxide by dispersing graphene

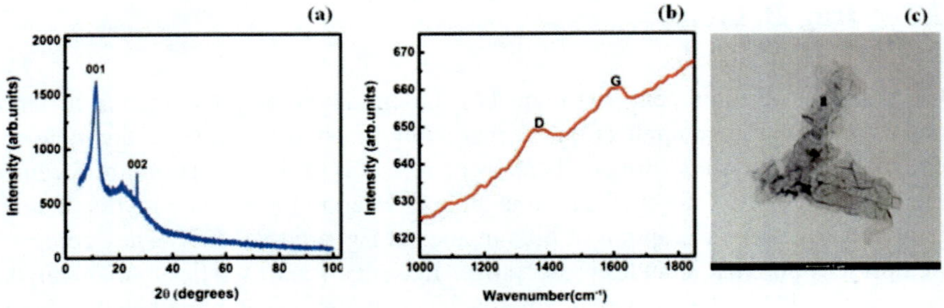

Fig. 1 **a** PXRD, **b** RAMAN, and **c** TEM micrographs of GO

oxide in reducing alkaline solution. They determined that this technique produces flakes of reduced graphene oxide having below oxygen and hydrogen content. The "Graphite oxide soot" is obtained by the combustion of graphene oxide at high temperature. Through electron microscopy, it was declared that this soot contains thin carbon multilayer hexagonal sheets (nearly 12.6 Å equal to 2–3 layers) acknowledged as graphene nanoplatelets. Meantime, graphite and its intercalated products had achieved attention because of their spectacular properties like they opened up the path for ion storage. In 1970, Blakely et al. found that graphene can be obtained from graphite by isolating carbon on a nickel surface [48]. At high temperatures, carbon layers get dissolved and get isolated into single or multilayers on the metal surface. Till 1995, certain terms like graphite layers, carbon layers or carbon sheets were adopted for graphene. After that IUPAC resolved one definition for graphene i.e. "A single carbon layer of the graphite structure, describing its nature by analogy to a polycyclic aromatic hydrocarbon of quasi-infinite size".

Previously it was known to scientists that a 2-D crystal of carbon atoms i.e. graphene exist but no one worked on extracting it from graphite. Until 2004, two investigators' prof. Andre Geim and prof. Kostya Novoselov confined, discovered, and characterized graphene in the University of Manchester. They had used the Scotch tape process for the extraction of graphene layers. In 2010 they were awarded with Nobel in Physics "for groundbreaking experiments regarding the 2-D material graphene".

3 Bio-functionalization of GO

The astonishing properties of GO inspired scientists to invent suitable GO derivatives with higher solubility and greater biocompatibility. GO has many functional groups on its edges (carboxylic acid groups) and basal planes (epoxy and hydroxyl groups). Figure 2 describes two different functionalization strategies such as covalent conjugation alongside noncovalent physical adsorption of molecules have been established till now.

3.1 Covalent Functionalization

Covalent functionalization accommodates polymer binding that demonstrate virtuous solubility and biocompatibility. Binding of polymers to the surface has occurred via a variety of methods such as approaches of trans-esterification, one pot synthesis, radical chemistry, and nitrene intermediate addition [49]. Among all polymers, binding of polyethylene glycol (PEG) with GO is most common and developed till now. Liu et al. analyzed the use of sheets of graphene for transporting drugs in cancer chemotherapy [50]. He treated GO with amine-terminated branched PEG, producing ultra-small sizes of PEGylated nano-GO (NGO-PEG) with substantial physiological stability [50]. Quite a few other polymers like polyethylenimine (PEI), poly (vinyl alcohol), dextran (DEX), polyacrylic acid (PAA), poly (allyl amine hydrochloride) (PAH), and chitosan (CH) [51–55] have also been tested for the purpose with promising results.

3.2 Non-covalent Functionalization

Coupling of GO to other molecules can be done by various forms of surface interactions that includes various forms of electrostatic bonding, coordinate conjugation, non bonding interactions and stacking [56]. Attachment of polycyclic molecules on graphene surface can be done via π-π stacking. The binding affinity in noncovalent interactions is not very strong and a change in the surrounding environment can alter this binding changing its stability and allowing it to release the coupled molecules. Therefore in drug-delivery systems, the functionalization of GO has become one of the furthermost used method [57].

A number of biomolecules are also analyzed which can act as good binder on GO surface through non covalent interactions. Binding of several proteins (like bovine serum albumin (BSA) etc.), phospholipid monolayer, protamine sulfate, and sodium alginate have been reported [58–60]. These findings indicate the possible paradigm of GO to work as a potent nano vector in cytoplasmic system.

4 Formulating Sustained-Delivery Systems

For the continuous delivery of drugs there are five extensive paths. The first path comprises attaching a specific molecule to the drug to enhance its retention time. Poly (ethylene glycol) or PEG is known as the captive of this method. Through generating an increased hydrodynamic radius surrounding the pharmacophore molecule, the bloodstream flow duration of a drug can be significantly enhanced by PEGylation [61]. The second wide path comprises of to enclose the drug of concern in a reservoir system that consists of a membrane that permits the drug to disperse

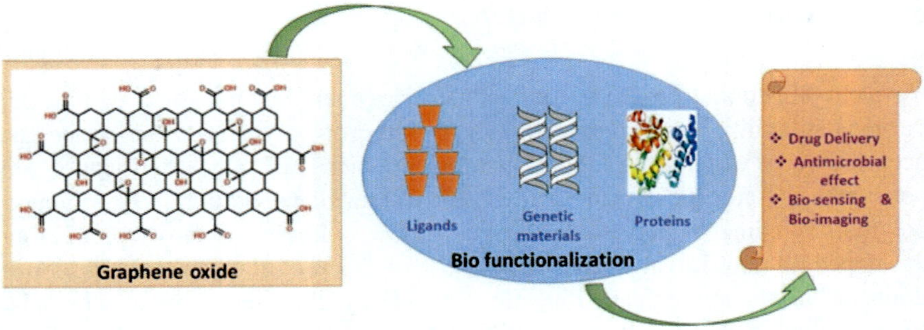

Fig. 2 Application of GO using bio-functionalization

through the device and into the adjacent tissue. This method comprised of Ocusert implant distributing pilocarpine and it depends on the origin of drug delivery scheme [62]. Diffusion-based systems pursue finite, usually small molecules with molecular weights less than 600 Da [63]. There are prodigious advantages of the membrane-based systems, like the facility to distribute several drugs over very long timescales. Although there is uncertainty, comprising the rift of the membrane, prominent to the impulsive supply of the whole reservoir [64].

The third pathway involves embedding the drug in a polymer matrix. Diffusion dictated the drug delivery along with release of nondegradable polymers [65]. The drug delivery profile along with degradable polymers, is verbalized by a sequence of diffusion and degradation actions that happen over time [65]. The tasks in the matrix-based methodologies incorporate upright involvement of the drug in the matrix to obtain twisting passageways along with encouraging connections among the drug and matrix. As the two accomplices approaching towards each other, another probable one will obtain very long (weeks to months to years) release. This interaction improves a portion of the route of developing drug delivery systems. The fourth path contains osmotic systems. This type of systems provides a matrix-promoted diifusion mechanism for drug release through adsorption and swelling of matrix water. Usually, osmotic systems exhibit a fast release often involving a burst phase. Researchers are mounting distorted membranes to regulate the distribution in a more constant manner over time [66]. However for long term delivery osmotic systems have very limited use. Fifth approach implicates pump systems that enable constricted regulator of drug delivery overtime for a vast collection of treatments. Pumps can either be incessant and submissive or vigorously engaged to distribute medicines, whereas in some cases pumps are reusable. Size is the main drawback of pumps. Generally, they have been large and needed more comprehensive embedding and assignment than some of the added materials, which are higher susceptible to delivered by minor needles.

5 Biocompatibility and Toxicity of Graphene Oxide-Related Drug Carriers

Biocompatibility and toxicity of GO are confirmed through a broad range of research works including intracellular studies and animal experiments, this study describes use of GO in the existent world via clinical applications. In contrast to the one dimensional rod like nano structure or usual spherical shaped particles, the 2-D graphene structures are exclusive, and their biocompatibility, chemical and physical factors for their cellular interfaces are unidentified [67]. Prodigious consideration is paid to describe the toxic effect, bio availability and biocompatibility of graphene-based materials, as the concerns around the security and noxious of these materials have not been entirely determined. Earlier reports graphene derivatives as drug carriers indicate low cytotoxic potential, however consequences achieved later are contradictory [68, 69]. This is similar to reports on polymer wrapping or PEG modification of CNTs, which shows enhanced biocompatibility and hydrophilicity after their functionalization [70]. These studies are proof that improved solubility of graphene will hike its biocompatibility. As huge number of hydrophilic groups, like carboxyl, hydroxy and epoxy groups are present at GO edge or basal planes, it shows improved biocompatibility and hydrophilicity as compare to graphene [71].

The interaction among isolated graphene or GO layers and cells has are well investigated by using cells from different organs including lung, fibroblasts and neurons [72–74]. Planar GO sheets were incorporated and confiscated in cytoplasmic, membrane-bound vacuoles through human lung epithelial or fibroblast cells and prompted cytotoxicity at microgram dosage after twenty four hours [75]. By calculating trans-endothelial electrical resistance, toxicity of graphene was verified across blood–brain barrier (BBB) by Hondroulis et al. As per the resistance dimensions were analogous to those of the control, the results demonstrate that graphene exhibited slight or not any toxicity near both distinct cell types. Thus, graphene is biocompatible for biomedical applications as it did not disturb the reliability of the BBB model. Liao et al. analyzed the bio-compatibility of graphene-associated materials to red blood cells and inferred that physiochemical parameters like size, physical state, surface charge greatly influences their biological/toxicological effects [76, 77]. Comparative biocompatibility of GO on mice and human fibroblast cells have also been explored, which indicates that the dosage higher than 50 $\mu g\ ml^{-1}$ could deliver evident cytotoxicity [78]. Zhang et al. used the radiotracer method and a sequence of biological assays to resolved the circulation and biocompatibility of GO in mice, and they establish that GO was mostly accumulated in the lungs, somewhere it was recollected for a longer duration [79]. Hu et al. reported slight apoptosis of A549 cells on treatment with GO but chemically reduced graphene oxide showed significantly higher cytotoxicity against A549 cells [80]. Liu et al. advanced a novel alteration method for the construction of gelatin functionalized graphene nanosheets via gelatin as a reducing reagent to enhance its biocompatibility [81]. The gelatin–graphene nanosheets exhibit significant biocompatibility and greater drug-loading ability. Duch et al. discovered hybrid graphene based material to enhance their biocompatibility in lung [82]. Their outcomes indicate pulmonary toxicity of graphene is mainly due to the oxidation, which can be significantly reduced by attaching block copolymer Pluronic. PEG (hydrophilic biocompatible polymer) to the non-oxidized graphene surface [83]. Hence, this process with PEG modifications is realistic for graphene and GO to gain the similar consequences. Various experiments revealed that GO, subsequently covering with PEG, illustrate insignificant cytotoxic effect to various cell lines, even at large concentrations [84, 85].

Yang et al. developed fluorophore tagged nanographene sheets (NGSs) and examined in vivo toxicity of PEGylated graphene in mice model through fluorescence imaging and revealed no appreciable toxic effects [86]. They similarly examine the long-term in vivo bio distribution of 125I-labeled NGS functionalized with PEG and thoroughly studied the probable toxicity of graphene over time. Their work established that PEGylated NGSs generally collect in the reticuloendothelial system containing liver and spleen later intravenous management, and can be progressively unoccupied beyond affecting any considerable toxicity to the treated mice over 3 months under experimental dosage condition [87]. Henceforth this study verified direct correlation of graphene toxicity to its functionalization that is similar to carbon nanotubes (CNTs) [88]. GO revealed extreme thrombogenic effect to mice model and capable of inducing aggregation in human platelets [89]. This thrombotoxic nature of GO somewhats confines its uses in medical field. It was observed that when amine-modified graphene (G-NH_2) introduced intravenously it had unconditionally no stimulatory influence on human platelets. This distinguished extremely with remarks achieved using unmodified GO and reduced GO. Dextran (DEX) (biocompatible polymer), was operated to covalently alter GO [90], exhibit amended constancy in physiological solutions and abnormally compact cell toxicity associated to basic un-functionalized GO. Fetal bovine serum (FBS) is a usually-engaged constituent in a cell culture medium, was similarly used to reduce the cytotoxicity of GO [91]. At low concentrations of FBS (1%), human cells were penetrating in the occurrence of GO and exhibit concentration-dependent cytotoxicity, although the cytotoxicity of GO was prominently diminished at 10% FBS. This research proposed that the cytotoxicity of GO nanosheets ascends after direct interactions amongst the cell membrane and GO nanosheets that induce physical destruction to the cell membrane, and this consequence is fundamentally mitigated when GO is protected with FBS because of tremendously huge protein adsorption capacity of GO. In end, all these foregoing biocompatibility investigations determined that graphene toxicity is strictly related to their surface functionalization. The functionalized graphene, GO and their derivatives are demonstrated in maximum circumstances to be expressively not as much of toxic as related with their non-functionalized counter-parts. It is imperative to manage appropriate long-standing toxicity research of graphene-based nanomaterials as animal tests executed on rodent models might have dissimilar outcomes from primates and humans. Therefore, numerous pre-clinical toxicity investigations are mandatory for all the graphene based nano constructs. Nonetheless, extra attempts are essential to additional check out their chronological toxic effects over a period of time, although primary outcomes from the available results offer an encouraging perspective for their fruitful use in medical and diagnostic field.

5.1 Gene and Cellular Delivery of GO

With some modification of functional groups present on GO, it could be used as a proficient gene delivery carrier [92, 93]. For an effective gene carrier, Kim et al. conjugated GO with branched polyethylenimine (BPEI) [94]. Actual molecular weight and subsequently DNA binding, condensation, transfection efficiency, gene delivery efficiency also cell viability of BPEI was enhanced when it was conjugated with GO. GO nanosheets were adhered to gold NPs/NRs by electrostatic interactions which was used as a gene vector for transferring pDNA into HeLa cells by Xu et al. [95]. Capping of the NPs/NRs with surfactant cetyltrimethylammonium bromide (CTAB) provided sufficient functional groups on NPs surface for an additional link of polyethylenimine (PEI). GO-encapsulated gold NPs (GOPEI-AuNPs) containing PEI were executed as a novel gene vector, as they delivered DNA into HeLa cells. In vitro, gene transfection experiments determined that viable PEI 25 kDa exhibit higher cytotoxicity and lower proficiency in HeLa cells as compared to GOPEI-AuNPs. GOPEI-AuNPs show analogous transfection proficiency (65% efficiency and 90% viability) for GOPEI, probably due to its tiny size and spherical shape that promote cellular uptake. Li et al. demonstrate the confined and effective gene delivery to many cell lines in a substrate-mediated manner, evolved by substrates of nano-GO [84]. The GO substrates revealed efficient gene transfection for multiple cell lines comprising stem cells along with accomplished biocompatibility. The approach comprised one step of patterning and two phases of transporting GO nanosheets.

Nano graphene oxide (NGO) was impregnated on acetate membrane via vacuum filtration and its true thickness was measured by volume-concentration calculation of the suspension. Moreover, the integral arrangement of the polydimethylsilcone (PDMS) imprint on the GO film was obtained by pressing the stamp beside the topmost of the GO film/acetate membrane. Previously, the arranged GO film/acetate membrane was repealed onto the new substrate, like a glass coverslip, and compelled instinctively with a heavy metal unit [96]. For myocardial therapy, Paul et al. developed a biocompatible and injectable hydrogel where vascular endothelial advance factor-165 (VEGF) pro-angiogenic genes for distributed in a nano complex of GO [97]. The nonviral gene delivery structure by polyethylenimine (PEI) functionalized GO nanosheets (fGO) complexed with DNAVEGF. For organized and confined gene therapy, DNAVEGF was framed and combined in the low-modulus methacrylated gelatin (GelMA) hydrogel. The distribution of bioactive molecules inside cellular structure is of immense uses in biological field. Hung et al. demonstrated a method via intrinsic GO to distribute hydrophilic molecules through co-incubation in culture [98]. Cellular delivery of a diversity of hydrophilic small molecules in culture has increased by GO co-incubation. Molecules that induced sedimentation and adsorbed strongly were reflected up to 13-fold enrichment. Alkyls, hydroxyls, phosphates, and amines have reinforced the adsorption, although triazoles interrupt the interactions. GO with an adjacent measurement of 150 nm accomplished inherently advanced cellular delivery contrast to its 1500 nm equal autonomous of sedimentation. Disinfected nano complexes have co-incubation concede for a 100-fold enlarges in cargo incubation concentration.

5.2 *Graphene Oxide for Biomolecule Delivery*

Gene delivery and protein delivery in vivo is playing an important role in cancer therapy. As numerous proteins are elaborate in cellular signaling through activating or deactivating the analogous pathways, thus the transfer of proteins into cells may vary cellular functions. GO-PEG was used for packing and delivering the cytoplasmic proteins [99]. To study the behavior of the GO-PEG/protein complex, GO-PEG was encumbered with BSA and results indicated that GO-PEG protects the proteins from degradation by enzymes and helps them to retain their apoptotic function. GO was used to carry DNA and RNA aptamers by Wang et al. for delivering nucleoside triphosphates (ATP, GTP) in living cells, ensuring good biocompatibility and protecting DNA/RNA from enzymatic degradation [100]. Delivery of siRNA has emerged as an important technological advancement in drug delivery in the recent years. Yang et al. utilized FA-PEG-GO for the the targeted delivery of hTERT siRNA in cells [101].

Yin et al. considered the initiation of an indicator transducer and activator of transcription 3 (Stat3) that endorses existence in an inclusive range of human cancers [102]. Here, plasmid-based Stat3-specific siRNA was used instead of GO-PEI-PEG. In a mouse melanoma model, this showed a substantial reversion in tumor progression and weight loss. After administration, blood parameters and histology revealed no substantial side effects. Huang et al. used the GO-PEI to deliver Si-RNA along with a biomarker in an invasive breast cancer cell line [103]. Additional biomolecules, like microRNA (miRNA), have been used for delivery in diabetes and cancer. Ryoo et al. in his experiment showed the use of a nano-graphene Oxide (NGO) assembly for visualizing diverse miRNA stages in living cells [104]. For this function, NGO was bound to a peptide nucleic acid (PNA) containing a fluorescent dye. The dye quenched when the PNA was attached to NGO but then again it was reinstated on combination of miRNA to PNA. In an attempt for the meticulous discharge of biomolecules, dual-polymer GO-PEG-PEI was utilized by Tao et al. for the effective transfer of unmethylated cytosine-phosphate-guanine (CpG) that represented nucleic acid having therapeutic effects with active immune-stimulatory actions [105]. The research revealed that GO-PEG-PEI can boost the immune stimulatory consequence of CpG. In vivo assay determined that on irradiation with laser, GO-PEG-PEI-CpG complex caused synergistic photo thermal and immunological effects, which exhibited the maximum efficacy in reduction of the tumor volume. This signifies the exceptional therapeutic efficiency of the GO-PEG-PEI-CpG complex against cancer.

5.3 Antibacterial Activity or Bacteria Filtration of GO

Diverse form of metal nanoparticles adhered with GO are being used as effective antibacterial agents [106–110]. Tang et al. constructed nanocomposites of GO-Ag with altered AgNPs to GO proportions and explored their antibacterial activities in both the Gram-negative (G−) bacteria *Escherichia coli* (*E. coli*) and the Gram-positive (G+) bacteria *Staphylococcus aureus* (*S. aureus*) [111]. At moderately low dosage (2.5 μg/mL), GO-Ag nanocomposite was extra capable and exhibited stable antibacterial activities as compared to AgNPs. Bactericidal activity of the GO-Ag construct was more prominent against *E. coli* compared to *S. aureus*. This nano construct appears to work as a better antibacterial agent against gram negative bacteria by adequately inhibiting cell division by disturbing bacterial cell wall integrity. Ocsoy et al.demonstrated DNA-assisted silver nanoparticles developed on GO displays an exceptional antibacterial activity in contrast to bacterial advert produced through Xanthomonas perforans, which is a main infection of tomatoes, caused 10–50% reduction in the production of tomatoes [112]. The actual small concentration (16 ppm) of Ag@dsDNA@GO, composites exhibited distinguished antibacterial facility in in vitro conditions with substantial adsorption. The SERS tags method has piqued an enormous awareness as optical nanoprobes in the application of biomedical imaging and phototherapies. Invigorated by the part of GO as a simulated receptor for several dyes, Lin et al. established as in situ procedure for GO enveloped gold nanocluster SERS tags with a tris (2,2′-bipyridyl)ruthenium-(II) chloride (Rubpy)/GO nanohybrid as a complex Raman reporter [113]. The GO implemented systematic results for monitoring distinct main limitations of SERS tags, like reproducibility, sensitivity, colloidal and signal immovability. Additionally for optical classification and photo thermal ablation of bacteria multifunctional, aldehyde group adjoined Au@Rubpy/GO SERS tags were formulated that exhibit acceptable photo thermal killing efficiency for both types of bacteria gram-positive and gram-negative.

Wang et al. reported ZnO/GO complexes with efficient antibacterial activities and low cytotoxicity against the bacterium *E. coli* and HeLa cells [114]. The antibacterial activity of ZnO/GO composites was generally because of the synergism between ZnO and GO. As ZnO NPs were augmented on the GO matrix, the possibility of bacteria contact to Zn was enhanced. The close addition of the *E. coli* cells and ZnO NPs on the GO nanosheets reinforce the penetrability of the bacterial membrane by Zinc ions. Filtration is an additional technique to eliminate the bacteria from the solutions and economically existing filters are not capable ample to take away them finally. Musico et al. improved the economically feasible membrane filters by the distribution of poly (*N*-vinyl carbazole) (PVK), graphene, poly (*N*-vinyl carbazole)-graphene, GO and poly (*N*-vinyl carbazole)-GO (PVK-GO) in demand to induce antibacterial properties [115]. These improvised membranes exhibited major enhancement in the antibacterial properties against *E. coli* and *Bacillus subtilis*. The improved antimicrobial property of the filter was ascribed to the fabrication of reactive oxygen species by the nanomaterials. Amongst the nanomaterials used, the PVK-GO-modified membrane filter demonstrated the finest elimination of *B. subtilis* and *E. coli.* Periodontal diseases and Dental caries have an inherent correlation with microbes such as *Streptococcus mutants*, *Porphyromonas gingivalis,* and *Fusobacterium nucleatum.* He et al. established that GO can be efficient in killing dental microbes. GO nanosheets were very proficient in inhibiting the development of dental pathogens [116]. From the study, it was found that GO-treated *S. mutants* died in higher numbers compared to dead the control. This outcome opens an avenue for GO nanosheets to be used as a potential antibacterial agent.

Xie et al. developed the biomimetic GO/chitosan supports with classified arrangements. First, GO/CS frameworks with huge micro-5pores (~300 μm) provide great mechanical strength due to strong bonding among the oxygen-enclose functional groups of GO and the amine groups of CS [117]. Additionally, another molecule octacalcuim phosphate (OCP) with porous structures was bio mimetically impregnated on GO/CS scaffolds (OCP-GO/CS). The classified microporous structures of OCP-GO/CS frameworks delivered an auspicious atmosphere for cell progress and adhesion. Bone morphogenetic protein-2 (BMP-2)-enclosed bovine serum albumin (BSA) nanoparticles and Ag-NPs were adsorbed in the frameworks for improvement of osteoinductivity and antibacterial properties, correspondingly. Antibacterial assessments exhibited that the frameworks revealed great antibacterial properties against selective gram positive and gram negative stains. In vitro and tests discovered that the frameworks had virtuous biocompatibility, amended bone marrow stromal cells production and variation, and encouraged bone tissue rejuvenation. Sametband et al. described the usage of GO derivatives to prevent herpes simplex virus type-1 (HSV-1) infections [118]. From his work it appeared that cell-to-cell escalation was severely affected by the application of the nano-materials and the main inhibition mechanism was viral attachment blocking. Composition of the culture media seems to greatly influence antiviral properties of GO composites and in the end, solubility was found to decline with time.

6 GO in Chemotherapy

Chemotherapy is only the leading remedial path in terminal cancer treatments. Majority of the cancer drugs used in chemotherapy are water insoluble which makes them difficult to be given orally due to their poor intestinal absorption. The maximum significant research issue in this area is, to develop a facility to deliver competently non soluble cancer drugs [119]. Liu et al. delivered non soluble chemo agents via GO as nanocarriers. Here, a nano-graphene oxide-polyethylene glycol (NGO-PEG) system was non covalently encumbered with SN38, a camptothecin (CPT) to form NGO-PEG-SN38 [120]. The efficiency of this system to entrap drug was 10% and the toxicity of NGO-PEG-SN38 in water and SN38 in DMSO were similar with advanced efficacy than irinotecan (CPT-11). At a similar period, Sun et al. performed the binding study of doxorubicin (DOX) for NGO PEG/DOX drug-delivery system

[121]. In GO-based nanocarriers, different drugs were loaded using chemotherapy and also a share of resolvable polymer-GO systems was discovered.

For stacking and distribution of a low water-soluble antioxidant and anticancer drug, ellagic acid (EA), Kakran et al. reported hydrophilic, covalently bonded and biocompatible Pluronic F38 (F38), Tween 80 (T80), and maltodextrin (MD) to GO [122]. Loading was 100% for GO-F38, 122% for GO-T80, and 114% for GO-MD. The EA was verified in human breast carcinoma cells (MCF-7) and human colon adenocarcinoma cells (HT29) which concludes that the EA discharge rate was pH-dependent and showed enhanced cytotoxic activity as compared to free EA dissolved in DMSO. GO loaded Adriamycin (ADR) nanocarrier exhibited the earliest approach of resistance reversal in cells as demonstrated by Wu et al. [123]. ADR was encumbered adjacent the GO surface (ADR-GO), with a loading of 93.6%, and exhibited release in response to change in pH. It was exhibited that GO expressively improved the uptake of ADR in MCF-7/ADR cells (an ADR-resistant breast cancer cell line) and displayed considerably advanced cell toxicity as compared to ADR used alone, proposing that ADR-GO could efficiently reverse the drug resiatance, with 8.35 reversal index. Others drugs like, paclitaxel (PTX) was coupled to GO through π stacking interaction [124], developing the GO-PEG/PTX system which exhibited an 11.2% loading of PTX. This system showed good targeting and significant cytotoxicity against A549 lung cancer and MCF-7 breast cancer cell line.

Few reports not only discovered covalent bonds for some compounds but also represent the procedure of stacking drugs through noncovalent interactions. Lu et al. has developed a drug molecule for successful use in chemotherapy against brain tumors with malignancy, characterized with small life and low thermal stability via PAA-modified GO for loading 1,3-bis(2-chloroethyl)-1-nitrosourea (BCNU) [125]. The drug-loading ability attained up to 19.8%. A 24 h increase in the half-life of targeted BCNU with enhanced cellular uptake was observed with the use of this nanocarrier in GL261 cancer cell. The in vitro anticancer efficiency of PAA-GO/BCNU was revealed through a 30% growth in DNA cross-linking and enhanced cytotoxic potential. Tian et al. established an NGO-PEG nanocarrier for a potent delivery of cis-diammine dichloro platinum (II) (cisplatin, CDDP) by simple covalent attachment of platinum (II) atom and carboxylic group on GO surface [126]. This system showed a 58% loading efficiency for the drug with the release mechanism was found to be strongly pH dependent. This synthetic system exhibited excellent cytotoxic effect against MCF-7 breast cancer cell line CAL 27 oral carcinoma cell lines in in vitro experiments.

7 Application of Nano-medicine as Drugs

The surface of graphene having a plethora of delocalized π electrons is an excellent candidate for conjugation to aromatic drugs like doxorubicin (DOX) and camptothecin (CPT) (which contain aromatic moiety) by stacking and other non-covalent interactions. In graphene, every atom is disclosed on its two sides and its extremely large surface area is permitted for ultra-high drug-loading efficacy. DOX works as a broad spectrum antibiotic for treating diverse classes of cancers through DNA intercalation mechanism. Though, its conventionally intravenous supervision may prompt its irregular circulation along with poor performance in intercellular membrane crossing. To overwhelm this problem, a drug carrier is urgently demanded which can immobilize drug molecules and transference them over cellular barriers. GO has a largely hydrophobic basal plane along with hydrophilic groups on the edges. Quinine portion of DOX show hydrophobic effect with GO via π-π stacking interaction, due to the huge π conjugated structure of GO. Furthermore, the hydroxyl and amino groups in DOX could produce an efficient network of electrostatic interaction with hydroxyl and carboxyl groups present on the GO sheets. GO is an ultimate exporter for loading and delivering DOX because they are fairly miscible in aqueous solutions. Chen's group explored the extensive interaction and release kinetics of DOX on GO sheets [127]. The Stacking of DOX on GO sheets was completed by simple mixing in an aqueous solution with the utility of sonication. Pan et al. reported that the loading efficiency and release kinetics of DOX-GO system is strongly pH dependent because of extremely extensive Hydrogen bonding network that exists between the drug and GO surface.

Close to adverse, the exemption of DOX is extra effective underneath basic and acidic environments than beneath neutral conditions. After 30 h at pH 10 and 2, the total bound DOX could be released 25 and 71% correspondingly. Depan et al. also revealed as a matter of fact, stacking and discharge of DOX were primarily be governed by pH and hydrogen-bonding interaction amongst GO and DOX [128]. Essentially DOX was encumbered on to GO surface through non- covalent π-π loading interaction, monitored through enclosing of GO with folic acid conjugated chitosan (CHI-FA). They establish that the nanohybrid arrangement displayed that at pH 5.3 higher drug release (35%) was observed as compared to much lower drug release of 11% at pH slightly higher than 7, which was accredited to the condensed interaction among DOX and the drug carrier. Under neutral conditions, the hydrogen-bonding interaction amongst DOX and GO was extra conspicuous, causing in a controlled release. The quantity of released DOX from GO was considerably advanced during the protonation of the amine ($-NH_2$) groups of DOX, which lead to the fractional alienation of hydrogen-bonding interaction beneath acidic conditions. Hu et al. accomplished the stacking of DOX on pluronic F127 and GNS (F127/GNS) at various ratio of DOX and F127/GNS. It was observed that at 0.9 mg/mL concentration of DOX provided highest stacking efficiency of 29% [129].

Later, the in vitro cytotoxicity of the non-covalent DOX/F127/GNS nanohybrid in MCF-7 human breast cancer cells was investigated. In acidic conditions the drug release amount was advanced in contrast to underneath neutral and basic conditions. After loading DOX with F127/GNS the cytotoxicity of MCF-7 was enhances as associated with free F127/GNS. To improve the cytotoxicity, Wang et al. designed reduced GO (rGO) by attaching a cluster of gold nanoparticles, commonly known as GNC through physical adhesion [130]. It was observed that DOX-loaded GNC-rGO repressed HepG2 cell growth extra firmly than the free drug or the matrix, which suggests the more efficient delivery of the drug loaded in the matrix to the cellular cytosol compared to free drug. To increase the consequence of embattled drug delivery, Yang et al. synthesized a GO-Fe_3O_4 composite and folic acid (FA) was conjugated onto Fe_3O_4 nanoparticles alongside chemical attachment to the amino groups of 3-aminopropyl triethoxysilane (APS)-to derive a chemically functionalized GO-Fe_3O_4 nanohybrid [131].

By π-π stacking, DOX was loaded onto a multi-functionalized GO surface. The in vitro tests designated that DOX-loaded GO-Fe_3O_4 nanohybrid showed considerably higher cytotoxicity against SK3 human breast cancer cells as compared to blank GO-Fe_3O_4, nonetheless the cytotoxicity was less as compared to GO-Fe_3O_4-FA- DOX at identical drug concentration. These outcomes proposed that it is probable to use GO as a perfect multi-functionalized drug carrier for tumor combination therapy. Furthermore, GO based material have correspondingly shown prodigious perspective in photothermal treatment. Zhang et al. established a PEG-GO-DOX for the antitumor consequence in vitro and in vivo by the amalgamation of photo thermal and conventional chemotherapy in one scheme [132]. The outcomes indicate, cumulative effect of the combined treatment procedure leads to a synergism effect that causes improved cytotoxicity as compared to either of the treatments when used alone.

7.1 Co-delivery of Multi-drug

By a detailed scheme of structural understanding, delivery schemes based GO template can also comprehend the co-delivery of several pharmacophores and biomolecules like DNA and RNA, to achieve an improved therapeutic efficiency. Zhang et al. executed simultaneous stacking of two chemotherapeutic molecules DOX and CPT, on the FA-GO through non-covalent interactions [133]. Co-delivery of these molecules drugs via FA-GO showed enhanced targeted delivery with greater cytotoxic potential as compared to GO loaded with either of two drugs only. Cationic polymers like PEI are generally adopted regularly in gene delivery, as the phosphate moieties of nucleic acids establish a strong electrostatic interaction with the polymer. Nonetheless, limited biocompatibility and intense cytotoxicity of PEI are the foremost complications for using it in drug delivery [134]. Both Zhang et al. and Liu et al. constructed PEI functionalized GO (PEI-GO) used in transfection of circular DNA or plasmids (PDNA) via a covalent link and non-covalent methods. Correspondingly GO-PEI exhibited lesser cytotoxicity and greater transfection efficacy compared to native PEI. These results lead to the presumption that GO-PEI constructs can effectively load drugs with aromatic moieties in the planar sheet via non covalent interaction. Additionally these GO-PEI constructs seemed to be capably of binding electrostatically to the side chains of nucleic acids. Latterly, Zhang et al. reported an in vitro cellular delivery of siRNA and DOX through a GO-PEI nano construct [135]. This co-delivery ominously increased therapeutic effectiveness and bring about the synergistic result in cancer cell killing. The authors similarly established CS-GO as a nanocarrier to distinctly distribute CPT and pRL-CMV into human cancer cell lines [136]. Even though most of the results with GO-polymer constructs are still in initial stage, it appears to have a beneficial chemotherapeutic efficiency and demand a significant requirement of forthcoming research in this field.

8 Limitations

In drug delivery in vitro and in vivo cytotoxic analysis is extremely important to eliminate the risk potent complexity while designing a drug carrier. However many physical parameters of GO like loading ratio, eluent fraction, surface morphology are sometimes hard to be organized, which might lead to the aggregation and establishment of micro gel. Furthermore, heating treatment was frequently used, which might affect the biological activity of the biopolymer, like heparin or enzyme. Frequently for the same delivery systems, all the significant mechanisms of absorption, distribution, metabolism, and excretion in the body are not quantified. There are also several systems existing and verified operative maximum of the time in vitro, however after the essential consequences in vivo, leaving these improvements waiting to take the succeeding phase and can become a regular weapon in future cancer treatment.

9 Future Prospective

The future of graphene based materials is tremendously auspicious for drug delivery. These materials can be beneficial both in single dosage and combinatorial therapy. Their cost-effectiveness can be a vast improvement over current expensive anticancer drugs. Usage of them maybe lessen the possible side effects of painful chemotherapeutic agents. Combination of these graphene based materials with other biopolymers therapeutic agents can develop new mechanisms for killing cancer cells. Toxicity investigations illustrate functionalized GO to be biocompatible. Several works establish GO derivatives are stimulating nanocarriers for the loading and delivery of therapeutic drugs. Graphene may come up as an exceptional nanoparticle to be used in the biomedical study by effective association with several branches of science.

10 Conclusions

The astonishing features of GO have prepared this material a tremendous stage to embolden innovative applications in diverse fields of science and technology. In the past several years the application of GO in drug delivery has grown rapidly. One of the main reason is its interesting surface structure and chemical composition which allows a number of diverse functionalization. In the initial phase of the enhancement of nanocarriers founded on GO, the difficultly of their stumpy solubility below physiological situations restricted their usage to biomedical applications. This difficulty was resolved through the covalent binding of polymers like PEG to their structures. The noncovalent type interfaces permit molecules to bind through electrostatic or π-π interactions and indorsed the foundation of a zone of volatile advancement in GO applications for drug delivery. Earlier research evidently specifies that substantial and stimulating growth has been completed in this field and that these developing biomaterials establish a prodigious perspective for biomedical applications.

Though initial preclinical research is inspiring the arena is still far away from clinical applications, which need to mark about continuing experiments. GO apparently exhibit various benefits equated with additional drug delivery systems. By using very simple preparation procedures GO has the capability to deliver immense drug loading capacity of multifold diverse drugs and beneficial molecules. As compared to other nanomaterials GO has elasticity and ability to enterprise composite multifunctional drug delivery systems used for collective treatments. One of the furthermost significant deliberations of these materials is a thoughtful considerate of GO collaboration with living cells (tissues and organs), exclusively the cellular uptake instruments. This information will surely simplify the upcoming enlargement of graphene-based nanocarriers, particularly for anti-cancer treatment wherever these materials displayed significant prospective. Biocompatibility and toxicity is the furthermost precarious concern for their regular use as therapeutic drugs. An analysis of present works has published the mainstream of current studies show that GO is bio adaptable and small toxic material, while attention must be taken with differing and indistinct consequences achieved by diverse investigation groups. Though a contract has been progressively grasped that the functionalization of GO ominously enhance their biocompatibility. Hence, further toxicity research with in vivo animal replicas are essential to ascertain the biocompatibility of GO. In brief, it can be ascribed that graphene based drug delivery systems shows immense potential and wide scope to provide a significant number of clinical solutions in nearby future.

Acknowledgements SM, SK, and DG would like to thanks Amity University, Haryana for supporting this work. SM and DG would like to acknowledge the support provided under the DST-FIST Grant No.SR/FST/PS-I/2019/68 of Govt. of India.

References

1. Geim AK, Novoselov KS (2007) The rise of graphene. Nat Mater 6:183–191
2. Brumfiel G (2009) Graphene gets ready for the big time. Nature 458(7237):390–401
3. Sykes E (2009) Graphene goes undercover. Nature Chem 1:175–176
4. Li D, Kaner RB (2008) Materials science, graphene-based materials. Science 320(5880):1170–1171
5. Gupta V, Sharma N, Singh U, Arif M, Singh A (2017) Higher oxidation level in graphene oxide. Optik (Stuttg) 143:115–124
6. Boehm HP, Clauss A, Fischer G, Hofmann U (1962) Surface properties of extremely thin graphite lamellae. In: Proceedings of the fifth conference on carbon, pp 73–80

7. Boehm HP, Setton R, Stumpp E (1994) Nomenclature and terminology of graphite intercalation compounds. Pure Appl Chem 66(9):1893–1901
8. Bolotin KI, Sikes KJ, Jiang Z, Klima M, Fudenberg G, Hone J (2008) Ultrahigh electron mobility in suspended graphene. Solid State Commun 146:351–355
9. Lee C, Wei X, Kysar J, Hone J (2008) Measurement of the elastic properties and intrinsic strength of monolayer graphene. Science 321:385–388
10. Bonaccorso F, Sun Z, Hasan T, Ferrari AC (2010) Graphene photonics and optoelectronics. Nat Photo 4:611–622
11. Balandin AA, Ghosh S, Bao W, Calizo I, Teweldebrhan D, Miao F (2008) Superior thermal conductivity of single-layer graphene. Nano Lett 8:902–907
12. Peigney A, Laurent C, Flahaut E, Bacsa RR, Rousset A (2001) Specific surface area of carbon nanotubes and bundles of carbon nanotubes. Carbon N Y 39:507–514
13. Stoller M, Park S, Yanwu Z, An J, Ruoff RS (2008) Graphene-based ultracapacitors. Nano Lett 8:6–10
14. Bao Q, Loh KP (2012) Graphene photonics, plasmonics, and broadband optoelectronic devices. ACS Nano 6:3677–3694
15. Zhang L, Xia J, Zhao Q, Liu L (2010) Functional graphene oxide as a nanocarrier for controlled loading and targeted delivery of mixed anticancer drugs. Small 6:537–544
16. Liu Z, Robinson JT, Tabakman SM, Yang K, Dai H (2011) Carbon materials for drug delivery and cancer therapy. Mater Today 14:316–323
17. Novoselov KS, Geim AK, Morozov SV, Jiang D, Zhang Y, Dubonos SV, Grigorieva IV, Firsov AA (2004) Electric field effect in atomically thin carbon films. Science 306(5696):666–669
18. Kim KS, Zhao Y, Jang H, Lee SY, Kim JM, Kim KS, Ahn JH, Kim P, Choi JY, Hong BH (2009) Large-scale pattern growth of graphene films for stretchable transparent electrodes. Nature 457(7230):706–710
19. Eizenberg M, Blakely JM (1979) Carbon monolayer phase condensation on Ni (111). Surf Sci 82:228
20. Wintterlin J, Bocquet ML (2009) Graphene on metal surfaces. Surf Sci 603:1841–1852
21. Berger C, Song Z, Li X, Wu X, Brown N, Naud C, Mayou D, Li T, Hass J, Marchenkov AN, Conrad EH, First PN, de Heer WA (2006) Electronic confinement and coherence in patterned epitaxial graphene. Science 312(5777):1191–1196
22. Land TA, Michely T, Behm RJ, Hemminger JC, Comsa G (1992) Scanning tunneling microscopy i: general principles and applications. Surf Sci 264:261
23. Yang M, Hou Y, Kotov NA (2012) Graphene-based multilayers: critical evaluation of materials assembly techniques. Nano Today 7:430–447
24. Singh RK, Kumar R, Singh DP (2016) Graphene oxide: strategies for synthesis, reduction and frontier applications. RSC Adv 6:64993–65011
25. Chen K, Song S, Liu F, Xue D (2015) Structural design of graphene for use in electrochemical energy storage devices. Chem Soc Rev 44:6230–6257
26. Guo S, Dong S (2011) Graphene nano sheet: synthesis, molecular engineering, thin film, hybrids, and energy and analytical applications. Chem Soc Rev 40:2644–2672
27. Liu Y, Dong X, Chen P (2012) Biological and chemical sensors based on graphene materials. Chem Soc Rev 41:2283–2307
28. Fan H, Wang L, Zhao K, Li N, Shi Z, Ge Z, Jin Z (2010) Fabrication, mechanical properties, and biocompatibility of graphene-reinforced chitosan composites. Biomacromolecules 11(9):2345–2351
29. Lee DY, Khatun Z, Lee JH, Lee YK, In I (2011) Blood compatible graphene/heparin conjugate through noncovalent chemistry. Biomacromolecules 12(2):336–341
30. Wang Y, Zhang P, Fang LC, Zhan L, Fang LY, Huang CZ (2012) Graphene–inorganic nanocomposites. RSC Adv 2(6):2322–2328
31. Depan D, Girase B, Shah JS, Misra RDK (2011) Structure–process–property relationship of the polar graphene oxide-mediated cellular response and stimulated growth of osteoblasts on hybrid chitosan network structure. Acta Biomater 7(9):3432–3445

32. Hu W, Peng C, Lv M, Li X, Zhang Y, Chen N, Fan C, Huang Q (2011) Protein corona-mediated mitigation of cytotoxicity of graphene oxide. ACS Nano 5(5):3693–3700
33. Shen J, Yan B, Shi M, Ma H, Li N, Ye MJ (2011) One-step hydrothermal synthesis of TiO_2-reduced graphene oxide sheets. Colloid Interf Sci 356(2):543–549
34. Shen J, Shi M, Yan B, Ma H, Li N, Hu Y, Ye M (2010) Covalent attaching protein to graphene oxide via diimide-activated amidation. Colloids Surf B 81(2):434–438
35. Tang L, Wang Y, Liu Y, Li J (2011) DNA-directed self-assembly of graphene oxide with applications to ultrasensitive oligonucleotide assay. ACS Nano 5(5):3817–3822
36. Liao KH, Lin YS, Macosko CW, Haynes CL (2011a) Cytotoxicity of graphene oxide and graphene in human erythrocytes and skin fibroblasts. ACS Appl Mater Interf 3(7):2607–2615
37. Murthy N, Chang I, Stayton P, Hoffman A (2001) pH-sensitive hemolysis by random copolymers of alkyl acrylates and acrylic acid. Macromol Symp 172(1):49–56
38. Robinson JT, Tabakman SM, Liang Y, Wang H, Sanchez CH, Vinh D, Dai HJ (2011) Ultrasmall reduced graphene oxide with high near-infrared absorbance for photothermal therapy. Am Chem Soc 133(17):6825–6831
39. Pan Y, Bao H, Sahoo NG, Wu T, Li L (2011) Water-soluble poly (N-isopropylacrylamide)–graphene sheets synthesized via click chemistry for drug delivery. Adv Funct Mater 21(14):2754–2763
40. Lavik EB, Kuppermann BD, Humayun MS (2012) Drug delivery. In: Retina, 5th edn. Elsevier Inc., pp 1
41. Friedmann T (1992) A brief history of gene therapy. Nat Genet 2:93–98
42. Frid A, Hirsch L, Gaspar R (2010) New injection recommendations for patients with diabetes. Diabetes Metab 36(2):3–18
43. Kim F, Cote LJ, Huang J (2010) Graphene oxide: surface activity and two-dimensional assembly. Adv Mater 22:1954–1958
44. Eda G, Chhowalla M (2010) Chemically derived graphene oxide: towards large-area thin-film electronics and optoelectronics. Adv Mater 22:2392–2415
45. Hummers WS, Offeman RE (1958) Preparation of graphitic oxide. J Am Chem Soc 80:1339–1416
46. Park S, Rodney SR (2009) Chemical methods for the production of graphenes. Nat Nanotechnol 4:217–224
47. Li X, Zhang G, Bai X, Sun X, Wang X, Wang E, Dai H (2008) Highly conducting graphene sheets and Langmuir-Blodgett films. Nat Nanotechnol 3:538–542
48. Khani M, Mehdipour E, Faghani A, Guday G, Donskyi IS, Unger WES, Adeli M (2016) Preparation of graphene oxide by cyanuric chloride as an effective and non-corrosive oxidizing agent. RSC Adv 6:115055–115057
49. Layek RK, Nandi AK (2013) A review on synthesis and properties of polymer functionalized graphene. Polymer (Guildf) 54:5087–5103
50. Liu Z, Robinson JT, Sun X, Dai H (2008a) PEGylated nanographene oxide for delivery of water-insoluble cancer drugs. J Am Chem Soc 130:10876–10877
51. Zhang L, Lu Z, Zhao Q, Huang J, Shen H, Zhang Z (2011) Enhanced chemotherapy efficacy by sequential delivery of siRNA and anticancer drugs using PEI-grafted graphene oxide. Small 7:460–464
52. Sahoo NG, Bao H, Pan Y, Pal M, Kakran M, Cheng HKF (2011) Functionalized carbon nanomaterials as nanocarriers for loading and delivery of a poorly water-soluble anticancer drug: a comparative study. Chem Commun 47:5235–5237
53. Zhang S, Yang K, Feng L, Liu Z (2011a) In vitro and in vivo behaviors of dextran functionalized graphene. Carbon N Y 49:4040–4049
54. Kurapati R, Raichur AM (2012) Graphene oxide based multilayer capsules with unique permeability properties: facile encapsulation of multiple drugs. Chem Commun 48:6013–6015
55. Depan D, Shah J, Misra RDK (2011a) Controlled release of drug from folate-decorated and graphene mediated drug delivery system: synthesis, loading efficiency, and drug release response. Mater Sci Eng C31:1305–1312

56. Zhang Q, Wu Z, Li N, Pu Y, Wang B, Zhang T (2017) Advanced review of graphene-based nanomaterials in drug delivery systems: synthesis, modification, toxicity and application. Mater Sci Eng C 77:1363–1375
57. Yang X, Wang Y, Huang X, Ma Y, Huang Y, Yang R (2011) Multi-functionalized graphene oxide based anticancer drug-carrier with dual-targeting function and pH-sensitivity. J Mater Chem 21:3448–3454
58. Shen H, Liu M, He H, Zhang L, Huang J, Chong Y (2012a) PEGylated graphene oxide-mediated protein delivery for cell function regulation. ACS Appl Mater Inter 4:6317–6323
59. Liu J, Guo S, Han L, Wang T, Hong W, Liu Y (2012) Synthesis of phospholipid monolayer membrane functionalized graphene for drug delivery. J Matter Chem 22:20634–20640
60. Xie M, Zhang F, Liu L, Zhang Y, Li Y, Li H (2018) Surface modification of graphene oxide nanosheets by protamine sulfate/sodium alginate for anti-cancer drug delivery application. Appl Surf Sci 440:853–860
61. Harris JM, Chess RB (2003) Effect of PEGylation on pharmaceuticals. Nat Rev Drug Discov 2:214–221
62. Stewart RH, Novak S (1978) Introduction of the ocusert ocular system to an ophthalmic practice. Ann Ophthalmol 10:325–330
63. Langer RS, Peppas NA (1981) Present and future applications of biomaterials in controlled drug delivery systems. Biomaterials 2:201–214
64. Langer R (1983) Implantable controlled release systems. Pharmacol Ther 21:35–51
65. Langer R (1990) New methods of drug delivery. Science 249:1527–1533
66. Philip AK, Pathak K (2006) Osmotic flow through asymmetric membrane: a means for controlled delivery of drugs with varying solubility. AAPS PharmSciTech 7:56
67. Sanchez VC, Jachak A, Hurt RH, Kane AB (2011) Biological interactions of graphene- family nanomaterials: an interdisciplinary review. Chem Res Toxicol 25:15–34
68. Chang YL, Yang ST, Liu JH, Dong E, Wang YW, Cao AN (2011) In vitro toxicity evaluation of graphene oxide on A549 cells. Toxicol Lett 200:201–210
69. Ryoo SR, Kim YK, Kim MH, Min DH (2010) Behaviors of NIH-3T3 fibroblasts on graphene/carbon nanotubes: proliferation, focal adhesion, and gene transfection studies. ACS Nano 4:6587–6598
70. Bianco A, Kostarelos K, Prato M (2005) Applications of carbon nanotubes in drug delivery. Curr Opin Chem Biol 9:674–679
71. Dreyer DR, Park S, Bielawski CW, Ruoff RS (2010) The chemistry of graphene oxide. Chem Soc Rev 39:228–240
72. Shen J, Shi M, Li N, Yan B, Ma H, Hu Y (2010) Facile synthesis and application of Ag-chemically converted graphene nanocomposite. Nano Res 3:339–349
73. Li XY, Huang XL, Liu DP, Wang X, Song SY, Zhou L (2011) Synthesis of 3D hierarchical Fe3O4/graphene composites with high lithium storage capacity and for controlled drug delivery. J Phys Chem C 115:21567–21573
74. Liu HW, Hu SH, Chen YW, Chen SY (2012) Characterization and drug release behavior of highly responsive chip-like electrically modulated reduced graphene oxide–poly (vinyl alcohol) membranes. J Mater Chem 22:17311–17320
75. Misra SK, Kondaiah P, Bhattacharya S, Rao CNR (2012) Graphene as a nanocarrier for tamoxifen induces apoptosis in transformed cancer cell lines of different origins. Small 8:131–143
76. Hondroulis E, Zhang ZQ, Chen CY, Li CZ (2011) Impedance based nanotoxicity assessment of graphene nanomaterials at the cellular and tissue level. Anal Lett 45:272–282
77. Liao KH, Lin YS, Macosko CW, Haynes CL (2011b) Cytotoxicity of graphene oxide and graphene in human erythrocytes and skin fibroblasts. ACS Appl Mater Interf 3:2607–2615
78. Wang K, Ruan J, Song H, Zhang J, Wo Y, Guo S (2011) Biocompatibility of graphene oxide. Nanoscale Res Lett 6:1–8
79. Zhang XY, Yin JL, Peng C, Hu WQ, Zhu ZY, Li WX (2011) Distribution and biocompatibility studies of graphene oxide in mice after intravenous administration. Carbon 49:986–995

80. Hu W, Peng C, Luo W, Lv M, Li X, Li D (2010) Graphene-based antibacterial paper. ACS Nano 4:4317–4323
81. Liu KP, Zhang JJ, Cheng FF, Zheng TT, Wang CM, Zhu JJ (2011) Green and facile synthesis of highly biocompatible graphene nanosheets and its application for cellular imaging and drug delivery. J Mater Chem 21:12034–12040
82. Duch MC, Budinger GRS, Liang YT, Soberanes S, Urich D, Chiarella SE (2011) Minimizing oxidation and stable nanoscale dispersion improves the biocompatibility of graphene in the lung. Nano Lett 11:5201–5207
83. Liu Z, Tabakman S, Welsher K, Dai H (2009) Carbon nanotubes in biology and medicine: in vitro and in vivo detection, imaging and drug delivery. Nano Res 2:85–120
84. Liu Z, Robinson JT, Sun XM, Dai HJ (2008b) PEGylated nanographene oxide for delivery of water-insoluble cancer drugs. J Am Chem Soc 130:10876–10887
85. Sun XM, Liu Z, Welsher K, Robinson JT, Goodwin A, Zaric S (2008a) Nano-graphene oxide for cellular imaging and drug delivery. Nano Res 1:203–212
86. Yang K, Zhang SA, Zhang GX, Sun XM, Lee ST, Liu ZA (2010) Graphene in mice: ultrahigh in vivo tumor uptake and efficient photothermal therapy. Nano Lett 10:3318–3323
87. Yang K, Wan JM, Zhang SA, Zhang YJ, Lee ST, Liu ZA (2011) In vivo pharmacokinetics, long-term biodistribution, and toxicology of PEGylated graphene in mice. ACS Nano 5:516–522
88. Ali Boucetta H, Bitounis D, Raveendran-Nair R, Servant A, Van den Bossche J, Kostarelos K (2013) Purified graphene oxide dispersions lack in vitro cytotoxicity and in vivo pathogenicity. Adv Healthc Mater 2:433–441
89. Singh SK, Singh MK, Kulkarni PP, Sonkar VK, Gracio JJA, Dash D (2012) Amine- modified graphene: thrombo-protective safer alternative to graphene oxide for biomedical applications. ACS Nano 6:2731–2740
90. Zhang SA, Yang K, Feng LZ, Liu Z (2011b) In vitro and in vivo behaviors of dextran functionalized graphene. Carbon 49:4040–4049
91. Hu W, Peng C, Lv M, Li X, Zhang Y, Chen N (2011) Protein corona-mediated mitigation of cytotoxicity of graphene oxide. ACS Nano 5:3693–3700
92. Hollanda LM, Lobo AO, Lancellotti M, Berni E, Corat EJ, Zanin H (2014) Graphene and carbon nanotube nanocomposite for gene transfection. Mater Sci Eng C 39:288–298
93. Ricci R, Leite NCS, da-Silva NS, Pacheco-Soares C, Canevari RA, Marciano FR, Webster TJ, Lobo AO (2017) Graphene oxide nanoribbons as nano- material for bone regeneration: effects on cytotoxicity, gene expression and bactericidal effect. Mater Sci Eng C 78:341–348
94. Kim H, Namgung R, Singha K, Oh IK, Kim WJ (2011) Graphene oxide–polyethyle-nimine nanoconstruct as a gene delivery vector and bioimaging tool. Bioconjug Chem 22:2558–2567
95. Xu C, Yang D, Mei L, Lu B, Chen L, Li Q, Zhu H, Wang T (2013) Encapsulating gold nanoparticles or nanorods in graphene oxide shells as a novel gene vector. ACS Appl Mater Interf 5:2715–2724
96. Li K, Feng L, Shen J, Zhang Q, Liu Z, Lee ST, Liu J (2014) Patterned substrates of nano-graphene oxide mediating highly localized and efficient gene delivery.ACS Appl. Mater Interf 6:5900–5907
97. Paul A, Hasan A, Kindi HA, Gaharwar AK, Rao VTS, Nikkhah M, Shin SR, Krafft D, Dokmeci MR, Shum-Tim D, Khademhosseini A (2014) Injectable graphene oxide/hydrogel-based angiogenic gene delivery system for vasculogenesis and cardiac repair. ACS Nano 8:8050–8062
98. Hung AH, Holbrook RJ, Rotz MW, Glasscock CJ, Mansukhani ND, MacRenaris KW, Manus LM, Duch MC, Dam KT, Hersam MC, Meade TJ (2014) Graphene oxide enhances cellular delivery of hydrophilic small molecules by co-incubation. ACS Nano 8:10168–10177
99. Shen H, Liu M, He H, Zhang L, Huang J, Chong Y (2012b) PEGylated graphene oxide-mediated protein delivery for cell function regulation. ACS Appl Mater Inter 4:6317–6323
100. Wang Y, Li Z, Weber TJ, Hu D, Lin CT, Li J (2013) In situ live cell sensing of multiple nucleotides exploiting DNA/RNA aptamers and graphene oxide nanosheets. Anal Chem 85:6775–6782

101. Yang X, Niu G, Cao X, Wen Y, Xiang R, Duan H (2012) The preparation of functionalized graphene oxide for targeted intracellular delivery of siRNA. J Mater Chem 22:6649–6654
102. Yin D, Li Y, Lin H, Guo B, Du Y, Li X (2013) Functional graphene oxide as a plasmid-based Stat3 siRNA carrier inhibits mouse malignant melanoma growth in vivo. Nanotechnology 24:105102
103. Huang YP, Hung CM, Hsu YC, Zhong CY, Wang WR, Chang CC (2016) Suppression of breast cancer cell migration by small interfering RNA delivered by polyethylenimine-functionalized graphene oxide. Nanoscale Res Lett 11:247
104. Ryoo SR, Lee J, Yeo J, Na HK, Kim YK, Jang H (2013) Quantitative and multiplexed microRNA sensing in living cells based on peptide nucleic acid and nano graphene oxide (PANGO). ACS Nano 7:5882–5891
105. Tao Y, Ju E, Ren J, Qu X (2014) Immunostimulatory oligonucleotides-loaded cationic graphene oxide with photothermally enhanced immunogenicity for photothermal/immune cancer therapy. Biomaterials 35:9963–9971
106. Tegou E, Magana M, Katsogridaki AE, Ioannidis A, Raptis V, Jordan S, Chatzipanagiotou S, Chatzandroulis S, Ornelas C, Tegos GP (2016) Terms of endearment: bacteria meet graphene nanosurfaces. Biomaterials 89:38–55
107. Ristic BZ, Milenkovic MM, Dakic IR, Todorovic-Markovic BM, Milosavljevic MS, Budimir MD, Paunovic VG, Dramicanin MD, Markovic ZM, Trajkovic VS (2014) Photodynamic antibacterial effect of graphene quantum dots. Biomaterials 35:4428–4435
108. Fathalipour S, Pourbeyram S, Sharafian A, Tanomand A, Azam P (2017) Biomolecule-assisted synthesis of Ag/reduced graphene oxide nanocomposite with excellent electrocatalytic and antibacterial performance. Mater Sci Eng C 75:742–751
109. Shen J, Li T, Shi M, Li N, Ye M (2012) Polyelectrolyte-assisted one-step hydrothermal synthesis of Ag-reduced graphene oxide composite and its antibacterial properties. Mater Sci Eng C 32:2042–2047
110. Govindaraju S, Samal M, Yun K (2016) Superior antibacterial activity of GlcN-AuNP-GO by ultraviolet irradiation. Mater Sci Eng C 69:366–372
111. Tang J, Chen Q, Xu L, Zhang S, Feng L, Cheng L, Xu H, Liu Z, Peng R (2013) Graphene oxide–silver nanocomposite as a highly effective antibacterial agent with species-specific mechanisms. ACS Appl Mater Interf 5:3867–3874
112. Ocsoy I, Paret ML, Ocsoy MA, Kunwar S, Chen T, You M, Tan W (2013) Nanotechnology in plant disease management: DNA-directed silver nanoparticles on graphene oxide as an antibacterial against Xanthomonas perforans. ACS Nano 7:8972–8980
113. Lin D, Qin T, Wang Y, Sun X, Chen L (2014) Graphene oxide wrapped SERS tags: multifunctional platforms toward optical labeling, photothermal ablation of bacteria, and the monitoring of killing effect. ACS Appl Mater Interf 6:1320–1329
114. Wang A, Cao Y, Jiang X, Zhang JH, Liu Y, Liu H, Wang (2014) Superior antibacterial activity of zinc oxide/graphene oxide composites originating from high zinc concentration localized around bacteria. ACS Appl Mater Interf 6:2791–2798
115. Musico YLF, Santos CM, Dalida MLP, Rodrigues DF (2014) Surface modification of membrane filters using graphene and graphene oxide-based nanomaterials for bacterial inactivation and removal. ACS Sustain Chem Eng. 2:1559–1565
116. He J, Zhu X, Qi Z, Wang C, Mao X, Zhu C, He Z, Li M, Tang Z (2015) Killing dental pathogens using antibacterial graphene oxide. ACS Appl Mater Interf 7:5605–5611
117. Xie C, Lu X, Han L, Xu J, Wang Z, Jiang L, Wang K, Zhang H, Ren F, Tang Y (2016) Biomimetic mineralized hierarchical graphene oxide/chitosan scaffolds with adsorbability for immobilization of nanoparticles for biomedical applications. ACS Appl Mater Interf 8:1707–1717
118. Sametband M, Kalt I, Gedanken A, Sarid R (2014) Herpes simplex virus type-1 at-tachment inhibition by functionalized graphene oxide. ACS Appl Mater Interf 6:1228–1235
119. Corrie PG (2007) Cytotoxic chemotherapy: clinical aspects. Medicine 36:24–28
120. Liu Z, Robinson JT, Sun X, Dai H (2008c) PEGylated nanographene oxide for delivery of water-insoluble cancer drugs. J Am Chem Soc 130:10876–10887

121. Sun X, Liu Z, Welsher K, Robinson JT, Goodwin A, Zaric S (2008b) Nano-graphene oxide for cellular imaging and drug delivery. Nano Res 1:203–212
122. Kakran M, Sahoo N, Bao H, Pan Y, Li L (2012) Functionalized graphene oxide as nanocarrier for loading and adriamycin can reverse drug resistance in breast cancer cells. Nanotechnology 23:355101
123. Wu J, Wang YS, Yang XY, Liu YY, Yang JR, Yang R (2011) Graphene oxide used as a carrier for delivery of ellagic acid. Curr Med Chem 18:4503–4512
124. Xu Z, Wang S, Li Y, Wang M, Shi P, Huang X (2014) Covalent functionalization of graphene oxide with biocompatible poly(ethylene glycol) for delivery of paclitaxel. Appl Mater Interf 6:17268–17276
125. Lu YJ, Yang HW, Hung SC, Huang CY, Li SM, Ma CCM (2012) Improving thermal stability and efficacy of BCNU in treating glioma cells using PAA-functionalized graphene oxide. Int J Nanomed 7:1737
126. Tian L, Pei X, Zeng Y, He R, Li Z, Wang J (2014) Functionalized nanoscale graphene oxide for high efficient drug delivery of cisplatin. J Nanopart Res 16:2709
127. Yang X, Zhang X, Liu Z (2008) High-efficiency loading and controlled release of doxorubicin hydrochloride on graphene oxide. J Phys Chem C 112:17554–17558
128. Depan D, Shah J, Misra RDK (2011b) Controlled release of drug from folate-decorated and graphene mediated drug delivery system: synthesis, loading efficiency, and drug release response. Mater Sci Eng C 31:1305–1312
129. Hu HQ, Yu JH, Li YY (2012) Engineering of a novel pluronic F127/graphene nanohybrid for pH responsive drug delivery. J Biomed Mater Res A 100A:141–148
130. Wang CS, Li JY, Amatore C (2011) Gold nanoclusters and graphene nanocomposites for drug delivery and imaging of cancer cells. Angew Chem Int Ed 50:11644–11648
131. Yang XY, Wang YS, Huang X (2011) Multi-functionalized graphene oxide based anticancer drug-carrier with dual-targeting function and pH-sensitivity. J Mater Chem 21:3448–3454
132. Zhang W, Guo ZY, Huang DQ (2011) Synergistic effect of chemo-photothermal therapy using PEGylated graphene oxide. Biomaterials 32:8555–8561
133. Zhang LM, Xia JG, Zhao QH (2010) Functional graphene oxide as a nanocarrier for controlled loading and targeted delivery of mixed anticancer drugs. Small 6:537–544
134. Jager M, Schubert S, Ochrimenko S (2012) Branched and linear poly(ethylene imine)-based conjugates: synthetic modification, characterization, and application. Chem Soc Rev 41:4755–4767
135. Zhang LM, Lu ZX, Zhao QH (2011) Enhanced chemotherapy efficacy by sequential delivery of siRNA and anticancer drugs using PEI-Grafted graphene oxide. Small 7:460–464
136. Bao H, Pan Y, Ping Y (2011) Chitosan-functionalized graphene oxide as a nanocarrier for drug and gene delivery. Small 7:1569–1578

Application of Graphene-Based Biopolymer Nanocomposites for Automotive and Electronic Based Components

Partha Pratim Das and Vijay Chaudhary

Abstract For polymer composites as a filler material, Graphene is gaining tremendous importance in the recent year. Biopolymer nanocomposites reinforced with graphene as a filler (GRMs) or carbon nanotubes (CNTs) have been extensively explored for use as engineering materials in various applications, as they have low mass density exceptional mechanical properties. The most relevant sectors are the automotive and aerospace industries, in which various components have to be built by considering light-weight materials and along with this they have to have high strength, low cost and one of the important things is their recyclability One of the other important sector in which graphene plays an important role is in electronic components. They are used in manufacturing of photovoltaic cells, biosensors, Electromagnetic Interference shielding (EMI) and many more. This present chapter provides a better understanding of how graphene acts a better option for improving the properties of biopolymer nanocomposites used in various automotive and electronic components.

Keywords Graphene · Low mass density · Biopolymer · Automotive · Recyclability

1 Introduction

Nanocomposites are materials where at least one phase displays nanometre dimensions (1 nm = 10^{-9}) Nanocomposite materials have emerged as suitable alternatives for overcoming micro composite limitations. It is a high-performance material exhibiting rare combinations of properties and special possibilities for design. With an expected average growth rate of about 25% and the highest market in engineering plastics and elastomers, their ability is so striking that they are useful in a range of

P. P. Das (✉) · V. Chaudhary
Department of Mechanical Engineering, Amity University, Uttar Pradesh, Noida 201313, India
e-mail: das1.parthapratim@gmail.com

V. Chaudhary
e-mail: vijaychaudhary111@gmail.com

B. Sharma and P. Jain (eds.), *Graphene Based Biopolymer Nanocomposites*, Composites Science and Technology,
https://doi.org/10.1007/978-981-15-9180-8_17

areas ranging from construction, automotive, electronics to biomedical applications [1]. In present scenario, glass and carbon fibers are often used among the different synthetic fibers due to a higher enhancement of the epoxy resin's mechanical properties, ideal for high-performance applications. Such synthetic fibers, however, are of expensive in nature, non-recyclable, non-renewable, non-biodegradable, and harmful chemicals are emitted during their processing [2]. And these fibers create major environmental hazards. Natural fibers are cheaper than synthetic fibers, lower in diameter, biodegradable, and when added after chemical changes, the properties of different polymers are improved by electrical, mechanical, and water absorption.

Graphene provide attractive opportunities in automotive applications because they are lightweight and simple to process, although their stiffness and impact strength must be enhanced simultaneously for use in many structural components to maximize stiffness, the surrounding flexible polymer matrices allow for great durability, so that the proper configuration of the components determines the strength [3]. However, in this case the interfacial adhesion between the filler and the matrix is a prerequisite for efficient load sharing and mechanical stability. Thus, over the past few years, significant work has centred on how to improve interfacial adherence between elements, and how to enhance them on the basis of mechanical properties. Although different nano fillers are now available now a days, graphene is emerging as a possible filler material used in biopolymer nanocomposites to achieve better mechanical properties, making the composite ideal for different applications [4]. In polymer nanotechnology [1], the synthesis of polymer nanocomposites is an important part. Through adding the inorganic nanometre compounds, the properties of polymers improve and this also has too many applications based on the inorganic content present in the polymers. One of the simplest approaches for the production of polymer nanocomposites is solvent casting and is less time consuming. Graphene is a carbon allotrope in the form of a single layer of atoms in a hexagonal two-dimensional lattice in which each vertex is formed by one atom which is becoming a potential filler material in recent years. Nowadays technology is largely based on nanoparticles and the two-dimensional graphene sheets have emerged as a topic of tremendous scientific interest due to their exceptional transportation of electrons, mechanical properties and high region. Therefore, graphene also appears in the domain of electronics as an emerging filler material.

With its fascinating tensile strength, elastic modulus, electrical and thermal conductivity, thermal stability, gas barrier and flame retardance [4], the Graphene discovery has opened exciting windows to develop new light composites while enhancing trade-offs between light weighting and safety problems [5]. Table 1 shows the various properties of the family of graphene. The incorporation of graphene into biopolymer nanocomposites and the resulting improvements in the mechanical strength as well as the electrical conductivity of the composites have been documented by various research groups [6]. Such nanocomposites usually often exhibit improved thermal stability and electrochemical behaviour, as well as gas barrier properties. The use of graphene in these composites improves the mechanical/electrical properties of biomaterials, and increases the cell attachment and growth on the biomaterial surface. Besides that, the main advantage of these nanocomposites is that only

Table 1 Showing the various properties of graphene [5]

Properties	Graphene	Graphene oxide	Reduced graphene oxide
Young's modulus (TPa)	1.0	0.208	0.25
Tensile strength (MPa)	130×10^3	76.8	0.9×10^5
Fracture toughness (MPa)	4–5	–	2.8–3
Optical transmittance	97.7%	–	60–90%
Charge carrier concentration (cm^{-2})	1.4×10^{13}	–	–
Room-temperature mobility (cm^{-2}/V s)	200,000	2–200	17–2000
Thermal conductivity (W/mK)	1500–5000	2000	30/250
Energy gap	0.26	–	0.35
Surface area (m^2/g)	560.6	2630	–

with a small amount of graphene filler and in the presence of a variety of polymer matrices leads to the positive changes in its properties.

In recent years, a number of research groups have developed and investigated the use of supercapacitor graphene nanocomposite [7] in various automotive applications [8]. Rajesh et al. [9] performed in situ chemical reactions of the synthesized reduced graphene oxide and/Zn nanoparticle/polycarbazole (rGO/Zn/PCz) nanocomposite. The developed nanocomposites were investigated by scanning electron microscopy-energy-dispersive X-ray analysis (SEM-EDX) and transmission reflectance attenuated by Fourier infrared spectroscopy (FTIR-ATR) transfer [9].

2 Graphene as Potential Filler Material in Biopolymer Nanocomposites

Owing to the presence of biopolymers, biopolymer nanocomposites have many advantages, such as mechanical-robust, safe, biocompatible, biodegradable, and have multifunctional characteristics, over traditional polymer-based products [10]. Based on the particular applications, synthetic components with controlled dispersion and highly structured organization can be incorporated into the biopolymer matrix. The filler materials are used to enhance the performance of the manufactured composites [11]. Graphene is much stiffer and heavier than the carbon nanotube (CNT), a special monolayer hexagonal lattice. William et al. [12] explained that composite materials exhibited improved properties compared to metal and polymeric materials which produced composites used as structural constituents. Figure 1 demonstrates the interactions between nanocomposites of the graphene and biopolymers. A variety of series production models have been released on the market in recent years, demonstrating the technological viability of applying the multi-material Body-in-White approach to lower production volume and high-performance parts of vehicles [13].

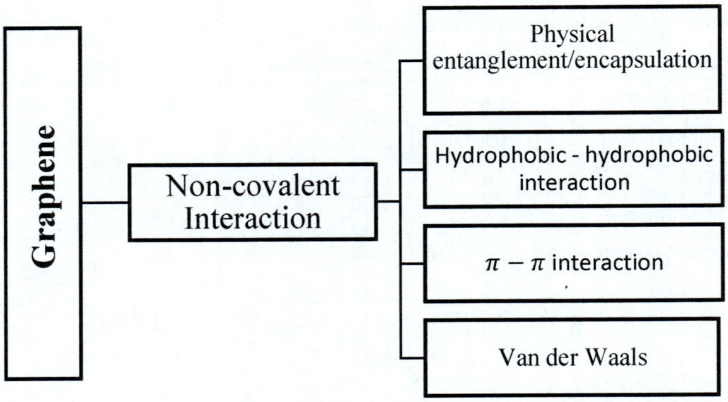

Fig. 1 Shows the interactions between graphene and biopolymers nanocomposites

Some of companies which also adopt the body in white approach such as The Porsche Panamera, the Audi TT, the new Daimler E-class and the BMW5, i3, and i8 series are examples [14]. In general, thermosetting polymer resins are commonly used in various applications such as in aerospace structures, coatings of automobile parts, automotive parts, and adhesives in various joining processes which also play an important role in electronic devices.

The primary composite aircraft structures could be lighter if the current resin durability and low conductivity limitations were also improved. Due to incorporation of graphene, it shows a significantly increase in rigidity, hardness and conductivity of polymer resins in thermosetting [15]. Galpaya et al. [16] compared mechanical-property graphene, functional graphene and graphene oxide to other nanofillers such as carbon nanotubes (CNTs), silicon dioxide (SiO_2), titanium dioxide (TiO_2) and aluminium oxide (Al_2O_3), respectively. They found that there is an increase in fracture toughness of 65% with the addition of just 0.125% of graphene which was produced by thermal annealing process. To achieve a comparable change, a 120-fold higher SiO_2 fraction of weight will allow a 30-fold higher Al2O3 fraction of weight and a 60-fold higher TiO_2 fraction of weight [16, 17]. The highest reported improvement in fracture toughness for CNTs is only 48%, and this is accomplished only with a 4-fold weight fraction higher than that resulting in an increase of 65% in the corresponding graphene epoxy nanocomposites [18].

The growing trend towards replacing traditional steel and casting irons for lightweight purposes in vehicles contributes to the production of automotive parts with lighter weight materials. Various traditional materials such as aluminium alloy, magnesium alloy and ultra-high resistance steel are used in engine blocks, cylinder heads, transmission, cases, valve bodies as well as channel plates [19]. Because of this graphene, as a possible material used by filler because of its exceptional mechanical properties, a filler material is continuously used to enhance the properties of the nanocomposites [20].

3 Application of Graphene-Based Biopolymer Nanocomposites

Graphene nanocomposite has numerous applications which including engineering science, electronics, biomedical, designing, household design and many more [21]. Figure 2 depicts some use of nanocomposites with biopolymers.

3.1 Applications of Graphene-Based Nanocomposites in Automotive Components

Graphene based nanocomposites for automotive applications find an effective way to incorporate graphene as polymer reinforcements for energy efficient and safe vehicles (EESVs) in composite materials [22]. The idea is based on the concept of introducing light-weight structures which aims to combine with new multifunctional materials to enhance the properties. Automotive industry is the one which has made extensive use of the natural-fiber reinforced composites (NFRCs). In Europe, plant fiber composites are mainly used by the automotive industry. The total reported use of natural fibers was around 4.0 ktonne in 1996 and this had increased to more than 21 ktonne by 1999. Shipping and aerospace industry to make commercial household products such as furniture, door and windows panel etc. Especially flax fiber processors have high hope in the thermal insulation. One area of concern to graphene is the automotive sector [23]. Today's scientists and engineers are discovering a variety of ways to intentionally create nanoscale materials and benefit from their enhanced properties.

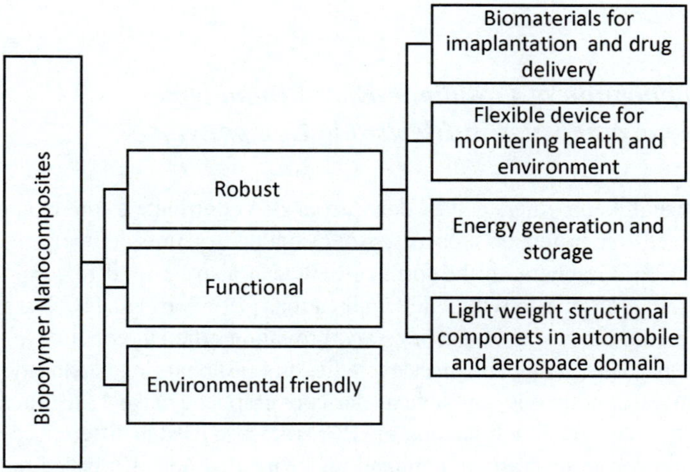

Fig. 2 Shows the application of biopolymer nanocomposite

Lin [24] and his team demonstrated the high potential graphene has for applications in automotive electronics. Modern vehicles have a variety of radar systems for protecting various viewing angles. The use of conventional fibre-reinforced plastics such as glass, carbon and aramid-reinforced thermoplastic and thermoset resins is considered important because of environmental problems [25]. Often called synthetic fibers, they are less biodegradable and made from non-renewable materials. Yet naturally occurring fibers may also bring environmental benefits as compared to synthetic fibres.

In automotive applications where high strength, lower cost and environmentally friendly features are needed, more interest was seen in investigating the suitability of natural fiber composites with graphene as a filler material. The criteria considered by a manufacturer are ultimate breaking force and elongation, flexural properties, impact resistance, fogging properties, flammability, odour, acoustic absorption, temperature and dwell time suitability for manufacturing, dimensional stability, water absorption and crash behaviour. Additionally, the most important variables for moving into new materials are length, quality, internal energy and the Head Injury Criterion (HIC) [26]. One way to boost the performance of cars is to reduce vehicle weight, or light weighting. Architectural upgrades and material substitution are the key strategic approaches in the Body-in-White (BIW) for weight loss. Table 2 demonstrates the use of biopolymer nanocomposites by different composite materials in the manufacture of cars in automotive components.

Now these interfaces find their way into the dashboards of many vehicles where the user interface provided to the driver is just as relevant as the vehicle's performance [27]. Another exceptional graphene property, its versatility, has been shown to be important in making screens that can fit over curved surfaces. This means that without the conventional restrictions associated with the dashboard form the entire vehicle can become a space for viewing material.

3.2 Application of Graphene-Based Biopolymer Nanocomposites in Electronic Components

Graphene and its polymers coated derivatives also contribute to the manufacture of various electronic products, such as sensors, condensers, tires, helmets, photovoltaic cells etc. [28]. Graphene in the form of balls can also be used for improving fast charging and high-rate activity in EV applications [29]. Son et al. [30] has shown that by suppressing adverse side reactions and providing efficient conductive pathways, the graphene-ball coating enhances cycle life and fast charging capability. Since the high electrical conductivity and optical clarity of graphene make it a useful material in user interface design, Touch screens and LCDs are also used in different applications. Figure 3 demonstrates various applications of nanocomposites based on graphene biopolymers in electronic components.

Table 2 shows the model application in automotive parts by various manufacturing companies

Company	Automotive parts manufactured	Fibre used	Ref
BMW	3, 5 and 7 series • Door panels • Headliner Panels • Boot lining • Seat back	Sisal and flax	[52]
AUDI	A3, A4, A6, Roadstar • Seat back • Side and Back panels • Boot lining • Hat Rack • Spare tire lining	Flax/sisal with thermoset resin	[52, 53]
Fiat	• Punto • Brava • Marea • Alfa Romeo 146, 156	Hemp, flax, Bamboo	[53]
FORD	Mondeo CD 162 • Door Panels • Boot Liner	Flax Wheat, hemp. Rice husk	[54]
Toyota	• Instrument panels/air conditioning vent • Floor mats/spare tire cover • Interior component • Radiator end tank	Kenaf, sugarcane fibres, banana fibres	[55]
Honda	• Floor area parts • Interior fabrics	Wood, flax	[55]
Nissan	• Car interior floor mats	Jute, hemp, nettle	[55, 56]
Volkswagen	• Interior door panel	Flax sisal	[57]
Mercedes-Benz	• Door Panels • Seat backrest • Floor panels • Under floor protection	Flax, sisal, coconut, wood	[58]

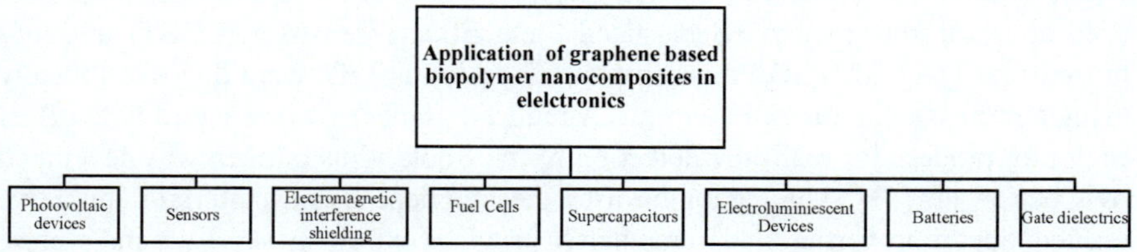

Fig. 3 Shows various applications of graphene-based biopolymer nanocomposites in electronic components

Graphene and its polymer composites have excellent properties in electroluminescent applications, dielectric gateways and touch screen devices because of their high power, thermal and electrical conductivity, low temperature processing conditions, and simple manufacturing process [31]. In 2018 Andrea Mura and his colleagues at Politecnico di Torino introduced a realistic application of graphene as a grease additive. In particular, grease applied to the graphene nanoplatelets has been evaluated as an improved spline coupling efficiency lubricant [32, 33]. With the developments in nanotechnology, engineers and researchers are in the process of enhancing radiator thermal performance by adding nanoparticles. In energy storage applications, polymer graphene composites outperform traditional rechargeable lithium batteries, as they have superior capacity and power density. Composites made from graphene are commonly known as supercapacitors [34]. Guo et al. [35] used graphene to improve a polymer cathode's electrochemical efficiency at 60% charge. This is very high concentration and the enhanced electrochemical efficiency at low graphene loading must be achieved in order to produce these products commercially. Graphene supercapacitors can be charged very fast, yet they can also store a large amount of electricity and are lightweight. Recently, Lu et al. [36] investigated the effect of graphene nanoplatelets nanoparticle suspension on the thermal efficiency of an automotive radiator. The study revealed that as volumetric concentration and Reynold number increase, the Nusselt number improves. This upgrade could elevate the efficiency of an automotive cooling system, resulting in a lower radiator and reduced engine fuel consumption. Nanocomposites based on graphene have wide range of application in Electromagnetic Interference shielding (EMI) [37].

Graphene mixed with polymer improves sensor efficiency in terms of selectivity, responsiveness and reaction time, sensors for various automotive parts, clinical diagnostics, metal ion speciation, battery and capacitor formations [38, 39]. Several study groups have also fabricated graphene nanocomposite for applications with biosensors in recent years. The strategy also emphasizes on the production of advanced Graphene-based automotive materials, helping to facilitate the introduction of new power-efficient and safe automobiles on the marketplace. Proa et al. [40] formed an experiment on a homogeneously decorated multi-layered reduced graphene oxide with platinum nanoparticles (reduced graphene oxide/Pt). The nanocomposite was used as a transducer of the Back-gated Field Effect Transistor (GFET) in nanobiosensors. The CS2, rGO/Pt, and protein G detecting IgG, were deposited locally using a professional microplotter unit. A reduced graphene oxide-based field-effect transistor process for real-time detection of the Ebola virus antigen was developed by Chen et al. [41]. The technique uses the graphene-based material's attractive semiconductor properties, and immediately provides extremely sensitive and precise detection of Ebola glycoprotein. This procedure is being checked for clinical use in point-of-care technologies for the feasibility of using Ebola glycoprotein suspended in diluted PBS solution, human serum, and plasma. Such results show the successful fabrication of a promising field-effect biosensor transistor for EVD diagnosis. Phua et al. [42] developed epoxy composites containing graphene nanoplatelets (from 0.2 to 5 vol.%), using simple heat-assisted bath sonics to promote the dispersion and exfoliation of nanofillers in the polymer matrix. The mechanical properties and the conduct of thermal deformation of the resulting composites were analyzed in detail. More specifically, the fracture's flexural strength and durability was found to expand to 0.4 vol.% of the filler filling. At the same time, the bulk electrical conductivity increased marginally in the presence of graphene nanoplatelets.

Several other energy related application areas battery and supercapacitors where graphene-based polymer nanocomposite can be used. Li et al. [43] demonstrated the manufacture of a Graphene oxide and polyaniline based flexible solid-state supercapacitors with a specific capacitance of 1095 Fg^{-1} at 1 Ag^{1}. Graphene-based polymer nanocomposites have become an appropriate option for the Li-ion batteries as an electrode material. It is also shown that protective coating [44] is another growing field where graphene-based polymer nanocomposites can be used, because graphene has excellent chemical resistance, gas impermeability, antibacterial properties, mechanical power, lubricity and thermal stability. For the synthesis of self-cleaning and corrosion protection coatings, it can be modified and combined with different forms of polymers. Graphene based nanocomposite materials are used in window electrodes for bulk heterojunction polymer organic photovoltaic cells [45]. Graphene's superior electrical conductivity is the major factor which contributes for the use in sensing technology.

In addition, graphene-based polymer nanocomposites are used in the specific fields of transparent flexible electrodes, mechanical devices, energy storage, memory storage, sensors, and in organic electronics. For their use in variable atmospheres, such as temperature shift, light strength, heat, magnetic and electric field or solutions, the shape memory alloy of polymer nanocomposite is very crucial. Graphene-based polymer nanocomposites can be used for this purpose, which can be used in robotics, orthopedic surgery, self-repairing structural components such as automotive fenders [46] where dents are fixed by temperature application, biomedical applications, etc. Graphene has recently become a favourite source for scientists as polymer matrix nanofiller [47]. Graphene sheets will several times boost the value of electrical conductivity of polymers at low concentration, more than carbon black. Graphene conductive composites can also be used extensively in large-scale applications such as: electromagnetic insulation, electrical modules, condensers, rechargeable battery electrodes, sensors and actuators, and aircraft ice protection systems [48]. Graphene-enhanced epoxy composites have attracted researchers 'interest in the field of aerospace, especially in solving some aircraft problems such as: lightning strike and ice accumulation on the surface of the aircraft [49]. Ming et al. [50] conducted an experiment using high graphene foam concentration (around 80 wt.%) to adjust the epoxy resin to improve its electrical conductivity. There is another dispersion method used for achieving high values of electrical conductivity, such as ball milling and mechanical stirring. And in the case of mechanical stirring, both thermal and electrical conductivities increased. Table 3 shows the stages leading to the development of advanced graphene composites.

Table 3 Shows the stages leading to the development of advanced graphene composites

Year	2017	2018	2019	2020
	Graphene based Composites	Hybrid Composites	Gels/Aerogels	Advanced graphene Hybrid Composites
	Viscous and thermal polymer that are blended with Graphene products	Graphene bases product compounded with other nanomaterials	Blende with binding substrates at low temperatures. Aerogels are made by removing liquid fully from a graphene gel	Advanced combination of graphene-based composites and gel technologies. These products make use multiple unique characteristics of graphene and other nanomaterials in functional programmable matrixes
Applications	Paints, cement-admixes, coating, epoxies, resins and polymers	Filtration, Sensors, electronics, high performance inks, power storage and generations	Aerospace, cosmetics, and power storage	Antennas, sensors, smart composites, light weight automotive parts, drones, building materials and nanorobotics

4 Limitations of Using Graphene as Filler Material

Although graphene has a wide range of application in various automotive and electronic components due to their low mass density and exceptional properties, there are also some limitations to the use of graphene as a filler material. Due to addition of graphene, biodegradability of the composites gets reduced which influence the biodegradation which make decomposition difficult. There should a limit of adding the graphene as a filler so as to get desired properties for desired application, otherwise the properties will get imbalance. As we also know that, interfacial adhesion is also important as it help in proper bonding between the fibre and the polymer used for the composites. If inadequate amount of graphene is added as required, it will affect the interfacial adhesion hence results in decrease in its strength.

5 Conclusion

In Recent, graphene and polymer-based nanocomposites have potential application in various automotive and electronics components. Reinforcement in graphene in nanocomposites also enhances the performance of nanocomposites. There are various renowned companies like BMW, Mercedes, Fiat, etc. who manufactured various automotive components of nanocomposites [51]. Several points regarding the applications of graphene-based nanocomposites in mechanical and electronics mentioned below:

- Graphene-based biopolymer nanocomposites have developed multiple applications in the years to come. Light weighting is an essential consideration for vehicle energy efficiency.
- Other applications of graphene polymer composites include: lightweight gas tanks, plastic containers, fuel-efficient aircraft and car parts, more grounded wind turbines, restorative embeds, and sports equipment.
- Various Composites that are manufactured are: door panels, headliner panels, boot lining, spare tire lining and many more.
- The other application such as biomedical devices is ultra-scaled down ease sensors of blood and pee inspection. When part of an electrochromic gadget array, guiding polymer/graphene composites may also be used.
- Graphene base polymer nanocomposites have gained lots of opportunity in Electromagnetic interference shielding applications.
- Graphene-based polymer nanocomposites are also gaining a lot of importance in the fields of transparent flexible electrodes, mechanical parts, energy storage, memory storage, sensors and as well as organic electronics.
- In Field effect transistor logic applications, graphene based nanocomposites also plays and important role.
- The adaptable polymer/graphene terminal has several market applications in LEDs, leading straightforward coatings for sun powered cells and showcases. Another promising potential that is directly related to polymer composite materials based on graphene is the recent developments in additive or 3D manufacturing and other innovative methods. In the context of material design these advances will lead to various perspectives and practical applications

References

1. Paul DR, Robeson L (2008) Polymer nanotechnology: nanocomposites. Polymer 49(15):3187–3204. https://doi.org/10.1016/j.polymer.2008.04.017
2. Pratim Das P, Chaudhary V (2019) Tribological and dynamic mechanical analysis of biocomposites: A Review. Mater Today: Proc. https://doi.org/10.1016/j.matpr.2019.08.233
3. Chaudhary V, Rajput A, Bajpai P (2017) Today: Proc 4(9):9893–9897. https://doi.org/10.1016/j.matpr.2017.06.289
4. William JJ (2012) J Mater 64(9):1032–1038. ISSN 1047-4838. http://dx.doi.org/10.1007/s11837-012-0424z
5. Kumar A, Sharma K, Dixit AR (2008) J Mater Sci 1–35. http://dx.doi.org/10.1007/s10853-018-03244-3
6. Sharma B, Malik P, Jain P (2018) Biopolymer reinforced nanocomposites: a comprehensive review. Mater Today Commun 16:353–363

7. Lawal A (2019) Biosens Bioelectron 141:111384. https://doi.org/10.1016/j.bios.2019.111384
8. Sharma B, Shekhar S, Gautam S, Sarkar A, Jain P (2018) Nanomechanical analysis of chemically reduced graphene oxide reinforced poly (vinyl alcohol) nanocomposite thin films. Polym Testing 70:458–466
9. Rajesh UC, Jinfeng WC, Takuya TS, Rawat DS (2015) ACS Sustain Chem. Eng 3(1):9–18. https://doi.org/10.1021/sc500594w
10. Jung YH, Chang TH, Zhang H, Yao C (2015) Nat Commun. https://doi.org/10.1038/ncomms8170
11. Ling S, Qin Z, Li C, Huang W, Kaplan DL, Buehler MJ (2017) Nat Commun 23–30. https://doi.org/10.1038/s41467-017-00613-5
12. Williams G, Seger B, Kamat PV (2008) ACS Nano 2:1487–1491. https://doi.org/10.1021/nn800251f
13. Elmarakbi A, Safty S, Martorana B, Azoti W (2016) Int J Automot Compos 2(2):155. https://doi.org/10.1504/ijautoc.2016.082043
14. Gundolf K, Elmar B, Roland S, Alexander K, Philipp S, Michael K (2012) Procedia—Soc Behav Sci 350–362. ISSN 1877-0428. http://dx.doi.org/10.1016/j.sbspro.2012.06.1015
15. Tapas K, Sambhu B, Dahu Y, Nam HK, Saswata B, Joong HL (2010) Prog Polym Sci 35(11):1350–1375. http://dx.doi.org/10.1016/j.progpolymsci.2010.07.005
16. Galpaya D, Wang M, Yan C, Liu M, Motta N, Waclawik E (2013) Proc SPIE Int Soc Opt Eng 87938793. https://doi.org/10.1117/12.2026759
17. Li Z, Zhou X, Pei C (2011) Int J Polym Sci 1–7. https://doi.org/10.1155/2011/803428
18. Chrysler C (2012) Sustainability report. Technical report, p 08
19. Stoller MD, Park S, Zhu Y (2008) Graphene-based ultracapacitors. Nano Lett 8:3498–3502. https://doi.org/10.1021/nl802558y
20. Chen J, Huang Z, Lv W, Wang CC (2017) Polym Compos. https://doi.org/10.1002/pc.24433
21. Ramkumar K (2019) Review article; Curr Graphene Sci 3:1–34. https://doi.org/10.2174/2452273203666191104150025
22. Arifurrahman F, Budiman BA, Aziz M (2018) Int J Sustain Transp Technol 1:21–29. http://dx.doi.org/10.31427/IJSTT.2018.1.1.4
23. Zhong Y, Zhen Z, Zhu H (2017) Graphene: Flat. Chem 4:20–32. http://dx.doi.org/10.1016/j.flatc.2017.06.008
24. Lin Y-M, Dimitrakopoulos C, Jenkins KA (2010) Sci J 327(5966):662. http://dx.doi.org/10.1126/science.1184289
25. Holberyand J, Houston D (2006) J Miner Metals Mater Soc 58(11):80–86. https://doi.org/10.1007/s11837-006-0234-2
26. Takuya N (2006) J Soc Automot Eng Japan 60(1):100–104. http://dx.doi.org/10.4028/www.scientific.net/AMM.564.3
27. Avouris P, Dimitrakopoulos C (2012) Mater Today 15:86–97. https://doi.org/10.1016/S1369-7021(12)70044-5
28. Deepalekshmi P, Kishor KS (2015) Springer Series on Polym Compos Mater. https://doi.org/10.1007/978-3-319-13875-6_1
29. Liu S, Wang A, Li Q (2018) Joule 2:184–193. https://doi.org/10.1016/j.joule.2017.11.004
30. Son IH, Park JH, Park S (2017) Nat Commun 8(1):1561. https://doi.org/10.1038/s41467-017-018237
31. Barber P, Balasubramanian S, Anguchamy Y, Gong S, Wibowo A, Gao H, Ploehn HJ, Zur Loye HC (2009) Materials 2(4):1697–1733. https://doi.org/10.3390/ma2041697
32. Vaia RA, Giannelis EP (2001) MRS Bull 26(5):394–401
33. Winey KI, Vaia RA (2007) Polymer nanocomposites. MRS Bull 32:314–322
34. Wu Q, Xu Y, Yao Z, Liu A, Shi G (2010) ACS Nano 4:1963–1970. https://doi.org/10.1021/nn1000035
35. Guo W, Yin YX, Xin S, Guo YG, Wan L (2012) J Energy Environ Sci 5:5221–5225. https://doi.org/10.1039/C1EE02148F
36. Lu Y, Niu Z, Chen J (2018) Adv Energy Mater 81702469. https://doi.org/10.1002/aenm.201702469

37. Gelves GA, Al-Saleh MH, Sundararaj U (2011) J Mater Chem. https://doi.org/10.1039/C0JM02546A
38. Pervaiz M, Panthapulakkal SKC, Sain B, Tjong M (2016) J Mater Sci Appl 07(01):26–38. https://doi.org/10.4236/msa.2016.71004
39. Elmarakbi A, Safty S, Martorana B, Azoti W (2016) Int J Automot Compos 2(2):155. https://doi.org/10.1504/ijautoc.2016.082043
40. Sergio Proa C, Jorge R, Vargas G, Arturo MR, Salvador MA (2018) Thin Solid Films 658:54–60. https://doi.org/10.1016/j.tsf.2018.05.032
41. Chen Y, Ren R, Pu H, Guo X, Chang J, Zhou G, Mao S, Kron M, Chen J (2017) Sci Rep 7(1):10974. https://doi.org/10.1038/s41598-017-11387-7
42. Phua JL, The PL, Ghani SA (2016) Int J Polym Sci. http://dx.doi.org/10.1155/2016/9767183
43. Li H, Song J, and Wang L., (2017). 193–200. Nanoscale. 2017;9(1) https://doi.org/10.1039/C6NR07921K
44. Zhu Y, Murali S, Cai W (2010). Adv Mater 22(35):3906–3924. https://doi.org/10.1002/adma.201001068
45. Eda G, Lin YY, Miller S, Chen CW, Su WF, Chhowalla M (2008) Appl Phys Lett 92(23):233305. https://doi.org/10.1063/1.2937846
46. Bera M, Maji PK (2017) MOJ Poly Sci 1(3):94–97. https://doi.org/10.15406/mojps.2017.01.00013
47. Cheng QF (2010) Carbon 48(1):260–266. http://dx.doi.org/10.1016/j.jes.2015.11.001.1001-0742
48. Teng X, Jiankai D, Haowen C, Yiqiang J, Yang Y (2018) Appl Therm Eng 11(4):371–398
49. Qin W (2014) Polym Compos 37(5):1549–1556. https://doi.org/10.1002/pc.23325
50. Ming P, Zhang Y, Bao J, Liu G, Li Z, Jiang L, Cheng Q (2015) RSC Adv 5(28):22283–22288. https://doi.org/10.1039/C5RA00233H
51. Maji P (2017) MOJ Polym Sci 1(3). https://doi.org/10.15406/mojps.2017.01.00013
52. Xu Y, Kawata S, Hosoi K, Kawai T, Kuroda S (2009) EXPRESS Polym Lett 3(10):657–664. https://doi.org/10.3144/expresspolymlett.2009.82
53. Vilay V, Mariatti M, Mat Taib R, Todo M (2008) J Compos Sci Technol 631–638. https://doi.org/10.1016/j.compscitech.2007.10.005
54. Mohd Yuhazri Y, Phongsakorn T, Sihombing H (2010) Int J Basic Appl Sci 10(3):63–66. https://doi.org/10.3144/expresspolymlett.2009.82
55. Norhidayah MH, Hambali A, Mohd Yuhazri Y, Zolkarnain M, Saifuddin HY (2014) Appl Mech Mater 564:3–7. https://doi.org/10.4028/www.scientific.net/AMM.564.3
56. Bledzki AK, Faruk O, Sperber VE (2006) J Macromolecular Mater Eng 29:449–457. https://doi.org/10.1002/mame.200600113
57. Karthi N, Kumaresan K, Sathish S, Gokulkumar S, Prabhu L, Vigneshkumar N (2020) An overview: natural fiber reinforced hybrid composites, chemical treatments and application areas. Mater Today: Proc
58. Threepopnatkul P, Kaerkitcha N, Athipongarporn N (2009) J Compos: Part B 40:628–632. https://doi.org/10.1016/j.compositesb.2009.04.008
59. Chaudhary V, Bajpai P, Maheshwari S (2018) Fibers Polym 19(2):403–415. https://doi.org/10.1007/s12221-018-7759-6